# Green Energy and Technology

Climate change, environmental impact, and the limited natural resources urge scientific research and novel technical solutions. The monograph series Green Energy and Technology serves as a publishing platform for scientific and technological approaches to "green" – i.e., environmentally friendly and sustainable – technologies. While a focus lies on energy and power supply, it also covers "green" solutions in industrial engineering and engineering design. Green Energy and Technology addresses researchers, advanced students, technical consultants, and decision-makers in industries and politics. Hence, the level of presentation spans from instructional to highly technical. **Indexed in Scopus**.

More information about this series at http://www.springer.com/series/8059

Ibrahim Dincer • Mehmet Akif Ezan

# Heat Storage: A Unique Solution For Energy Systems

Springer

Ibrahim Dincer
Department of Automotive
Mechanical and Manufacturing
Engineering
University of Ontario
Oshawa, ON, Canada

Mehmet Akif Ezan
Department of Mechanical Engineering
Dokuz Eylül University
Izmir, Turkey

ISSN 1865-3529 ISSN 1865-3537 (electronic)
Green Energy and Technology
ISBN 978-3-030-06323-8 ISBN 978-3-319-91893-8 (eBook)
https://doi.org/10.1007/978-3-319-91893-8



Printed on acid-free paper

This Springer imprint is published by the registered company Springer International Publishing AG part of Springer Nature.
The registered company address is: Gewerbestrasse 11, 6330 Cham, Switzerland

# Preface

Sustainable development is recognized as one of the most significant domains of society which is primarily related to the future of a country and how the country will sustain its progress without negative implications. It depends primarily on energy, environment, resources, economy, and social, cultural, and ethical dimensions. Energy sustainability appears to be the most significant tool in achieving a sustainable society. Energy sustainability, of course, requires sustainable resources, sustainable systems, and sustainable outputs to meet the needs of the society. The criticality is that there is a need for sustainable storage options between resources and energy systems, and between energy systems and energy services. There is a need to offset the mismatch between demand and supply by any means possible. Energy storage techniques can also be integrated into renewable-resources-based power plants, such as solar and wind, to overcome the intermittency of renewable sources and provide continuous power generation. Energy storage methods are also incorporated to produce electrical power when there is excessive demand and the source is not accessible. Different types of energy storage systems are currently used in diverse fields of engineering applications, such as *chemical*, *electrochemical*, *electrical*, *mechanical*, and *thermal energy storage*. Thermal energy storage (TES) is the storage of thermal energy at high (*heat storage*) or low (*cold storage*) temperatures. TES is an essential feature for using the conventional energy systems, and it is *sustainable*, *efficient*, *economical*, and *environmentally friendly*. TES is, therefore, a key technology in reducing the mismatch between the energy supply and demand for thermal systems. In addition to large-scale renewable-sourced power generation systems, TES methods are widely used in residential or commercial heating/cooling applications. TES systems not only provide a balance between supply and demand but also increase the *performance* and *reliability* of energy systems.

The book presents the essentials of energy storage techniques with some real-world applications and covers in-depth knowledge of heat storage systems. Different aspects of heat storage systems are illustrated, from *microscale* to *macroscale*. The book also covers material production, characterization, modeling, experimentation, and optimization of heat storage systems. It is intends to provide a new perspective to the researchers, scientists, engineers, technologists, students, policy-makers, etc. who wish to learn more about heat storage systems and applications.

Chapter 1 introduces the fundamental aspects of thermodynamics and heat transfer. The chapter includes some step-by-step solved illustrative examples of simple closed and open heat storage systems. Chapter 2 describes the importance and methods of energy storage. In Chap. 3, the essentials of sensible heat TES, latent heat TES, and thermochemical ES techniques are illustrated. Chapter 4 focuses on the implementation of TES units into buildings and solar power generation systems. Chapter 5 discusses the thermodynamics, heat transfer, and computational fluid dynamics analyses of TES systems. Chapter 6 addresses the fundamentals of the second law-based optimization of TES systems. Some illustrative examples provide an in-depth understanding of the importance of optimization to build a better system. Chapter 7 covers comprehensive information about case studies related to heat storage systems from microscale to macroscale applications.

This book, in closing, offers unique perspectives on fundamentals, systems, and applications of heat storage systems. The book follows the International System of Units (SI). At the end of each chapter, some useful references are provided to guide the readers for further knowledge.

Oshawa, ON, Canada — Ibrahim Dincer
Izmir, Turkey — Mehmet Akif Ezan
September 2018

# Contents

# Chapter 1
# Fundamental Aspects of Thermodynamics and Heat Transfer

## 1.1 Introduction

The application of energy storage systems requires an in-depth knowledge of the thermal-fluid sciences. These sciences generally cover thermodynamics, heat transfer, and fluid mechanics. Thermodynamics, which is also known as the *science of energy*, builds up the framework of the heat and work interactions of a system that operates at various design and working conditions. In thermodynamics analyses, system performance must be evaluated by considering both the *first law* and *second law* aspects. The first law of thermodynamics, also known as *conservation of energy*, states that the quantity of energy remains constant for a system that undergoes a process. The first law mainly examines the interactions between heat transfer and work. The second law of thermodynamics, on the other hand, considers the *quality* of energy and defines *work potential*. Second law analyses make it possible to combine economic and environmental aspects in the thermodynamics models. Heat transfer applications combine the laws of thermodynamics with mathematics to evaluate spatial or temporal variations of scalar, i.e., temperature or pressure, and vectorial, i.e., heat flux or velocity, quantities within a system.

The objective of the current chapter is to introduce the fundamental aspects of thermodynamics and heat transfer. As the book mainly focuses on *heat storage* and its applications, the current section excludes fluid mechanics and covers only the essentials of the thermal sciences. Authors encourage readers to refer to relevant textbooks (i.e., Cengel and Boles 2010; Cengel and Ghajar 2014; Dincer and Rosen 2011) for further reading on the topics presented and discussed in the following sections.

 
I. Dincer, M. A. Ezan, *Heat Storage: A Unique Solution For Energy Systems*, Green Energy and Technology, https://doi.org/10.1007/978-3-319-91893-8_1

## 1.2 Dimensions and Units

Dimensions characterize physical quantities, and units are used to assign numerical values to dimensions. There are seven fundamental dimensions, also known as *primary dimensions*, and the rest of the dimensions are derived from these seven. The primary dimensions and corresponding units according to the International System of Units (SI or metric unit system) are given in Table 1.1.

Some derived (*secondary*) dimensions are listed here:

| | | |
|---|---|---|
| Velocity | $V = x/t$ | (m/s) |
| Acceleration | $a = V/t$ | (m/s$^2$) |
| Force | $F = ma$ | (kgm/s$^2$) or (N) |
| Work (or energy) | $W = Fx$ | (Nm) or (J) |
| Power | $P = W/t$ | (J/s) or (W) |

Unit consistency is crucial to obtain meaningful results in thermal system analyses. Even in a simple problem, checking unit consistency can prevent possible mistakes. In this book, the SI unit system is used as it is an internationally accepted unit standard in engineering applications.

## 1.3 Thermodynamic Systems

An engineer should simplify a real-world device into a reduced mathematical model (or *system*) to consider the interactions between each component of the device and its surroundings. A *system* is a finite volume or mass in space that is selected for consideration. Anything outside of the system is called *surroundings*. The physical or artificial surface that separates the system from its surroundings is the *boundary*. As an illustrative example, a solid object that is immersed in water is shown in Fig. 1.1. Here the system (*solid object*), surroundings (*water*), and boundary (*fixed-real*) are indicated.

The identification of a system is an important step in thermal analyses since the definition of balance equations, either *mass*, *energy*, *entropy*, or *exergy*, has unique

**Table 1.1** Primary dimensions and SI units

| Dimension | Unit |
|---|---|
| Length | Meter (m) |
| Mass | Kilogram (kg) |
| Time | Second (s) |
| Temperature | Kelvin (K) |
| Electric current | Ampere (A) |
| Amount of light | Candela (cd) |
| Amount of matter | Mole (mol) |

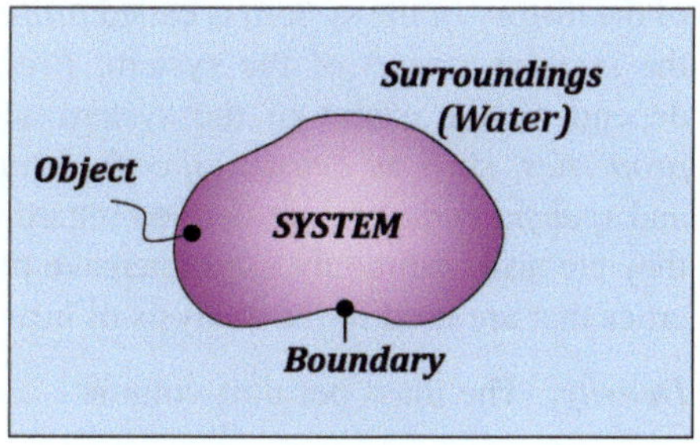


**Fig. 1.1** System concept

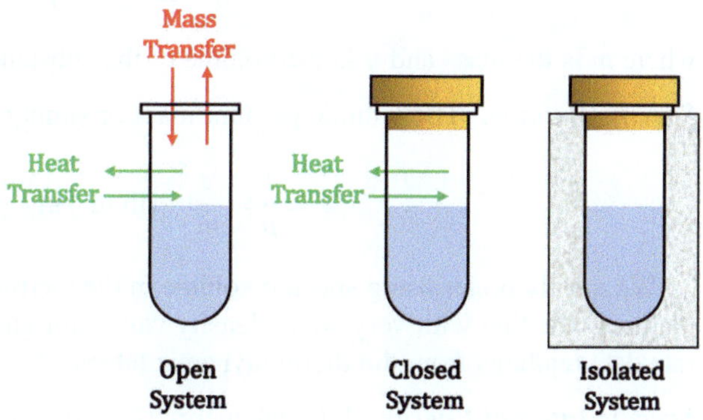


**Fig. 1.2** Thermodynamic systems

forms depending on the type of the system. In the thermodynamic point of view, there are three types of systems:

- ***Open system:*** Allows mass and energy (*work* or *heat*) transfer through the system boundaries.
- ***Closed system:*** Allows energy (*work* or *heat*) transfer through the system boundaries, but mass transfer does not take place.
- ***Insolated system:*** Neither energy transfer nor mass transfer is allowed.

The schematic representation of each thermodynamic system is shown in Fig. 1.2.

## 1.4 Thermodynamic Properties

Thermodynamics is the science of energy, and it deals with the variations between individual *states*. A state is a condition that is defined by *system properties.* In the case of heat storage, there are no external forces, such as magnetic or electrical, acting on the systems, so a *simple system postulate* is appropriate for such systems. To fix a state in a simple system, which corresponds to the evaluation of the thermodynamic properties, we do not need to know all its properties, instead two independent *intensive properties* are adequate. Any characteristic of a system which is independent of the history of the system is called *property*. An *intensive property* is independent of the size (or *extent*) of the system. *Pressure* and *temperature* of a system do not depend on the extent of the system and are mostly used to fix a state. *Specific properties*, such as *density*, *specific heat*, *specific volume*, *specific internal energy*, and *specific enthalpy*, are defined regarding the per unit volume or per unit mass, so they are also commonly used intensive properties. Some fundamental intensive properties that are used in the analysis of heat storage units are defined below.

***Density.*** The mass per unit volume

$$\rho = m/\forall(\text{in kg/m}^3) \tag{1.1}$$

where $m$ is the mass and $\forall$ is the volume of the substance.

***Specific Volume.*** The volume per unit mass or simply inverse of the density

$$v = \frac{1}{\rho} = \frac{\forall}{m} \quad (\text{in m}^3/\text{kg}) \tag{1.2}$$

We mostly prefer using specific volume in the thermodynamic analysis of power plants, as dealing with very small density values for gases may cause some inaccurate data readings from the thermodynamic tables.

***Specific Internal Energy.*** Internal energy ($U = mu$ in Joule) of a system is related to the microscopic forms of energy. It includes *sensible*, *latent*, *chemical*, and *nuclear* forms of energy. The sensible and latent forms of internal energy will be discussed in detail in the following section. Internal energy is used to calculate the energy variation of a closed system in which there is no boundary or flow work. In general, the internal energy of a system at a specific state can be evaluated from the thermodynamic tables. For incompressible liquids or solids without phase change, we can simply calculate the specific internal energy (per *unit mass*) variation of a system regarding temperature variations as follows:

$$du = c_v dT \quad (\text{in J/kg}) \tag{1.3}$$

where $c_v$ is the specific heat at constant volume (in J/kgK).

***Specific Enthalpy.*** Enthalpy ($H = mh$ in Joule) is commonly known as total energy since in addition to internal energy it also includes flow (or *boundary*) work. Specific enthalpy (per *unit mass*) is defined as

$$h = u + pv \quad (\text{in J/kg}) \tag{1.4}$$

In the case of incompressible liquids, we can evaluate enthalpy variation by using the following thermodynamic relation:

$$dh = c_p dT \quad (\text{in J/kg}) \tag{1.5}$$

where $c_p$ is the specific heat at *constant pressure* (in J/kgK).

Please note that, for incompressible substances, such as *liquids* or *solids*, the specific heat at constant volume ($c_v$) is identical to the specific heat at constant pressure ($c_p$). As we commonly deal with either solid or liquid phases of a substance in the case of thermal energy storage applications, we may simply use specific heat ($c$), which is defined as

$$c = c_p = c_v \quad (\text{in J/kgK}) \tag{1.6}$$

## 1.5 Sensible and Latent Heats

Internal energy has two parts: *sensible* and *latent*. Sensible heat is related to the temperature variation of a substance, and latent heat is linked to the phase change of a material. Suppose that in a piston-cylinder system, water is initially at $T_{int} = -10\ °C$ under atmospheric pressure. As the initial temperature is below the solidification temperature of water ($T_{melting} = 0\ °C$), the initial phase is *solid*. Heat energy is supplied to the piston-cylinder assembly to obtain water vapor in the final state. In Fig. 1.3, the variation in water temperature with respect to energy throughout the process is illustrated.

In this process, the sensible and latent heat regions and the corresponding energy variations for each region are defined as follows:

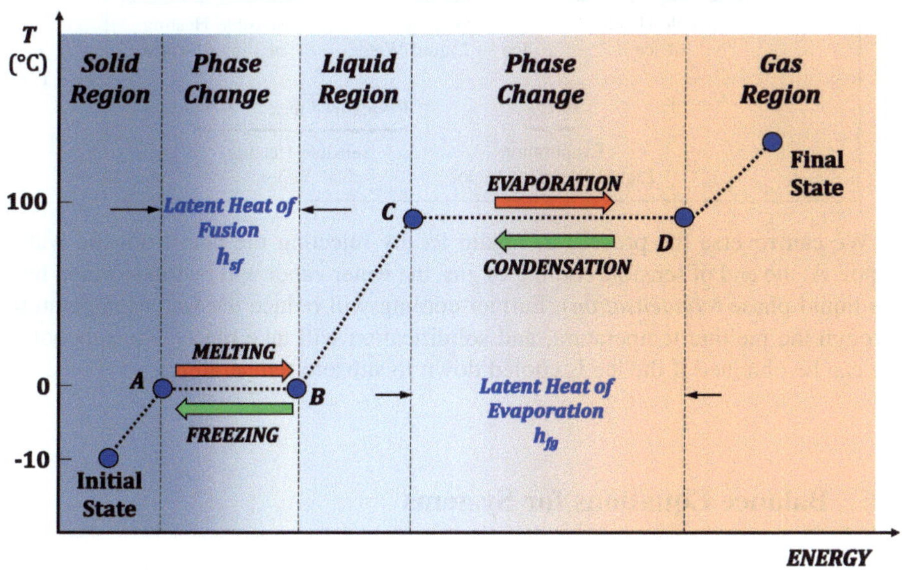


**Fig. 1.3** Process of water from solid to gas phase

### Sensible Heat

***Initial* → A:** *Heat transfer increases the temperature of ice until the melting temperature of water ($T_{melting} = 0$ °C). Variation of internal energy for the current region is $\Delta E_{int \to A} = mc_{solid}(T_A - T_{int})$.*

**B → C:** *Heat transfer increases the temperature of the liquid water until the boiling temperature of water ($T_{boiling} = 100$ °C). Variation of internal energy for the current region is $\Delta E_{B \to C} = mc_{liquid}(T_B - T_C)$.*

**D → *Final:*** *In the gas phase of water, heat transfer increases the temperature of water vapor. Under constant pressure, the variation of internal energy for the current region is $\Delta E_{D \to final} = mc_{p,gas}(T_{final} - T_D)$.*

### Latent Heat

**A → B:** *Melting region. Ice transforms into liquid water without any temperature change. The solid substance should absorb the latent heat of fusion to transform into liquid. The variation of internal energy is defined in terms of the latent heat of fusion ($h_{sf}$ in J/kg) as $\Delta E_{A \to B} = mh_{sf}$.*

**C → D:** *Evaporation region. Liquid water transforms into gas phase without any temperature change. Liquid should absorb the latent heat of evaporation to transform into gas. The variation of internal energy is defined in terms of the latent heat of evaporation ($h_{fg}$ in J/kg) as $\Delta E_{C \to D} = mh_{fg}$.*

Consequently, from the initial to the final state, the total internal energy variation can be written as follows:

$$\boxed{\begin{aligned}\Delta E = & \underbrace{mc_{solid}(T_A - T_{int})}_{\substack{\text{Sensible Heating}\\ \text{of Ice}}} + \underbrace{mh_{sf}}_{\substack{\text{Melting}\\ \text{Ice} \to \text{Liquid Water}}} + \underbrace{mc_{liquid}(T_B - T_C)}_{\substack{\text{Sensible Heating}\\ \text{of Liquid Water}}} \\ & + \underbrace{mh_{fg}}_{\substack{\text{Evaporation}\\ \text{Liquid Water} \to \text{Vapor}}} + \underbrace{mc_{p,gas}\left(T_{final} - T_D\right)}_{\substack{\text{Sensible Heating}\\ \text{of Vapor}}}\end{aligned}} \tag{1.7}$$

We can reverse the process to obtain ice by rejecting the heat from the water vapor. At the end of sensible cooling of gas, the water vapor will be transformed into the liquid phase (*condensation*). Further cooling will reduce the water temperature through the melting temperature, and solidification will take place. The subcooled ice can be obtained if the ice is cooled down to subzero temperatures.

## 1.6 Balance Equations for Systems

Mass, energy, entropy, and exergy balance equations should be considered to carry out complete thermodynamic analyses for a thermal system. In the following, the balance equations are given with simple worked examples.

### 1.6.1 Mass Balance

The following balance equation is derived from the conservation of mass principle for a transient and open system:

$$\sum \dot{m}_{in} - \sum \dot{m}_{out} = \frac{dm_{cv}}{dt} \quad (\text{in kg/s}) \tag{1.8}$$

The terms on the left-hand side represent the total mass flow rate at the inlet and outlet sections of a system, respectively. The term on the right-hand side is the variation of the total mass within the control volume.

**Example 1.1**
Waste heat recovery is a prevalent subject and is applicable for a broad range of thermal systems. In a recent review, Hepbasli et al. (2014) stated that in buildings we lose an enormous amount of useful heat from wastewater and it has a key role in energy conservation and environmental pollution. We can store warm water in a large tank and release the useful heat to various heating applications in the building.

Assume that we are collecting the wastewater of a building complex in a storage tank which has a total volume of 2500 m$^3$. An engineer designed the wastewater recovery unit in such a way that the water is supplied to the reservoir with a volumetric flow rate of 1.2 m$^3$/min, and from a pipeline ($D_{pipe}$ = 0.1 m) at the bottom of the reservoir, we reject the water with a mean velocity of 0.5 m/s. If the initial water volume is 1000 m$^3$, determine the volume of water at the end of 24 h.

**Solution:**
As a first step, we should draw a simple sketch to define the system boundary (or *control volume*) and shows the mass flows crossing the system boundary as shown in Fig. 1.4 as well as write the mass balance equation for this system.

(continued)

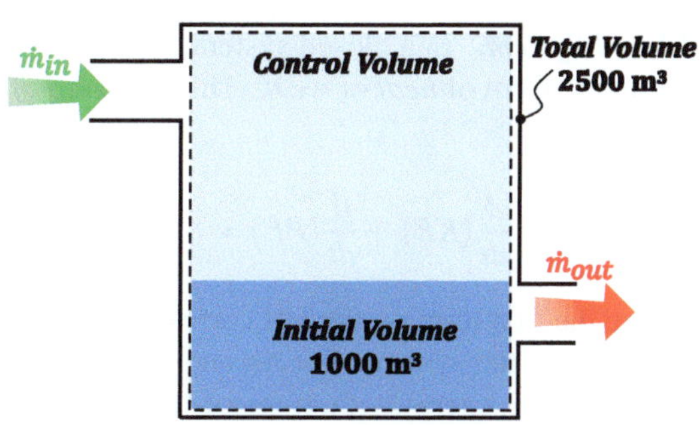


**Fig. 1.4** Sketch of a open system with single inlet/outlet

**Example 1.1** (continued)

$$\dot{m}_{in} - \dot{m}_{out} = \frac{dm_{cv}}{dt}$$

The mass flow rates at the inlet and outlet are constant throughout the process, hence the integration of the mass balance equation yields

$$\left[\dot{m}_{in} - \dot{m}_{out}\right]\Delta t = \left[m_{final} - m_{initial}\right]$$

The mass flow rate of a fluid is defined in terms of the density ($\rho$), the cross-sectional area ($A$), and the mean velocity ($V$) along the normal of the surface as $\dot{m} = \rho VA$. The relationship between the mass flow rate ($\dot{m}$ in kg/s) and the volumetric flow rate ($\dot{\forall}$ in m$^3$/s) is $\dot{m} = \rho\dot{\forall}$. For simplicity, we can assume that the density of water is constant as $\rho = 1000$ kg/m$^3$. So, we can reorganize the mass balance equation to yield

$$\left[\left(\rho\dot{\forall}\right)_{in} - (\rho VA)_{out}\right]\Delta t = \left[(\rho\forall)_{final} - (\rho\forall)_{initial}\right]$$

As the density is constant, we can drop the terms. The given parameters can be implemented to determine the final volume of water within the storage tank after 24 h:

$$\forall_{final} = \left[\left(\frac{1.2}{60}\right) - \left(0.5 \times \pi\frac{0.1^2}{4}\right)\right](24 \times 3600) + (1000) \quad \text{and hence}$$

$$\boxed{\forall_{final} = 2389\ \text{m}^3}$$

## 1.6.2 Energy Balance

Energy balance equations differ for closed and open systems, so it is essential to use them appropriately. In the following, general forms of balance equations are given first, then some reductions are proposed to obtain balance equations which are adequate for modeling heat storage systems.

***Closed System.*** In a closed system, the energy can pass through the system boundary in the form of *heat* or *work*. The rate form of the energy balance equation can be written as

$$\frac{d}{dt}(U) + \frac{d}{dt}(KE) + \frac{d}{dt}(PE) = \left[\dot{W}_{in} - \dot{W}_{out}\right] + \left[\dot{Q}_{in} - \dot{Q}_{out}\right] \quad \text{(in W)} \qquad (1.9)$$

where the first term on the left corresponds to the internal energy ($U = mu$ in Joule) variation of the system. The second and third terms on the left represent the kinetic and potential energy variations of the system. The two brackets on the right-hand side of the equation indicate the work and heat transfer interactions between the system and its surroundings, respectively. In the case of thermal storage applications, the components of the storage systems are mostly stationary, and the last two terms on the left-hand side of Eq. (1.9) can be dropped:

$$\frac{d}{dt}(U) = \left[\dot{W}_{in} - \dot{W}_{out}\right] + \left[\dot{Q}_{in} - \dot{Q}_{out}\right] \quad \text{(in W)} \tag{1.10}$$

Note that the kinetic and potential energy variations are omitted for stationary thermal system. That is, if we combine Eqs. (1.3) and (1.10), we can obtain the energy balance equation for incompressible substances in terms of the temperature:

$$\frac{d}{dt}(mcT) = \left[\dot{W}_{in} - \dot{W}_{out}\right] + \left[\dot{Q}_{in} - \dot{Q}_{out}\right] \quad \text{(in W)} \tag{1.11}$$

**Example 1.2**

One of the oldest forms of the thermal energy storage application is leaving rocks or bricks in the sunlight during daytime. After sunset, the stored energy within the system is used in some specific applications, such as heating small rooms or water reservoirs. Suppose that a rock bed contains 200 kg of rocks with a specific heat of 2000 J/kgK is exposed to solar radiation throughout the day. Due to the convective and radiative heat loss through the ambient, some portion of the incident solar radiation will be stored within the rocks. For simplicity, we can assume that 60% of the incident radiation is lost. If the daily average incident solar radiation is 150 W/m$^2$ and the initial temperature of the rock is $T_{initial} = 15$ °C, calculate the final temperature of the rock bed at the end of the day. Here, assume that the surface area of the rock bed is 5 m$^2$ and the lumped capacitance method is valid for the preliminary calculations.

**Solution:**

The forty percent of the incident solar radiation is stored in the rock bed throughout the day. Note that there is no work interaction for the current system, that is, the balance equation that is given in Eq. (1.11) is simplified in the following form (Fig. 1.5):

$$\frac{d}{dt}(mcT) = 0.4(I_{solar}A)$$

The right-hand side of the equation is constant for 24 h, so we can simply integrate the equation to end up with

$$mc\left(T_{final} - T_{initial}\right) = 0.4(I_{solar}A) \times (24 \times 3600)$$

(continued)

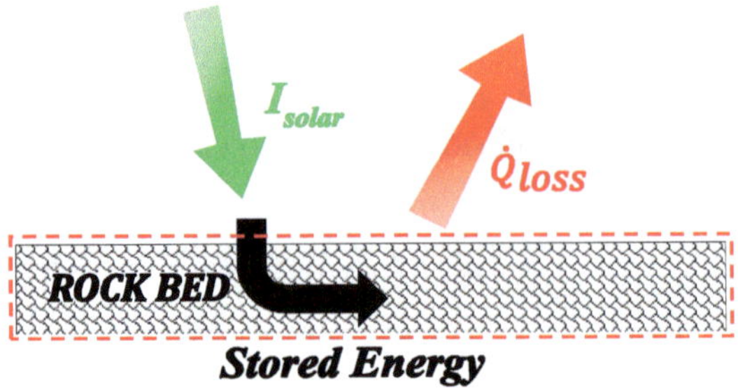


**Fig. 1.5** Energy balance for the rock bed

**Example 1.2** (continued)

Consequently, the final temperature of the rock bed is obtained by using the given parameters:

$$T_{final} = \frac{0.4(150 \times 5)}{200 \times 2000} \times (24 \times 3600) + 15 \quad \text{resulting in} \quad \boxed{T_{final} = 79.8\ ^{\circ}\text{C}}$$

**Example 1.3**

Electrical thermal energy storage (ETES) is an alternative and economical way of heating where natural gas is not accessible. ETES consists of a well-insulated storage unit in which high-dense solids are used with immersed electrical heaters. Thermal energy is stored in storage material (high-dense solid) during off-peak hours by electrical heaters for further usage. During off-peak hours, i.e., nighttime, the electric rate is considerably lower than the high-usage periods, hence the ETES provides economic advantages for end users.

Consider an ETES unit for residential usage. The system stores thermal energy during nighttime and releases thermal energy to the room during daytime. Electrical heaters are embedded into the storage unit to charge the thermal energy. At the end of the charging process, the temperature of the storage medium, i.e., magnetite, reaches up to 700 °C. During the discharge process, the indoor air circulates inside the storage medium and releases the thermal energy from the ETES. The temperature of the storage medium reduces with advancing time, that is, the rate of heat transfer from the storage medium to the air decreases with time. Assume that within a discharge duration of 16 h, the discharge rate reduces linearly from 600 W to 100 W. If the duration of off-peak hours is 8 h, evaluate the required electric heater

(continued)

**Example 1.3** (continued)
capacity to charge the ETES. Assume that there is no heat loss through the environment throughout the charging and discharging processes.

**Solution:**
For a well-insulated ETES unit, the energy stored at nighttime is discharged from the unit during daytime. The electrical input increases the internal energy of the system via temperature change. The total amount of discharged energy is evaluated from the average discharge rate and the discharge duration as follows:

$$\Delta E_{\text{discharged}} = \dot{Q}_{average} \times \Delta t_{\text{discharge}} \quad \text{which becomes} \quad \Delta E_{\text{discharged}} = 20,160\,\text{kJ}$$

For an idealized process without any heat loss through the environment, the same amount of energy should be supplied to the system in 8 h by the electrical heaters. So, the balance equation can be reduced to the following form to calculate the capacity of the electrical heater ($\dot{W}_{electric}$):

$$\Delta E_{\text{charged}} = \Delta E_{\text{discharged}} = \dot{W}_{electric} \times \Delta t_{\text{charge}} \quad \text{which becomes} \quad \boxed{\dot{W}_{electric} = 700\,\text{kW}}$$

**Example 1.4**
Phase change materials (PCMs) are substances that are used in thermal energy storage systems to store (via *melting*) or release (via *solidification*) thermal energy by means of latent heat. Instead of using a rock bed, an engineer wants to use PCM tank to store the same amount of solar energy as in Example 1.2. Assume that the initial and final temperature values of the PCM are identical as in the previous example. Calculate the required mass of PCM for the following thermal properties of PCM:

$$T_{melting} = 30^{\circ}\text{C}, h_{sf} = 200\ \text{kJ/kg}, c = 2000\ \text{J/kgK}$$

**Solution:**

The initial and final temperatures of the PCM are as follows: $T_{initial} = 15\ ^{\circ}\text{C}$ and $T_{final} = 79.8\ ^{\circ}\text{C}$. As the melting temperature of the PCM stands between these two boundaries, the PCM is initially in the solid phase and at the end of the process it turns into a super-heated liquid. The total energy variation of the PCM can be written as

$$\Delta E = \left\{ mc(T_m - T_{\text{initial}}) + mh_{sf} + mc\left(T_{final} - T_m\right)\right\}_{PCM}$$

(continued)

**Example 1.4** (continued)

The internal energy variation of the PCM will be the same as that in Example 1.2 as $\Delta E = 0.4(I_{solar}A) \times (24 \times 3600)$. Hence, we can evaluate the mass of PCM as follows:

$$m_{PCM} = \frac{0.4(150 \times 5) \times (24 \times 3600)}{2000(30 - 15) + 200E3 + 2000(79.8 - 30)}$$

resulting in $\boxed{m_{PCM} = 78.64 \text{ kg}}$

In this example, the total mass of storage material that is used in latent heat storage unit (i.e., PCM tank) is 60% lower than the sensible heat storage unit (*rock bed*).

***Open System.*** In an open system, along with heat and work, energy can also pass through the system boundaries by flow. That is, in addition to the simplified form of the closed system energy balance equation (Eq. 1.9), the flow energy terms, including enthalpy, kinetic energy and potential energy, should be included in the energy balance equation as follows:

$$\begin{aligned} \frac{dE_{cv}}{dt} = \left[\dot{W}_{in} - \dot{W}_{out}\right] + \left[\dot{Q}_{in} - \dot{Q}_{out}\right] + \sum\left[\dot{m}\left(h + \tfrac{V^2}{2} + gz\right)\right]_{in} \\ - \sum\left[\dot{m}\left(h + \tfrac{V^2}{2} + gz\right)\right]_{out} \quad \text{(in W)} \end{aligned} \tag{1.12}$$

By neglecting the kinetic and potential energy variations between the inlet and outlet sections of the system, the energy balance equation can be reduced to the following form:

$$\frac{dE_{cv}}{dt} = \left[\dot{W}_{in} - \dot{W}_{out}\right] + \left[\dot{Q}_{in} - \dot{Q}_{out}\right] + \sum\left[\dot{m}\,h\right]_{in} - \sum\left[\dot{m}\,h\right]_{out} \quad \text{(in W)} \tag{1.13}$$

**Example 1.5**

In a cooling system, there is an insulated pump, with an efficiency of 50%, which consumes 2 kW power to circulate the water inside the insulated pipeline. If the temperature at the inlet of the pipeline is 10 °C, determine the exit temperature for the mass flow rate of 1.5 kg/s. Vary the pump efficiency from 50% to 90% and discuss the results.

(continued)

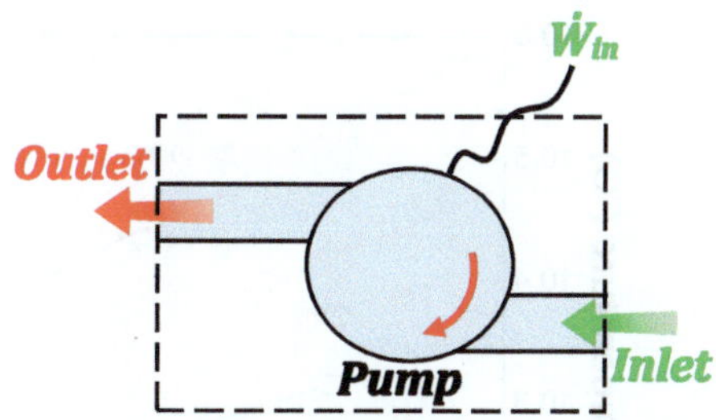


**Fig. 1.6** Energy balance for the pump

**Example 1.5** (continued)
**Solution:**

The pump's efficiency is defined as follows: $\eta_{pump} = \dot{W}_{flow}/\dot{W}_{in}$. Here the denominator is the power consumption of the pump, and the numerator is the portion of the supplied electrical power that is transferred to the liquid. In the current system, the pump can only transfer 50% of the supplied electrical power to the liquid. The rest of the consumed power will heat up the working fluid. The pipeline is completely insulated, and the system is working under steady-state condition (Fig. 1.6). Under these circumstances, the balance equation (Eq. 1.13) can be reduced to

$$0 = 0.5\dot{W}_{in} + \left[\dot{m}\,h\right]_{in} - \left[\dot{m}\,h\right]_{out}$$

For an incompressible fluid flow, water, the enthalpy variation between the inlet and outlet sections of the system can be determined by Eq. (1.5):

$$0.5\dot{W}_{in} = \dot{m}\,c(T_{out} - T_{in}) \text{ or}$$

$$T_{out} = \frac{0.5 \times 2000}{1.5 \times 4180} + 10 \quad \text{which becomes} \quad \boxed{T_{out} = 10.16^{\circ}\mathrm{C}}$$

If we vary the pump efficiency from 50% to 90% and repeat the same solution procedure that is given above, we can easily obtain the variation of outlet temperature against the pump efficiency as given in Fig. 1.7. Here we also varied the power consumption of the pump from 1 kW to 2.5 kW. It is clear that for the current flow and working conditions, the temperature variation between the inlet and the outlet sections of the pump is negligible.

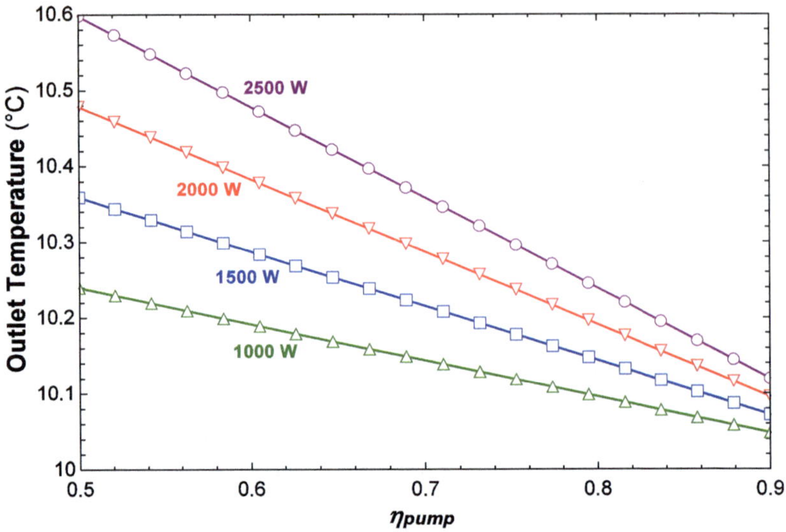


**Fig. 1.7** Outlet temperature of the pump as a function of power consumption and pump efficiency

### *1.6.3 Entropy Balance*

Entropy ($S = ms$ in J/K) is an extensive property, and it is related to the microscopic disorder of the system (Cengel and Boles 2010). There are two fundamental equations in thermodynamics to define the relationship between specific entropy ($s$ in J/kgK) and specific internal energy or specific enthalpy:

$$Tds = du + pdv \tag{1.14a}$$

$$Tds = dh - vdp \tag{1.14b}$$

From these equations, entropy variation of a system can be obtained as follows:

$$s_2 - s_1 = \int_{T_1}^{T_2} c_v(T)\frac{dT}{T} + R\ln\frac{v_2}{v_1} \quad \text{and} \quad s_2 - s_1 = \int_{T_1}^{T_2} c_p(T)\frac{dT}{T} - R\ln\frac{p_2}{p_1} \tag{1.15}$$

For incompressible substances with constant specific heat, entropy variation is defined as

$$s_2 - s_1 = c\ln\frac{T_2}{T_1} \tag{1.16}$$

In the case of phase change, i.e., melting or boiling, of an incompressible substance at a constant temperature, entropy variation should be written as

$$s_2 - s_1 = \frac{h_{sf}}{T_{melting}} \quad \text{or} \quad s_2 - s_1 = \frac{h_{fg}}{T_{boiling}} \tag{1.17}$$

where $h_{sf}$ and $h_{fg}$ indicate the latent heats for solid-liquid and liquid-gas phase change processes, respectively. The entropy balance equation for an open system is given in the following form:

$$\frac{dS_{cv}}{dt} = \sum_i \frac{\dot{Q}_i}{T_{b,i}} + \sum [\dot{m}\,s]_{in} - \sum [\dot{m}\,s]_{out} + \dot{S}_{gen} \tag{1.18}$$

where the left-hand side of the equation is the rate of total entropy variation of the system. The first term on the right-hand side represents entropy transfer by heat transfer through the $i^{th}$ boundary. Notice that the rate of entropy transfer by heat is related to the surface temperature ($T_b$) of the relevant boundary. The second and third terms on the right-hand side are entropy transfer by fluid flow. The last term, on the other hand, is entropy generation. By referring to the Clausius statement, we know that entropy generation cannot be less than zero ($\dot{S}_{gen} \geq 0$).

**Example 1.6**

Consider a building wall with dimensions of 3.5 m × 6 m that has a thickness of 35 cm. The wall is exposed to the outdoor, and under steady-state conditions, the temperatures on the inner and outer surfaces of the wall are measured to be 25 °C and 0 °C, respectively. If the rate of heat transfer through the wall is 1000 W, determine the rate of entropy generation within the wall.

**Solution:**

The balance equation for entropy (Eq. 1.18) is reduced into the following form for a system in which there is no flow across the boundaries at steady-state (Fig. 1.8):

$$0 = \sum_i \frac{\dot{Q}_i}{T_{b,i}} + \dot{S}_{gen}$$

(continued)

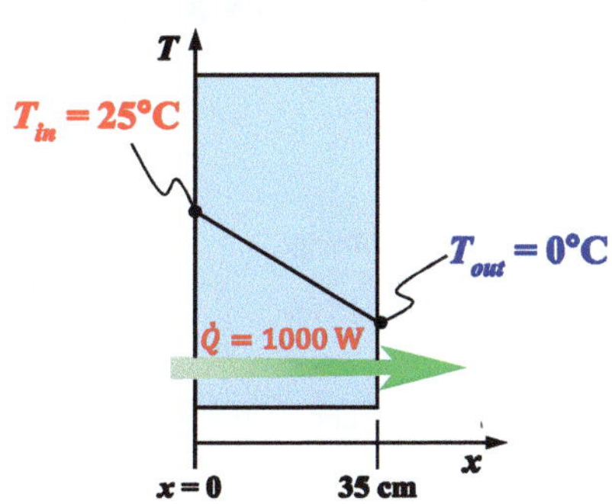


**Fig. 1.8** Steady-state heat transfer through a building wall

**Example 1.6** (continued)

The entropy transfer by heat on the inner and outer surfaces of the wall can be defined as the rate of heat transfer through the wall and the corresponding surface temperatures as

$$0 = \left(\frac{\dot{Q}}{T_{in}}\right) - \left(\frac{\dot{Q}}{T_{out}}\right) + \dot{S}_{gen} \quad \text{or} \quad 0 = \left(\frac{1000}{298}\right) - \left(\frac{1000}{273}\right) + \dot{S}_{gen}$$

The rate of entropy generation within the wall is obtained as

$$\boxed{\dot{S}_{gen} = 0.307\,\mathrm{W/K}}$$

## 1.6.4 Exergy Balance

Exergy ($Ex$ in Joule), or the availability, is the maximum theoretical work that can be obtained from a system when it passes to the dead state (see Fig. 1.9). It means that in the dead state, the exergy of a system is zero. Unlike energy, exergy is not conserved in all processes; rather, exergy of a real system reduces throughout the process due to irreversibilities. In general, the exergy balance equation is defined as

$$\frac{dEx_{cv}}{dt} = \sum \dot{E}x_{in} - \sum \dot{E}x_{out} - \dot{E}x_{dest} \tag{1.19}$$

The term on the left-hand side is the exergy variation of the system and is defined in terms of the difference between a system's properties at a given state and the dead state conditions:

$$Ex_{cv} = (U - U_0) + P_0(\forall - \forall_0) - T_0(S - S_0) + \frac{1}{2}mV^2 + mgz \tag{1.20}$$

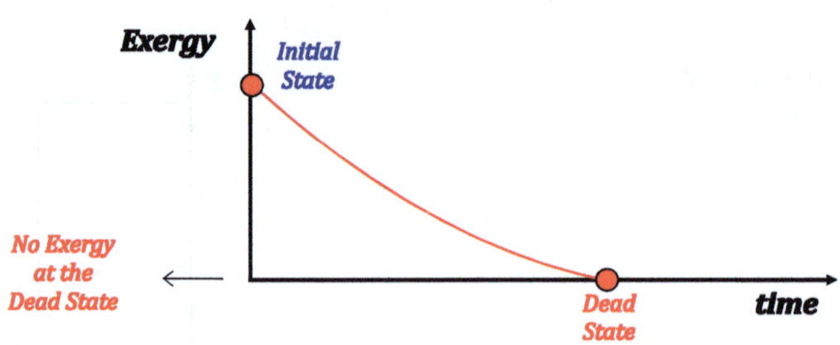


**Fig. 1.9** Exergy and dead state

**Table 1.2** Differences between energy and exergy concepts

| Energy concept | Exergy concept |
|---|---|
| Depends on the system properties or energy interactions with the surroundings and it is independent of environmental parameters | Depends on both the system properties and energy interactions with the surroundings and the environment parameters |
| Has a non-zero value (due to the Einstein's $mc^2$ equation) | If the system is in equilibrium (thermal + mechanical + chemical and so on) with the surroundings, exergy of the system is zero |
| In energetic analyses, the first law of thermodynamics should be satisfied | In exergetic analysis both first and second laws of thermodynamics should be satisfied |
| Conserved in all processes | Mostly destroyed in real processes |

Adapted from Dincer and Rosen (2011)

where subscript "0" denotes the system properties that are evaluated at the dead state (see Fig. 1.9). $U$ is the internal energy (in J), $P_O$ is the pressure at the dead state (in Pa), $\forall$ is the volume (in m$^3$), $T_O$ is the dead state temperature, and $S$ is the entropy (J/K). $V$ (in m/s) and $z$ (in m) are the velocity and elevation of the system, respectively. Exergy has the same unit as energy, so we can compare the energy and exergy concepts, as given in Table 1.2, to understand the differences between the first law and second law analyses in thermodynamics.

The exergy balance equation for an open system is given in the following form:

$$\frac{dEx_{cv}}{dt} = \underbrace{\sum \dot{E}x_{heat}}_{\substack{\text{Exergy Transfer}\\ \text{by HEAT}}} + \underbrace{\sum \dot{E}x_{work}}_{\substack{\text{Exergy Transfer}\\ \text{by WORK}}} + \underbrace{\sum \left[\dot{m}ex_{flow}\right]_{in} - \sum \left[\dot{m}ex_{flow}\right]_{out}}_{\substack{\text{Exergy Transfer}\\ \text{by FLUID FLOW}}} - \underbrace{\dot{E}x_{dest}}_{\substack{\text{Exergy}\\ \text{DESTRUCTION}}} \tag{1.21}$$

Each term in Eq. (1.21) is defined as

$$\dot{E}x_{\text{heat}} = \left(1 - \frac{T_0}{T}\right)\dot{Q} \tag{1.22}$$

$$\dot{E}x_{\text{work}} = \begin{cases} \dot{W} - P_0 d\forall/dt & \rightarrow \text{Boundary Work} \\ \dot{W} & \rightarrow \text{Other forms of Work} \end{cases} \tag{1.23}$$

$$ex_{flow,in} = (h_{in} - h_0) - T_0(s_{in} - s_0) + \frac{1}{2}V_{in}^2 + gz_{in} \tag{1.24a}$$

$$ex_{flow,out} = (h_{out} - h_0) - T_0(s_{out} - s_0) + \frac{1}{2}V_{out}^2 + gz_{out} \tag{1.24b}$$

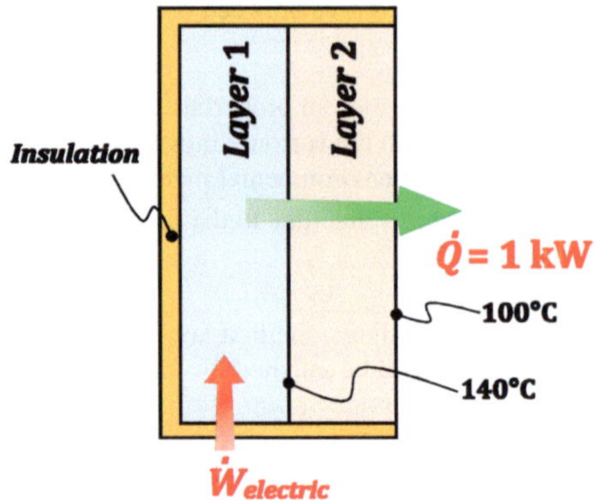


**Fig. 1.10** The steady-state heat transfer through a multilayer block

The exergy destruction can be determined from the balance equation (Eq. 1.21) if each of the terms for the system is calculated. As entropy generation is related to the internal/external irreversibilities of the system, the exergy destruction can also be evaluated in terms of entropy generation:

$$\dot{E}x_{dest} = T_0 \dot{S}_{gen} \tag{1.25}$$

**Example 1.7**
An electric heater is used in a multilayer storage tank to charge high-temperature thermal energy within the heavy mass. Assume that under steady-state condition, the supplied electrical energy passes through the layers of the tank. The rate of heat transfer and the interface temperatures are given in Fig. 1.10. Evaluate the exergy destructions in each layer of the tank and the overall exergy destruction within the composite block.

**Solution:**
The balance equation for exergy (Eq. 1.21) is reduced to the following form for a system in which there is no flow across the boundaries and steady-state:

$$0 = \sum \dot{E}x_{heat} + \sum \dot{E}x_{work} - \dot{E}x_{dest}$$

In Layer 1, the electrical input that is supplied from the heater turns into heat and passes through Layer 2. The exergy balance for Layer 1 includes the exergy terms associated with heat transfer and work:

$$0 = \dot{W}_{electric} - \left(1 - \frac{T_0}{T}\right)\dot{Q} - \dot{E}x_{dest,layer1}$$

(continued)

**Example 1.7** (continued)

Notice that the electrical input increases the exergy and the heat transfer from the system reduces the exergy. If the dead state temperature is assumed to be 300 K, the rate of exergy destruction is evaluated as

$$0 = 1000 - \left(1 - \frac{300}{413}\right)1000 - \dot{E}x_{dest,layer1}$$

resulting in $\boxed{\dot{E}x_{dest,layer1} = 726.4 \text{ kW}}$

For Layer 2, the exergy balance can be written to determine the rate of exergy destruction:

$$0 = \left(1 - \frac{300}{413}\right)1000 - \left(1 - \frac{300}{373}\right)1000 - \dot{E}x_{dest,layer2} \quad \text{and hence}$$

$$\boxed{\dot{E}x_{dest,layer2} = 77.9 \text{ kW}}$$

The total exergy destruction rate for the multilayer storage system is then calculated as

$$\boxed{\dot{E}x_{dest,system} = 804.3 \text{ kW}}$$

Alternatively, the total exergy destruction rate of the system can be evaluated by defining an overall balance equation. For the overall system, electrical power input passes through the system boundary as work, and the system loses heat through the outer surface, which is at 100 °C. Thus, the balance equation could be written as

$$0 = 1000 - \left(1 - \frac{300}{373}\right)1000 - \dot{E}x_{dest,system}$$

resulting in $\boxed{\dot{E}x_{dest,system} = 804.3 \text{ kW}}$

## 1.7 Heat Transfer Mechanisms

Heat transfer deals with the evaluation of *temperature distribution* and *rate of heat transfer* under steady-state or transient conditions. As shown in Fig. 1.11, there are three basic heat transfer mechanisms: *conduction*, *convection*, and *radiation*.

Even though in most real-world applications conductive, convective, and radiative heat transfers take place simultaneously, to develop mathematical models

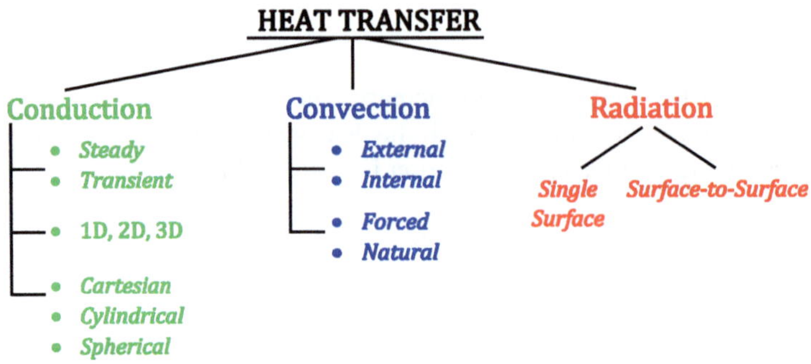


**Fig. 1.11** Mechanisms of heat transfer

with acceptable accuracy, engineers mostly make reasonable simplifications. Heat conduction problems are relatively simple for 1D and steady-state conditions. Heat conduction problems become complicated as the system becomes multidimensional and transient. The analytical solutions are limited for certain types of geometries and boundary conditions in transient and multidimensional situations. In thermal energy storage (TES) units, the heat transfer process within the storage medium, either solid or liquid, is commonly resolved by reducing the problem in transient and multidimensional heat conduction. Convective heat transfer, on the other hand, is characterized by considering the geometry (*internal/external flow*) and the formation mechanism (*forced/natural*). In TES units, thermal energy is commonly transported with working fluids, and the selection of the appropriate type of heat transfer coefficient has great importance to build a realistic mathematical model. The radiative heat transfer inside the TES units (surface-to-surface) is commonly neglected as the temperature difference and surface-to-surface interactions within the storage units are very small. Besides, the radiative heat exchange is quite important for the external surfaces of the TES units, such as receivers, tanks, and pipelines, which are considered as a single surface. Here we will only discuss the fundamentals of heat transfer mechanisms; for further information one can refer to the relevant textbooks (e.g., Cengel and Ghajar 2014).

### 1.7.1 Heat Conduction

Conductive heat transfer can take place inside both solids and fluids, but the driving mechanism may differ depending on the type of medium. Within a solid material, the heat conduction (also known as *diffusion*) occurs by the vibrations of the molecules. In the case of metals, the conductive heat transfer is primarily related to the mobility of the free electrons. Regardless of the driving mechanism, the heat conduction within the material is obtained from the Fourier's law. For a planar geometry (*see* Fig. 1.12), the Fourier's law is defined as

**Fig. 1.12** 1D heat conduction within a planar wall

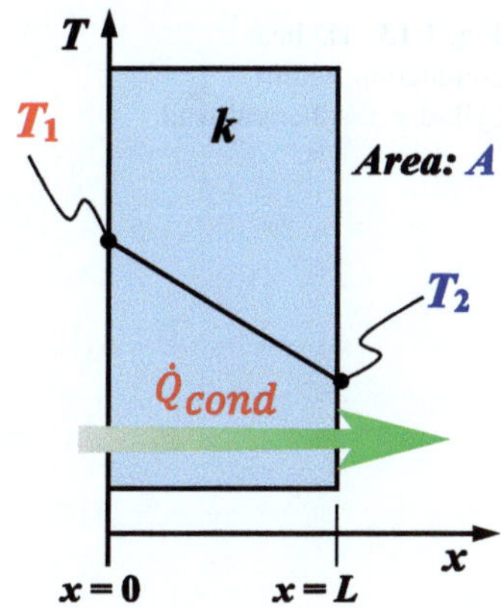


$$\dot{Q}_{cond} = -kA\frac{dT}{dx} \tag{1.26}$$

while $k$ is the thermal conductivity (in W/mK), $A$ is the cross-sectional area (in $m^2$) that is perpendicular to the heat transfer vector, and $dT/dx$ (in K/m) is the temperature gradient along the $x$-direction. Note that while the temperature is a scalar quantity, the rate of heat transfer is a vector and it has both a magnitude and a direction. As the direction of the heat transfer vector is along the decreasing temperature ($dT/dx < 0$) direction, to obtain the correct direction of heat flow, the negative sign is intentionally used in Fourier's law. Let's integrate the differential equation for constant thermal conductivity ($k = const.$) and a planar wall ($A = const.$) to obtain the algebraic form of Fourier's law:

$$\dot{Q}_{cond}\int_{x=0}^{L} dx = -kA\int_{T=T_1}^{T_2} dT \quad \text{or} \quad \dot{Q}_{cond} = kA\frac{T_1 - T_2}{L} \tag{1.27}$$

Equation (1.27) is limited to calculate the rate of heat transfer through a planar geometry. In the case of cylindrical or spherical coordinates, heat transfer through annular walls with constant temperatures at the inner and outer radii (*see* Fig. 1.13) can be calculated from the following equations:

$$\textit{Cylindrical wall}: \dot{Q}_{cond,\, cylinder} = 2\pi kL\frac{T_1 - T_2}{\ln r_2/r_1} \tag{1.28}$$

where $L$ is the length of the cylinder. Similarly, the heat rate from a spherical wall can be evaluated from

$$\textit{Spherical wall}: \dot{Q}_{cond, sphere} = 4\pi k\frac{T_1 - T_2}{1/r_1 - 1/r_2} \tag{1.29}$$

Equations (1.28) and (1.29) can be simplified into Eq. (1.27) if the wall thickness of cylinder or sphere is small enough regarding the radius of the sphere or cylinder.

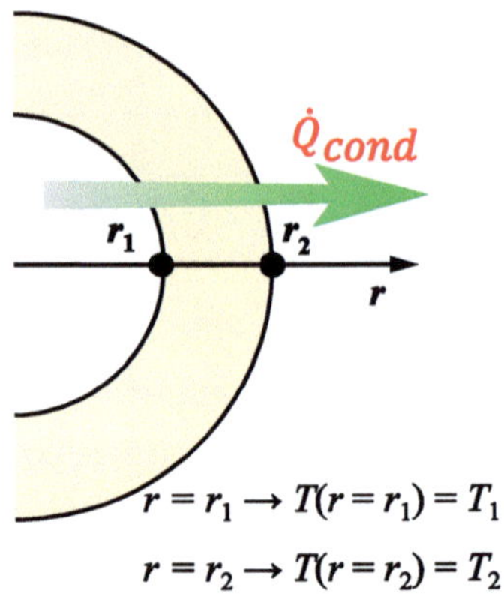


**Fig. 1.13** 1D heat conduction within cylindrical/spherical wall

**Table 1.3** Thermal conductivities and diffusivities of commonly used solid and liquid storage mediums

| Material | | Thermal conductivity (W/m$^2$K) | Thermal diffusivity (m$^2$/s) |
|---|---|---|---|
| Solid storage medium | Sand-rock minerals | 1.0 | 4.52E-07 |
| | Reinforced concrete | 1.5 | 8.02E-07 |
| | Cast iron | 37.0 | 9.18E-06 |
| | NaCl | 7.0 | 3.81E-06 |
| | Cast steel | 40.0 | 8.55E-06 |
| | Silica fire bricks | 1.5 | 8.24E-07 |
| | Magnesia fire bricks | 5.0 | 1.45E-06 |
| Liquid storage medium | Water | 0.58 | 1.39E-07 |
| | Mineral oil | 0.12 | 5.99E-08 |
| | Synthetic oil | 0.11 | 5.31E-08 |
| | Silicone oil | 0.10 | 5.29E-08 |
| | Nitrite salts | 0.57 | 2.08E-07 |
| | Liquid sodium | 71.0 | 6.43E-05 |
| | Nitrate salts | 0.52 | 1.74E-07 |
| | Carbonate salts | 2.0 | 5.29E-07 |

Adapted from Tian and Zhao (2013)

The spanwise or timewise temperature variation in a conductive medium depends on the boundary conditions and the thermal properties of the medium. For steady and transient heat conduction problems, the thermal behavior of the medium is related to the thermal conductivity and diffusivity of the material, respectively.

***Thermal Conductivity.*** Thermal conductivity is a measure of material to characterize the ability to conduct heat. Depending on the phase (i.e., solid, liquid, and gas) and type (i.e., pure, mixture, eutectic, or alloy), the thermal conductivity values may vary from an order of 0.01 to 10,000 W/mK.

***Thermal Diffusivity.*** Thermal diffusivity is a combined thermal property and characterizes the dynamic thermal response (*transient*) of a material to a sudden change. Thermal diffusivity ($\alpha$) of a material is the ratio of thermal conductivity ($k$) to volumetric heat capacity ($C = \rho c$) expressed as

$$\alpha = \frac{k}{\rho c} \quad \left(\text{in m}^2/\text{s}\right) \tag{1.30}$$

In Table 1.3, the thermal conductivities and diffusivities of commonly used materials are given.

**Example 1.8**

In latent heat thermal energy storage (LHTES) systems, phase change materials (PCMs) are mostly encapsulated inside spherical capsules. Assume that PVC ($k_{PVC} = 0.2$ W/mK) spherical balls with an inner diameter of 10 cm that are filled with paraffin ($\rho = 772$ kg/m$^3$, $h_{sf} = 200$ kJ/kg) with a melting temperature of 30 °C are heated by forced convection around the spherical capsules. During the melting period, the outer wall of the sphere is measured to be 40 °C, and the inner wall is at melting temperature. If the wall thickness of the material is 2 mm, (a) determine the rate of heat transfer, (b) determine how long it will take to achieve complete melting, (c) determine the rate of exergy destruction within the sphere wall, (d) vary the outer surface temperature of the sphere, and discuss the results according to the energetic and exergetic aspects. Assume that the environment is at 25 °C.

(Hint: *For this introductory example, one may assume that the inner and outer wall temperatures of the sphere are constant during the melting process.*)

**Solution:**

(a) The rate of heat transfer can be calculated from Eq. (1.29):

$$\dot{Q}_{cond,sphere} = 4\pi \times 0.2 \frac{40 - 30}{1/0.05 - 1/0.052}$$

which becomes $\boxed{\dot{Q}_{cond,sphere} = 32.67 \text{ W}}$

(b) We combine Eqs. (1.7) and (1.10) to obtain the following form of the energy balance equation:

$$\frac{d(mh_{sf})}{dt} = \dot{Q}_{in} \text{ or } \frac{d(mh_{sf})}{dt} = \dot{Q}_{cond,sphere}$$

If the rate of heat transfer is constant throughout the melting process, we can calculate the time for complete melting as

(continued)

**Example 1.8** (continued)

$$\Delta t_{melting} = \frac{\left(4/3\pi r_{sphere}^3 \rho\right) h_{sf}}{\dot{Q}_{cond,sphere}} \quad \text{resulting in} \quad \boxed{\Delta t_{melting} = 2474 \text{ s}}$$

(c) The exergy balance equation (Eq. 1.21) can be reduced to the following form for a closed system which undergoes a steady-state heat transfer:

$$0 = \sum \dot{E}x_{heat} + \dot{E}x_{dest}$$

The exergy by heat transfer (Eq. 1.22) has two terms for the heat transfers entering and leaving the spherical wall:

$$\dot{E}x_{heat} = \left(1 - \frac{T_0}{T_{outer}}\right)\dot{Q}_{cond,sphere} - \left(1 - \frac{T_0}{T_{inner}}\right)\dot{Q}_{cond,sphere} \quad \text{resulting in}$$

$$\boxed{\dot{E}x_{heat} = 1.026 \text{ W}}$$

We can end up with the same result if we consider Eq. (1.25). To do so, we can go through the entropy balance equation (Eq. 1.18) to compute entropy generation as

$$0 = \sum_i \frac{\dot{Q}_i}{T_{b,i}} + \dot{S}_{gen}$$

(d) We have varied the temperature difference between the melting temperature of the PCM and the outer wall of the sphere from 0 K to 20 K and obtained the required time for complete melting and also exergy destruction due to heat transfer. Figure 1.14 shows the variations of required time for complete melting and the rate of exergy destruction. The time for complete melting is obtained as 8373 s and 605 s for temperature differences of 3 K and 41 K, respectively. The speed of the melting process significantly drops as the temperature of the outer wall of the sphere improves.

However, in the point of view of the second law of thermodynamics, by increasing the temperature difference, we lose the available energy. As can be seen from Fig. 1.14, the rate of exergy destruction gradually increases with increasing temperature difference.

Considering the current findings in this simple problem, we can see that the perspective of the first law of thermodynamics is not enough to decide whether the design of a system is *efficient*. In the next chapters, we will discuss the methods that are used to optimize heat storage systems by considering first law and second law concepts simultaneously.

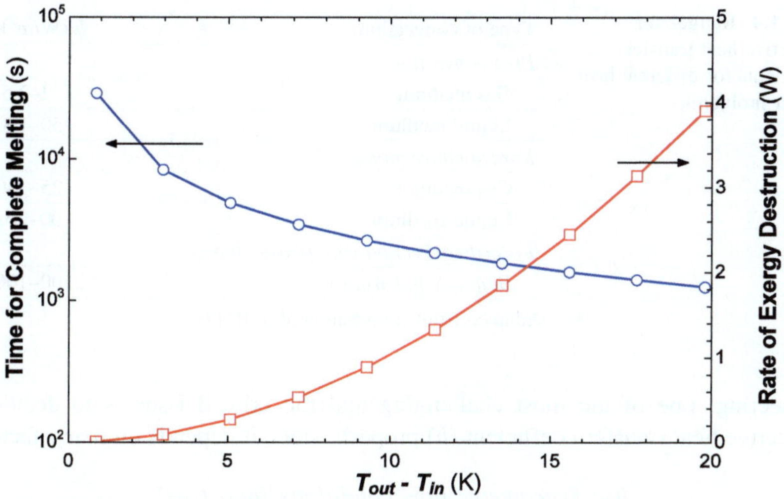


**Fig. 1.14** Variations of melting time and rate of exergy destruction as a function of temperature difference

**Fig. 1.15** Convective heat transfer from a flat plate

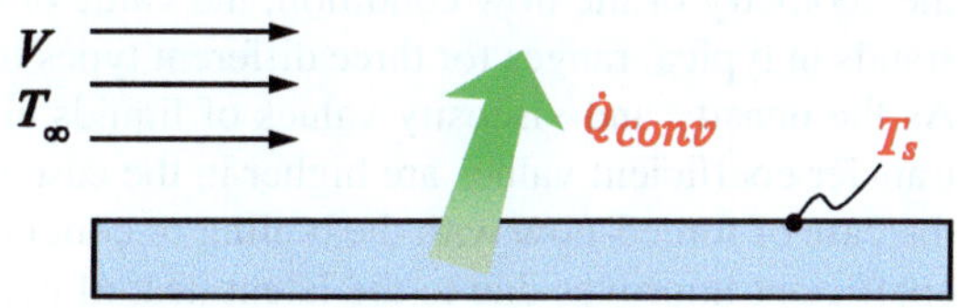


## *1.7.2 Convective Heat Transfer*

In engineering applications, we experience convection when a fluid flows on a solid body. There are two forms of heat convection: *free* and *forced*. In free (or natural) convection, the fluid moves due to the density gradient without any moving components, such as a pump or a fan. The density gradient is a result of temperature distribution within the fluid medium. In the forced mode of convection, on the other hand, there should be a fan or pump to push the fluid medium. Even though the driving mechanism differs, the resultant form of the rate equation is written in the following form for each mode of convective heat transfer (Fig. 1.15):

$$\dot{Q}_{conv} = hA(T_s - T_\infty) \tag{1.30}$$

where $h$ is the convective heat transfer coefficient (in W/m$^2$K), $A$ is the heat transfer surface area (in m$^2$) between solid and fluid, $T_s$ is the solid surface temperature, $V$ is the free stream velocity, and $T_\infty$ is the free stream temperature. In thermal

**Table 1.4** Ranges for convective heat transfer coefficients for different heat transfer problems

| Type of convection | $h$ (W/m$^2$K) |
|---|---|
| *Free convection* | |
| Gas medium | 2–2.5 |
| Liquid medium | 50–1000 |
| *Forced convection* | |
| Gas medium | 25–250 |
| Liquid medium | 100–20,000 |
| *Forced convection with phase change* | |
| *Boiling/condensation* | 2500–100,000 |

Adapted from Bergman et al. (2011)

engineering, one of the most challenging and the critical issue is to decide the convective heat transfer coefficient ($h$) properly since it depends on many factors:

$$h = f(geometry, flow\ conditions, fluid\ type)$$

The convective heat transfer coefficient hardly depends on the type of convection: *free*, *forced*, or *convection with phase change*. Experience shows that regardless of the geometry or the flow condition, the value of convective heat transfer coefficient stands in typical ranges for three different types of convection, as given in Table 1.4. As the density and viscosity values of liquids are quite higher than gases, the heat transfer coefficient values are higher in the case of the liquid medium. Moreover, in the case of forced-flow with the boiling or condensation, the convective heat transfer coefficient improves due to the latent heat of evaporation.

Convective heat transfer coefficient is evaluated from the dimensionless Nusselt (Nu) number. Nusselt number is defined as the ratio of convective heat transfer to the conduction heat transfer in a medium:

$$Nu = \frac{q_{convection}}{q_{conduction}} \tag{1.31}$$

From this definition, one can deduce that for a pure heat conduction problem, the Nusselt number is unity. Nusselt number is also defined in terms of the dimensionless temperature gradient on a solid surface. Both definitions are used to generalize the convective heat transfer. Nusselt correlations/equations are defined in terms of Reynolds (Re) and Prandtl (Pr) numbers for forced convection. Reynolds number characterizes the flowing fluid and is defined as the ratio of inertial forces to the viscous forces. In a generalized form, Reynolds is defined as follows:

$$\mathrm{Re}_{L_c} = \rho \frac{L_c V}{\mu} \tag{1.32}$$

where $L_c$ (in m) is the characteristic length of the solid body and $V$ (in m/s) is the bulk velocity of flowing fluid. $\rho$ (in kg/m$^3$) and $\mu$ (in kg/ms) stand for the density and dynamic viscosity of the fluid, respectively. For instance, for flow over a flat plate, the critical Reynolds number is $5 \times 10^5$ for the transition from the laminar to the turbulent boundary layer. On the other hand, for flow inside a tube, the critical Reynolds number is mostly assumed in the range of 2000 and 10,000 (Cengel and Ghajar 2014). In the case of natural convection, the Nusselt number is defined in terms of the dimensionless Grashof (Gr) and Prandtl number. Grashof is the ratio of the buoyancy force to viscous forces and is defined as follows:

$$Gr = \frac{g\beta(T_s - T_\infty)L_c^3}{\vartheta^2} \tag{1.33}$$

where $\vartheta$ (in $m^2/s$) is the kinematic viscosity ($=\mu/\rho$) and $\beta$ (in 1/K) is the volumetric thermal expansion coefficient. Thermal expansion coefficient is defined as the derivative of the density regarding the temperature:

$$\beta = \frac{1}{\rho}\frac{\partial \rho}{\partial T}\bigg|_T \tag{1.34}$$

For ideal gases, the volumetric thermal expansion coefficient is simply defined as the inverse of the mean fluid temperature. In a thermal system, the combined forced and natural convection effects could be observed. To determine the importance of forced and natural convection effects on the heat transfer rate, the following criteria are considered:

*Forced convection is dominant:* $Gr/Re^2 >> 1$
*Natural convection is dominant:* $Gr/Re^2 << 1$
*Mixed convection:* $Gr/Re^2 \approx 1$

***Forced Convection.*** Numerous correlations are derived to determine the heat transfer coefficients for various geometries such as flow over plates, cylinders, tube bundles, flow inside circular pipes, or rectangular channels. To represent both the dependent and independent variables in a compact form, the correlations are represented in dimensionless form. The general form of convective heat transfer correlations is as follows:

$$Nu = \frac{hL_c}{k} = CRe^m Pr^n \tag{1.35}$$

where Nu is the Nusselt number, and $L_c$ is the characteristic length. $L_c$ is characterized by the geometry and can be the length of the plate (i.e., $L_c = L$ *for flow over the flat plate*) or diameter of a tube (i.e., $L_c = D$ *for flow over the circular tube or flow inside the circular tube*) depending on the type of heat transfer domain. Nusselt number is characterized as the ratio of the convection to the conduction. That is, the average Nusselt number on a surface should not be less than unity. *C*, *m*, and *n* are the empirical or numerical equation constants. Re and Pr correspond to the dimensionless Reynolds and Prandtl numbers, respectively. In Table 1.5 Nusselt correlations/equations are given for flow over a flat plate, a cylinder, and a sphere for a wide range of Reynolds and Prandtl numbers.

**Table 1.5** Nusselt correlations/equations for forced external flow

| **Correlation *or* Equation** |
|---|
| ***Correlations for Flat plate in External Flow*** 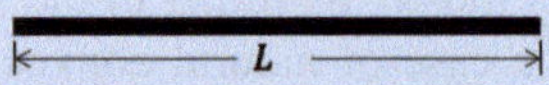 $\mathrm{Re} = \rho \frac{uL}{\mu}$ |
| $Nu = 0.332Re^{1/2}Pr^{1/3}$ ***for*** **laminar; Pr ≥ 0.6; local Nu; properties @ $T_{f,m}$** |
| $Nu = 0.664Re^{1/2}Pr^{1/3}$ ***for*** **laminar; Pr ≥ 0.6; average Nu; properties @ $T_{f,m}$** |
| $Nu = 0.565Re^{1/2}Pr^{1/2}$ ***for*** **turbulent ($Re \le 10^8$); Pr ≤ 0.05; local Nu; properties @ $T_{f,m}$** |
| $Nu = (0.037Re^{4/5} - 871)Pr^{1/3}$ ***for*** **mixed ($Re \le 10^8$); 0.6 < Pr < 60; average Nu; properties @ $T_{f,m}$** |
| ***Correlations for Circular cylinders in Cross-Flow***  $\mathrm{Re} = \rho \frac{uD}{\mu}$ |
| $Nu = cRe^{n}Pr^{1/3}$ ***for*** **$0.4 < Re < 4\times10^6$; Pr ≥ 0.7; average Nu; properties @ $T_{f,m}$**<br>***where***<br>$c = 0.989$ & $n = 0.330$ *for* $0.4 < Re < 4$<br>$c = 0.911$ & $n = 0.385$ *for* $4 < Re < 40$<br>$c = 0.683$ & $n = 0.466$ *for* $40 < Re < 4{,}000$<br>$c = 0.193$ & $n = 0.618$ *for* $4{,}000 < Re < 40{,}000$<br>$c = 0.027$ & $n = 0.805$ *for* $40{,}000 < Re < 400{,}000$ |
| $Nu = cRe^{s}(Pr_a/Pr_s)^{1/4}$ ***for*** **$1 < Re < 10^6$; 0.7 < Pr < 500; average Nu; properties @ $T_a$**<br>***where***<br>$c = 0.750$ & $n = 0.4$ *for* $1 < Re < 40$ — $s = 0.37$ *for* $Pr \le 10$<br>$c = 0.510$ & $n = 0.5$ *for* $40 < Re < 1{,}000$ — $s = 0.36$ *for* $Pr > 10$<br>$c = 0.260$ & $n = 0.6$ *for* $10^3 < Re < 2\times10^5$<br>$c = 0.076$ & $n = 0.7$ *for* $2\times10^5 < Re < 10^6$ |
| $Nu = 0.3 + [(0.62Re^{1/2}Pr^{1/3})/(1 + (0.4/Pr)^{2/3})^{1/4}][1 + (Re/28{,}200)^{5/8}]^{4/5}$<br>***for*** **RePr > 0.2, average Nu, @ $T_{f,m}$** |
| ***Correlations for Spheres in Cross-Flow***  $\mathrm{Re} = \rho \frac{uD}{\mu}$ |
| $Nu = 0.37Re^{0.6}Pr^{1/3}$ ***for*** **17 < Re < 70,000; average Nu; properties @ $T_{f,m}$** |
| $Nu = 2 + (0.4Re^{1/2} + 0.06Re^{2/3})Pr^{0.4}(\mu_a/\mu_s)^{1/4}$<br>***for*** **0.71 < Pr < 380; average Nu; properties @ $T_a$, $1 < (\mu_a/\mu_s) < 3.2$** |

Adapted from Dincer and Rosen (2011)

***Natural Convection.*** There is no external force acting on the fluid in natural convection. The fluid can freely move if there is density gradient that is caused by temperature variations within the fluid medium. Similar to forced convection, the heat transfer coefficient is characterized by using the dimensionless Nusselt number. The general form of convective heat transfer correlations is written as follows:

$$Nu = \frac{hL_c}{k} = CRa^m Pr^n \tag{1.36}$$

where Ra is the dimensionless Rayleigh number ($Ra = g\beta(T_s - T_\infty)L_c^3/\vartheta\alpha$),

$$Ra = GrPr \tag{1.37}$$

**Table 1.6** Nusselt correlations/equations for natural convection

| Correlation *or* Equation | |
|---|---|
| *General Definitions* | |
| $Nu = hL_c/k_f = cRa^n$ and $Ra = GrPr = g\beta(T_s - T_a)L_c/(\vartheta\alpha)$ | |
| where $n$ is ¼ for laminar and 1/3 for turbulent flow, | |
| $L_c$ is the height for vertical plates/pipes; diameter for horizontal pipes; radius of sphere | |
| properties @ $T_{f,m} = (T_s - T_a)/2$ | |
| *Correlations for Vertical plates (or Inclined plates $\theta \leq 60°$)* | |
| $Nu = [0.825 + 0.387Ra^{1/6}/(1 + (0.492/Pr)^{9/16})^{4/9}]^2$ for an entire range of Ra | |
| $Nu = 0.68 + 0.67Ra^{1/4}/(1 + (0.492/Pr)^{9/16})^{4/9}$ for $0 < Ra < 10^9$ | |
| *Correlations for Horizontal plates ($L_c = A_s/P$) As: Surface Area & P:Perimeter* | |
| For upper surface of heated plate and lower surface of cooled plate: | |
| $Nu = 0.54Ra^{1/4}$ for $10^4 \leq Ra \leq 10^7$ | |
| $Nu = 0.15Ra^{1/3}$ for $10^7 \leq Ra \leq 10^{11}$ | |
| For lower surface of heated plate and upper surface of cooled plate: | |
| $Nu = 0.27Ra^{1/4}$ for $10^5 \leq Ra \leq 10^{10}$ | |
| *Correlations for Horizontal cylinders* | |
| $Nu = hD/k = cRa^n$ | |
| where | |
| $c = 0.675$ and $n = 0.058$ for $10^{-10} < Ra < 10^{-2}$ | $c = 1.020$ and $n = 0.148$ for $10^{-2} < Ra < 10^2$ |
| $c = 0.850$ and $n = 0.188$ for $10^2 < Ra < 10^4$ | $c = 0.480$ and $n = 0.250$ for $10^4 < Ra < 10^7$ |
| $c = 0.125$ and $n = 0.333$ for $10^7 < Ra < 10^{12}$ | |
| $Nu = [0.60 + 0.387Ra^{1/6}/(1 + (0.559/Pr)^{9/16})^{8/27}]^2$ for an entire range of Ra | |
| *Correlations for Spheres* | |
| $Nu = 2 + 0.580Ra^{1/4}/(1 + (0.469/Pr)^{9/16})^{4/9}$ for $Pr \geq 0.7$ and $Ra \leq 10^{11}$ | |

Adapted from Dincer and Rosen (2011)

Gr number, Grashof, is the ratio of the buoyancy force to viscous forces. Nusselt correlations for natural convection on horizontal cylinders, spheres, and horizontal/vertical plates are listed in Table 1.6.

### *1.7.3 Radiative Heat Transfer*

Even though radiative heat transfer takes place inside both solid and fluid domains, in most engineering applications, we usually consider the radiative heat exchange between surfaces at different temperatures facing each other. Unlike the conductive or convective forms of heat transfer, radiative heat exchange can occur without requiring any medium between solid surfaces. It means that radiation occurs even in vacuum. Any matter can emit radiative energy if the temperature is higher than absolute zero. The Stefan-Boltzmann law defines the amount of radiation that is emitted from a blackbody as

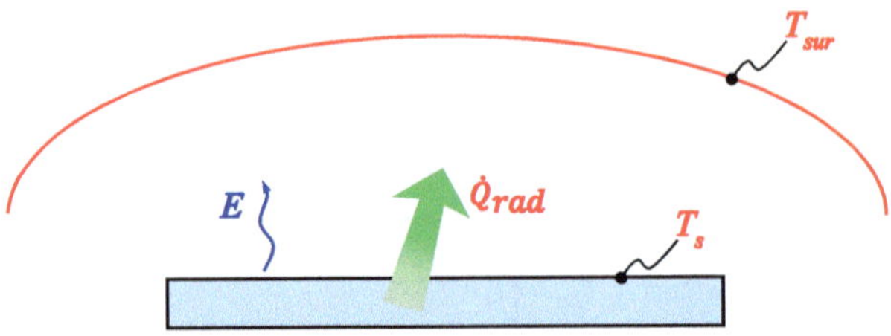


**Fig. 1.16** Radiative heat transfer between the surface and the surroundings

$$E_b = \sigma T_s^4 \tag{1.38}$$

where $\sigma$ is the Stefan-Boltzmann constant ($\sigma = 5.67 \times 10^{-8}$ W/m$^2$ K$^4$) and $T_s$ is the surface temperature in Kelvin. For a non-blackbody the emitted radiation should be defined in terms of the emissivity of the surface:

$$E = \varepsilon E_b = \varepsilon \sigma T_s^4 \tag{1.39}$$

where emissivity, $0 < \varepsilon \leq 1$, shows the ability of the surface to emit radiation. If an object is surrounded by a big surface, such as a furnace, a room, or the atmosphere, the net radiative heat exchange between the surface and the surroundings (see Fig. 1.16) can be defined as follows:

$$\dot{Q}_{rad} = \varepsilon \sigma A \left(T_s^4 - T_{sur}^4\right) \tag{1.40}$$

There are some other significant properties that characterize radiative heat transfer in objects: *absorptivity* ($\alpha$), *reflectivity* ($\rho$), and *transmissivity* ($\tau$).

For a system which is exposed to solar radiation, one should consider the absorptivity and transmissivity of properties of the materials to write down the correct energy balance equation. For the case of a solar collector, it is desired to have a transparent glass cover that permits solar radiation with very limited reflectivity or absorptivity. The air gap between the glass cover and the absorber acts as a thermal buffer and blocks the heat loss from the collector. The absorber of the collector, on the other hand, should possess high absorptivity to transfer the useful heat to the working fluid that flows inside the heat exchanger underneath the collector.

**Example 1.9**

Consider the vertical wall of a building that is exposed to a solar radiation of 800 W/m$^2$. Assume that the wall has a thickness of $W = 30$ cm and the thermal conductivity of the wall material is 5 W/mK. On the outer surface of the wall, the convective heat transfer coefficient of air is 25 W/m$^2$ K, and the outdoor temperature is 30 °C. The absorptivity and emissivity of the outer surface of the wall are given as $\alpha = \varepsilon = 0.8$. If the indoor wall temperature is $T_{s,in} = 20$ °C, determine the outer wall temperature ($T_{s,out}$). Assume that the sky temperature is −10 °C.

(continued)

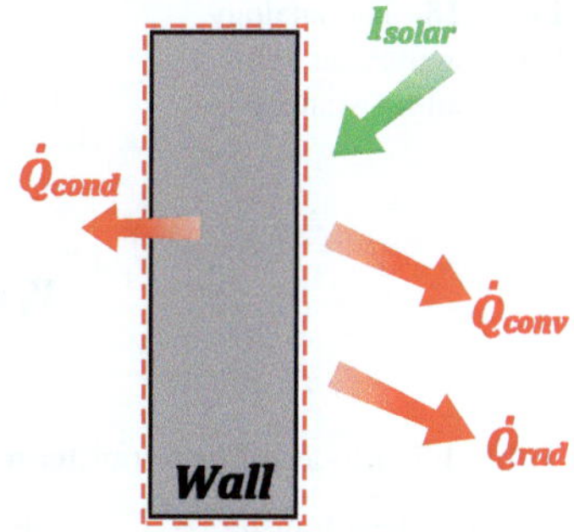


**Fig. 1.17** Energy balance for the building wall

**Example 1.9** (continued)
**Solution:**
Under steady-state condition, the energy balance equation for the wall can be written as follows:

$$\sum \dot{Q}_{in} = \sum \dot{Q}_{out}$$

The schematic diagram in Fig. 1.17 illustrates the heat transfer rates that are entering and leaving the wall. The conductive, convective, and radiative components can be written in the reduced balance equation that is given above:

$$(I_{solar}\alpha) = \dot{Q}_{cond} + \dot{Q}_{conv} + \dot{Q}_{rad}$$

Thus,

$$(I_{solar}\alpha) = k\frac{T_{s,out} - T_{s,in}}{W} + h(T_{s,out} - T_{\infty}) + \varepsilon\sigma\left(T_{s,out}^4 - T_{sky}^4\right)$$

We can determine the outer surface temperature as 31.2 °C by resolving the equation by following an iterative solution procedure. One of the following software can also be used to obtain the same result: MATLAB, MS Excel, or EES (Engineering Equation Solver).

### *1.7.4 The Concept of Thermal Resistance*

The complex heat transfer problems can be reduced into simple thermal resistance networks by using a concept similar to that used in electricity. The concept of thermal resistance is based on the analogy presented in Fig. 1.18.

We can define conductive heat transfer in terms of thermal resistance as follows:

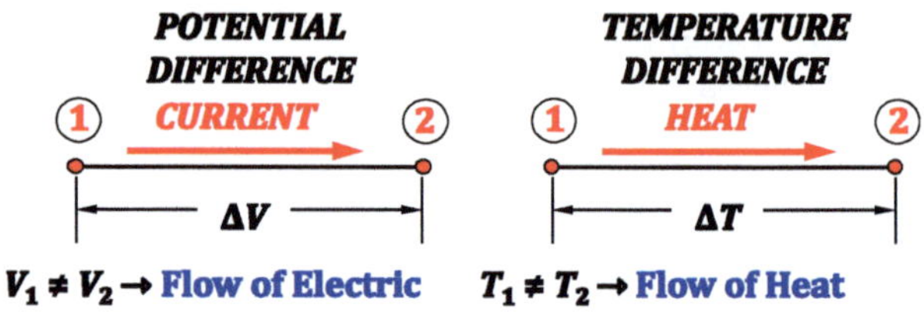


**Fig. 1.18** The analogy between thermal and electrical resistances

**Table 1.7** Modes of heat transfer and thermal resistances

| Mode of heat transfer | Rate of heat transfer (W) | Thermal resistance (K/W) |
|---|---|---|
| Conduction | $kA\frac{\Delta T}{L}$ | $\frac{L}{kA}$ |
| Convection | $hA\Delta T$ | $\frac{1}{hA}$ |
| Radiation | $\varepsilon\sigma A\Delta T$ | $\frac{1}{\varepsilon\sigma A\left(T_s^2\text{-}T_{sky}^2\right)\left(T_s+T_{sky}\right)}$ |

$$\text{Electrical Resistance: } I=\frac{\Delta V}{R_e}$$

$$\text{Heat Conduction: } \dot{Q}_{cond}=kA\frac{\Delta T}{L}\rightarrow\dot{Q}_{cond}=\frac{\Delta T}{^{L}/_{kA}}\Rightarrow\dot{Q}_{cond}=\frac{\Delta T}{R_{t,cond}}$$

where $R_e$ is the electrical resistance and $R_{t,cond}$ $=L/kA$ is the *thermal resistance for heat conduction*. In the case of convective heat transfer, the thermal resistance for convection can be obtained as

$$\text{Heat Convection: } \dot{Q}_{conv}=hA\Delta T\rightarrow\dot{Q}_{conv}=\frac{\Delta T}{^{1}/_{hA}}\Rightarrow\dot{Q}_{cond}=\frac{\Delta T}{R_{t,conv}}$$

where $R_{t,conv}$ $=1/hA$ is the *thermal resistance for convection*. The thermal resistance for radiative heat transfer can be evaluated as

$$\text{Heat Radiation: } \dot{Q}_{rad}=\varepsilon\sigma A\left(T_s^4-T_{sky}^4\right)\rightarrow\dot{Q}_{rad}=\frac{T_s-T_{sky}}{\frac{1}{\varepsilon\sigma A\left(T_s^2+T_{sky}^2\right)\left(T_s+T_{sky}\right)}}$$
$$\Rightarrow\dot{Q}_{rad}=\frac{\Delta T}{R_{t,rad}}$$

Here, the radiative heat transfer is alternatively defined in terms of the *radiation heat transfer coefficient* ($h_{rad}$)

$$\text{Heat Radiation: } \dot{Q}_{rad}=\varepsilon\sigma A\left(T_s^4-T_{sky}^4\right)\rightarrow\dot{Q}_{rad}=\frac{T_s-T_{sky}}{\frac{1}{h_{rad}A}}\Rightarrow\dot{Q}_{rad}=\frac{\Delta T}{R_{t,rad}}$$

where $h_{rad}=\varepsilon\sigma\left(T_s^2+T_{sky}^2\right)\left(T_s+T_{sky}\right)$. The *thermal resistance for radiation* is then defined as $R_{t,rad}$ $=1/h_{rad}A$. Table 1.7 summarizes the heat transfer mechanisms and the corresponding thermal resistances for each mode of heat transfer.

## 1.8 Closing Remarks

In this chapter, the focus is placed on the fundamental and basic aspects and concepts of two critical subjects, namely, thermodynamics and heat transfer. Numerous illustrative examples are presented to connect the basic concepts with some heat storage applications. This chapter is expected to serve as the base for the following chapters on various thermal energy storage systems and applications and their utilization in various sectors.

**Nomenclature**

| | |
|---|---|
| $a$ | Acceleration, m/s$^2$ |
| $A$ | Area, m$^2$ |
| $c$ | Specific heat, J/kgK |
| $c_p$ | Specific heat at constant pressure, J/kgK |
| $c_v$ | Specific heat at constant volume, J/kgK |
| $D$ | Diameter, m |
| $E$ | Energy, J |
| $Ex$ | Exergy, J |
| $F$ | Force, N |
| $g$ | Gravitational acceleration, m/s$^2$ |
| Gr | Grashof number |
| $h$ | Specific enthalpy, J/kg |
| $k$ | Thermal conductivity, W/mK |
| $L$ | Length, m |
| $m$ | Mass, kg |
| $\dot{m}$ | Mass flow rate, kg/s |
| Nu | Nusselt number |
| $p$, $P$ | Pressure, Pa, *or* power, W |
| $r$ | Radial position, m |
| $R$ | Ideal gas constant, J/kgK, or thermal resistance (K/W) |
| $\dot{Q}$ | Rate of heat transfer, W |
| Ra | Rayleigh number |
| Re | Reynolds number |
| $s$ | Specific entropy, J/kgK |
| $S$ | Entropy, J/K |
| $\dot{S}$ | Rate of entropy, W/K |
| $T$ | Temperature, K or °C |
| $u$ | Specific internal energy, J/kg |
| $U$ | Internal energy, J |
| $v$ | Specific volume, m$^3$/kg |
| $V$ | Velocity, m/s |
| $\forall$ | Volume, m$^3$ |
| $\dot{\forall}$ | Volumetric flow rate, m$^3$/s |
| $t$ | Time, s |
| $\dot{W}$ | Power, W |

| | |
|---|---|
| $x$ | Position, m |
| $z$ | Height, m |

**Greek Letters**

| | |
|---|---|
| $\alpha$ | Thermal diffusivity, m$^2$/s, or absorptivity |
| $\beta$ | Thermal expansion coefficient, 1/K |
| $\varepsilon$ | Emissivity |
| $\eta$ | Efficiency |
| $\mu$ | Dynamic viscosity, kg/ms |
| $\vartheta$ | Kinematic viscosity, m$^2$/s |
| $\rho$ | Density, kg/m$^3$, or reflectivity |
| $\sigma$ | Stefan-Boltzmann constant, W/m$^2$K$^4$ |
| $\tau$ | Transmissivity |

**Subscripts**

| | |
|---|---|
| 0 | Dead state |
| $b$ | Boundary or blackbody |
| $c$ | Characteristic |
| $cond$ | Conduction |
| $conv$ | Convection |
| $cv$ | Control volume |
| $dest$ | Destruction |
| $fg$ | Liquid to gas |
| $gen$ | Generation |
| $i$ | Index |
| $in$ | Inlet |
| $int$ | Initial |
| $rad$ | Radiation |
| $s$ | Surface |
| $out$ | Outlet |
| $sf$ | Solid to liquid |

# References


Bergman, T. L., Lavine, A. S., Incropera, F. P., & DeWitt, D. P. (2011). *Fundamentals of heat and mass transfer*. Hoboken: Wiley.

Cengel, Y. A., & Boles, M. A. (2010). *Thermodynamics: An engineering approach* (7th ed.). New York: McGraw Hill.

Cengel, Y. A., & Ghajar, A. J. (2014). *Heat and mass transfer: Fundamentals and applications* (5th ed.). New York: McGraw Hill.

Dincer, I., & Rosen, M. (2011). *Thermal energy storage: Systems and applications* (2nd ed.). Hoboken: Wiley.

Hepbasli, A., Biyik, E., Ekren, O., Gunerhan, H., & Araz, M. (2014). A key review of wastewater source heat pump (WWSHP) systems. *Energy Conversion and Management, 88*, 700–722.

Tian, Y., & Zhao, C. Y. (2013). A review of solar collectors and thermal energy storage in solar thermal applications. *Applied Energy, 104*, 538–553.

# Chapter 2
# Energy Storage Methods

## 2.1 Introduction

Energy storage is a key technology not only to maintain the sustainability of renewable energy resources but also to remove the mismatch between demand and supply. In the current chapter, first a global projection on the energy consumption is introduced and then the major advantages of energy storage techniques are discussed. The objective of this chapter is to present an overview of the fundamentals of *chemical*, *electrochemical*, *electrical*, *mechanical*, and *thermal energy storage* techniques.

## 2.2 Importance of Energy Storage

The increasing human population affects not only food or freshwater consumption but also has a crucial influence on worldwide energy consumption. Figure 2.1 is prepared according to the data from the World Population Data Sheet (2012) and US Energy Information Administration (2012). The figure illustrates the relationship between population and energy consumption for selected countries. It is certain that population is not the only factor to foresee the energy consumption of a country. On the contrary, industrial development and welfare level are the two main factors that influence the energy consumption of a country.

Since the beginning of the nineteenth century, humanity has been demanding more civilized daily life, and we are now facing the adverse effect of the uncontrolled growth of industrialization and urbanization. For the last 50 years, human-related emissions attracted attention of scientists, and there is an increasing concern about the projections for the near future. There are several local and global concerns, such as global warming, climate change, ozone layer depletion, and water/air pollution, that are all directly or indirectly related to the fossil fuel-based energy



I. Dincer, M. A. Ezan, *Heat Storage: A Unique Solution For Energy Systems*, Green Energy and Technology, https://doi.org/10.1007/978-3-319-91893-8_2

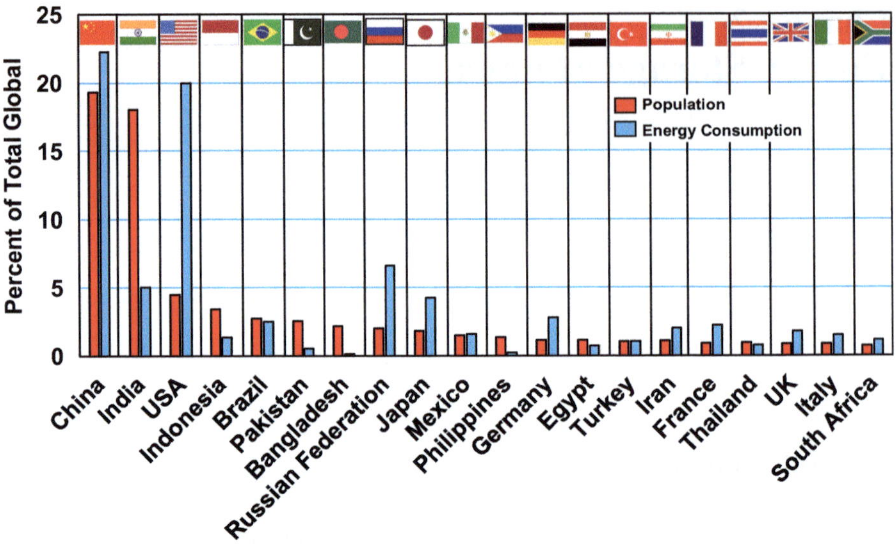


**Fig. 2.1** Relationship between the population, industrial development, and energy consumption (Data from World Population Balance 2017)

production strategies. Such concerns motivate humanity to shift from the consumption of fossil fuels to clean, efficient, and sustainable energy production methods. Renewable energy sources are promising options as they are clean and regenerated by nature. The following renewable energy sources are commonly used:

- Biomass
- Hydropower
- Geothermal
- Solar
- Wave
- Wind

There are two significant obstacles to widespread usage of renewable energy production systems: (*i*) *high capital costs* and (*ii*) *intermittent nature of the renewable resources*. According to the report of the International Renewable Energy Agency (IRENA) in June 2016, the investment costs and capacity factors of solar photovoltaic (PV), parabolic trough solar collector (PTSC), and onshore/offshore wind power are projected to 2050. The comparison between 2015 and 2025 regarding the costs and capacities is given in Table 2.1.

It is supposed that increasing technological improvements will help in the reduction of costs and tend to increase the capacity factors of the renewables. According to IRENA's predictions, the investment cost of the PV technology will reduce up to 50%. Here the technological challenge is twofold: (*i*) the development

**Table 2.1** Projections of the IRENA for investment costs and capacity factors for renewables

| Technology | Investment costs[a] (USD/kW) 2015 | 2025 | Percentage reduction (%) | Capacity factor (%) 2015 | 2025 | Percentage change (%) |
|---|---|---|---|---|---|---|
| Solar photovoltaic (PV) | 1810 | 790 | 57 | 18 | 19 | 8 |
| Parabolic trough solar collector (PTSC) | 5550 | 3700 | 33 | 41 | 45 | 8.4 |
| Onshore wind power | 1560 | 1370 | 12 | 27 | 30 | 11 |
| Offshore wind power | 4650 | 3950 | 15 | 43 | 45 | 4 |

[a]2015 financial data is used for pricing the systems
Adapted from IRENA (2016)

of high efficient power generation technologies and materials to obtain as much useful energy as possible from renewable sources such as wind or sun and (*ii*) improvement of energy storage technologies to provide continuous power generation to meet demands even in the case of intermittent renewable sources such as solar and wind. Without using an energy storage technique, solar and wind energy resources are non-dispatchable due to their unpredictable nature. Solar energy is not constant throughout the day; the insolation varies throughout daytime and is strongly influenced by weather conditions. For instance, for a PV-based electricity generation system, the power output will significantly reduce if there are dense clouds. To limit those fluctuations and maintain continuous power generation, energy storage systems should be implemented into energy systems.

Energy storage (ES) systems not only enable steady use of renewables, but they can also enable efficient use of conventional energy resources. ES systems are used in industrial applications, commercial facilities, and residential systems. As an example, we can consider a hybrid power generation system in which parabolic trough solar collector (PTSC) is used in a geothermal system. During daytime, PTSC absorbs solar energy and supplies thermal energy to the cycle and improves system efficiency. However, after sunset, a conventional hybrid system without a storage unit cannot work. Implementation of a thermal energy storage unit in such a hybrid system enables continuous use of solar energy throughout the day. A well-designed storage system with an electronic controller can store thermal energy during daytime to be used at nighttime. In commercial HVAC applications, trigeneration systems with absorption chillers are commonly in use to meet the heating and cooling loads. In such a system, there may be a dissimilarity between the heating/cooling demand and the electricity demand throughout a day. A wisely designed energy storage system can reduce the fuel consumption of the diesel engine by storing the electricity and thermal energy with a suitable controller system and improve overall system efficiency.

Dincer and Rosen (2011) and Dincer (1997) summarized the benefits of ES systems as follows:

- Reduced equipment size
- Reduced energy cost
- More efficient and effective utilization of equipment
- Increased flexibility of operation
- Improved indoor air quality
- Reduced initial and maintenance costs
- Reduced pollutant emissions ($CO_2$)

ES systems are being used intensively in power generation systems, industrial applications, residential or commercial buildings, and transportation. Research and development are ongoing to integrate advanced technologies and materials into the currently available ES methods to improve the success, capacity, and efficiency of storage systems. Nevertheless, the success of an ES system does not depend just on the system. The selection of a proper ES strategy for a specific application is also a crucial factor to realize the full potential of ES integration.

The increasing attention to ES methods is clear from the dramatic increase in the number of papers that are published in the scientific journals that are indexed by Web of Science (WOS). Figure 2.2 shows the variation in the total number of papers published yearly from 1979 to 2016 that are related to *energy storage*. In the last 10 years (from 2006 to 2016), the number of energy storage-related

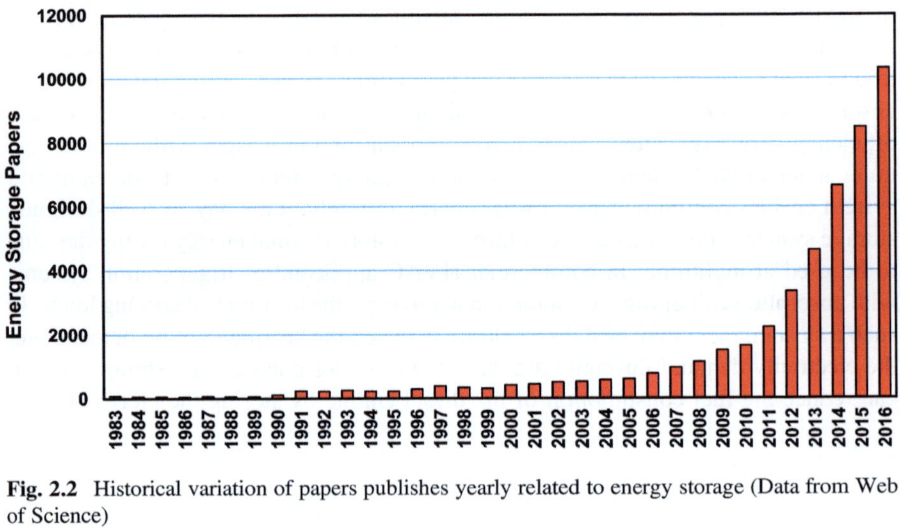


**Fig. 2.2** Historical variation of papers publishes yearly related to energy storage (Data from Web of Science)

scientific studies published yearly increased by more than 13 times. This chart is a key indicator to show the intense interest of the scientific community in energy storage technologies to resolve the current energy-related problems.

In the following section, we will review the commonly used energy storage strategies and give some detailed information about the (*i*) chemical, (*ii*) electrochemical, (*iii*) electrical, (*iv*) mechanical, and (*v*) thermal energy storage methods.

## 2.3 Energy Storage (ES) Methods

There are different types of energy storage systems that are currently used in diverse fields of engineering applications. Figure 2.3 shows energy storage methods based on the type of intermediate energy form. The thermochemical ES is intentionally highlighted under thermal energy storage since the thermochemical ES is considered as a chemical ES method in some classifications that are provided in the literature. Some other classifications are also possible by considering the energy input (i.e., *electricity or thermal*), energy output (i.e., *thermal energy, liquid fuels, or gaseous fuels*), or the conversion process (i.e., *power-to-power, power-to-gas, power-to-liquid, or power-to-thermal*).

In Fig. 2.4, an overview of the energy storage techniques based on discharge time and system power rating is presented. The discharge time varies in a wide range from seconds to hours. The supercapacitors and flywheels are suitable for short-term ES applications. On the other hand, the pumped hydro and compressed air technologies make it possible to store huge amounts of energy for a long term. The power rate also varies in range from kW to GW scale. The details and the recent advances in these technologies are discussed in the related subsections.

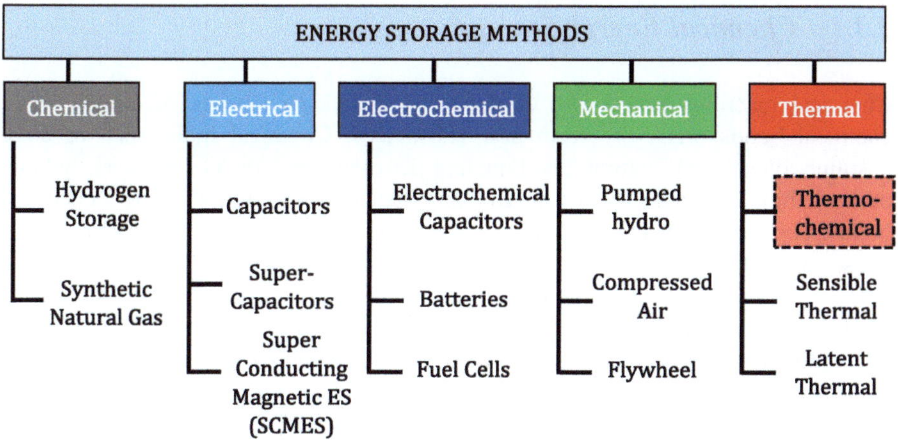


**Fig. 2.3** Commonly used energy storage (ES) methods

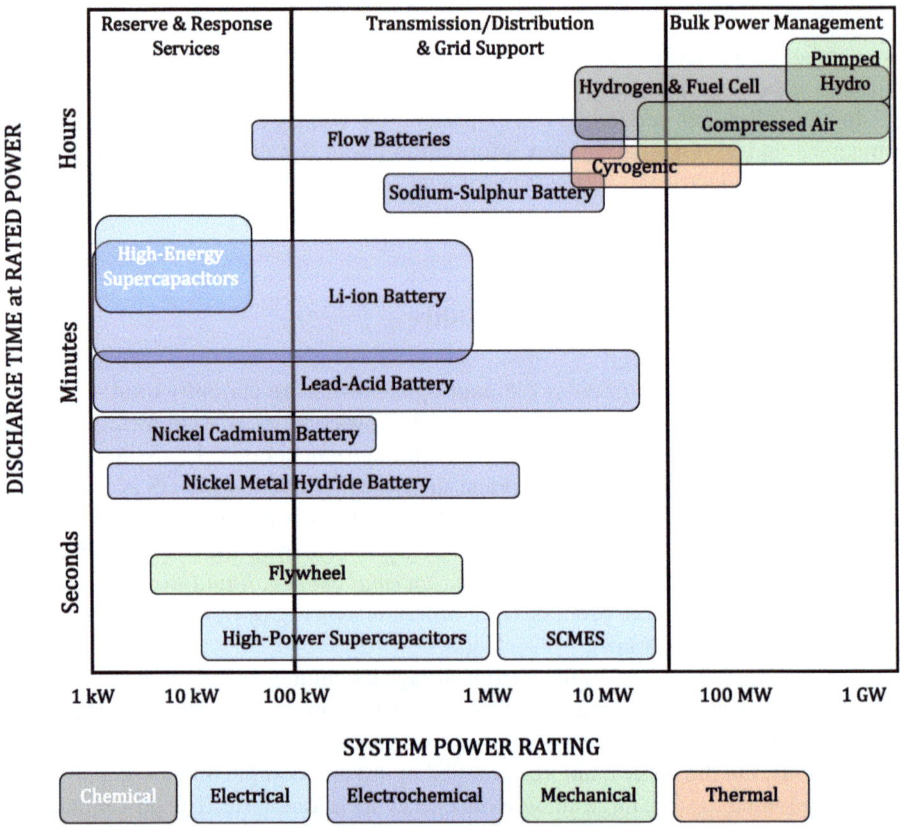


**Fig. 2.4** Discharge times and capacities of energy storage (ES) methods. (Data from EnergyStorageSense 2017)

## 2.3.1 *Chemical Energy Storage*

Chemical reactions can absorb or release a significant amount of energy when chemical bonds break or form new substances. Chemical fuels, such as coal, gasoline, diesel fuel, natural gas, liquefied petroleum gas (LPG), propane, butane, ethanol, biodiesel, and hydrogen, can be used to store energy in their chemical bonds. This process is known as power-to-gas or power-to-liquid, depending on the phase of the output chemical fuels. Chemical fuels are widely used to store large amounts of energy for longer durations or transport the energy to long distances to produce mechanical power or thermal energy in industrial, commercial, or residential applications. Figure 2.5 compares the energy densities (in MJ per unit volume) of commonly used chemical fuels and batteries.

Power-to-gas energy storage technology is a relatively new trend to convert the excessive electricity that is produced from renewables, i.e., wind or solar, to

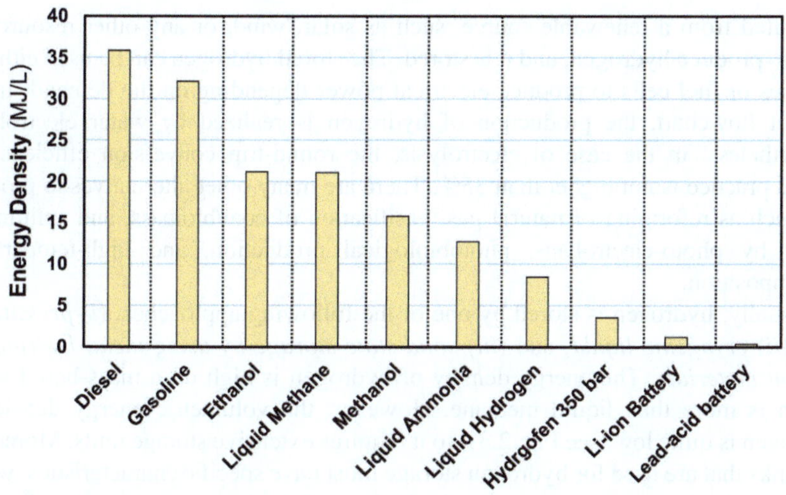


**Fig. 2.5** Comparison of the energy densities of chemical fuels (based on lower heating values) with batteries. (Data from DTU 2013)

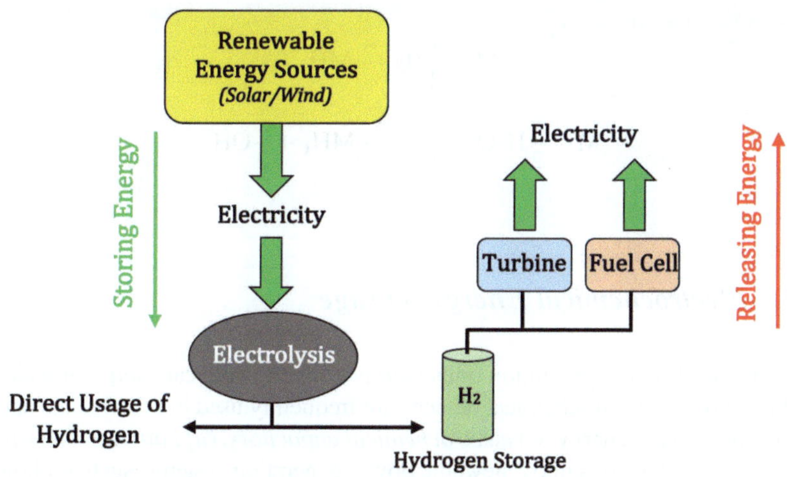


**Fig. 2.6** Hydrogen energy storage system. (Reproduced from Guney and Tepe 2017)

chemical fuels, such as hydrogen or methane, which can be used directly in vehicles, ships, or aircrafts. Storing electrical energy in chemical form, via electrolysis in the case of hydrogen, is compatible with the electrochemical storage since chemical fuels have higher energy densities. Hydrogen is not a direct energy source; instead, it is an intermediate energy form and is an ideal candidate as an energy carrier or storage medium. A conceptual flowchart of a hydrogen energy storage unit is illustrated in Fig. 2.6. Here, the excessive electrical power that is generated from a renewable source, such as solar, wind, or any other resource, is used to produce hydrogen, and it is stored. The stored hydrogen can be used either in turbines or fuel cells to produce electrical power depending on the demand. In the current flowchart, the production of hydrogen is realized by water electrolysis. Nevertheless, in the case of electrolysis, the round-trip conversion efficiency of such a practice is not higher than 35%. There are many other alternatives to produce $H_2$, such as reforming of natural gas, gasification of coal/biomass, and splitting of water by photo-electrolysis, photobiological production, and high-temperature decomposition.

Usually, hydrogen is stored by one of the following approaches: (*i*) *pressurized gas*, (*ii*) *cryogenic liquid*, and (*iii*) *solid-state storage by using metal hydrides or carbon materials*. The energy density of hydrogen is high on a mass-based scale, which is more than liquid methane. However, the volumetric energy density of hydrogen is quite low (see Fig. 2.3), so it requires extensive storage units. Moreover, the tanks that are used for hydrogen storage must have specific characteristics, which makes their investment costs quite high. Instead, hydrogen can be stored within metal hydrides (MHx) with a high volumetric energy density. Two commonly used methods for hydrating a metal (M) are: (*i*) *direct dissociative chemisorption* and (*ii*) *electrochemical splitting of water*. The reactions are as follows (Sakintuna et al. 2007):

$$M + \frac{x}{2}H_2 \leftrightarrow MH_x \tag{2.1}$$

$$M + \frac{x}{2}H_2O + \frac{x}{2}e^- \leftrightarrow MH_x + \frac{x}{2}OH^- \tag{2.2}$$

### 2.3.2 *Electrochemical Energy Storage*

Electrochemical power generation units merely convert chemical energy into electricity. Three types of electrochemical devices are frequently used in broad range applications to store/release energy: (*i*) *electrochemical capacitors*, (*ii*) *batteries*, and (*iii*) *fuel cells*. Integrating batteries into renewable power generation systems, such as photovoltaic or wind power, can provide storage of the excessive amount of energy that is produced during off-demand periods. The stored energy can be smoothly discharged from the batteries and supplied to the consumer when demand is higher than production. In comparison to mechanical energy storage methods, such as pumped hydro or compressed air, batteries are compact, affordable, and readily applicable to electrical power generation systems. Moreover, due to mechanical losses in the mechanical storage strategies, efficiencies drop as low as 50–70%. In the case of batteries, on the other hand, efficiencies stand in the range of 70–80%. One of the most important disadvantages of batteries, however, are their limited capacity. A typical storage facility composed of only electrochemical storage units, i.e., batteries, can meet the demand for

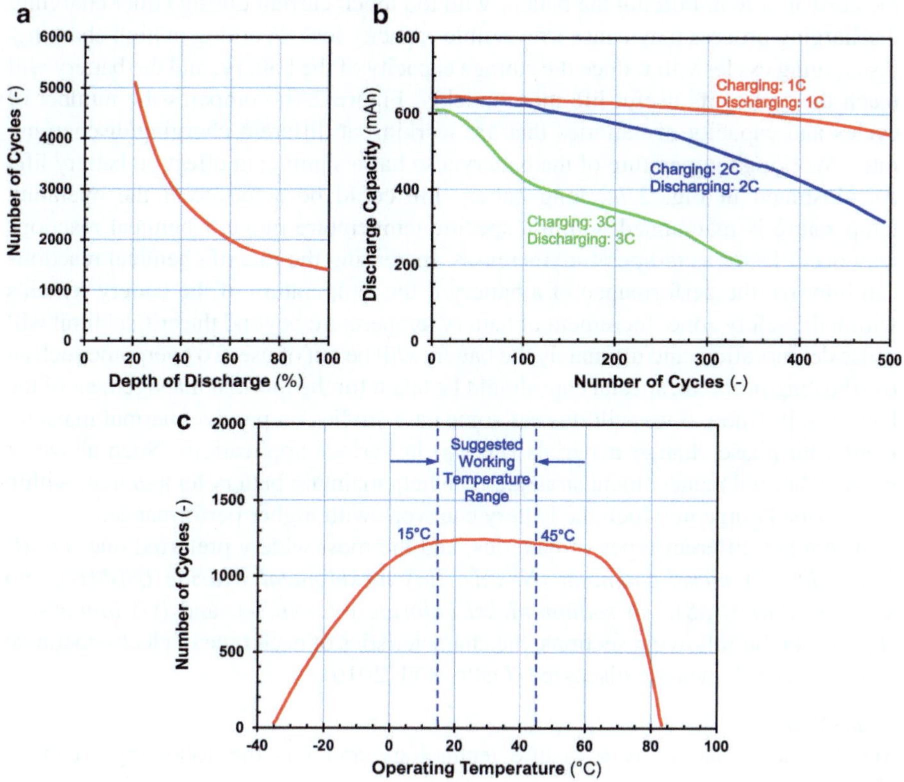


**Fig. 2.7** Influences of depth of discharge, charging/discharging rate, and operating temperature on the number of cycles of a battery (Data from Choi and Lim 2002)

only 3–5 h. On the other hand, mechanical energy storage systems allow a significant amount of energy storage. For instance, a compressed air energy storage unit can meet the required energy for as much as 8–12 h to generate electricity.

To determine the life cycle of a battery, one should consider three parameters: (*i*) *depth of discharge* (DOD), (*ii*) *discharge and charge rates*, and (*iii*) *ambient temperature*. It is recommended to keep the DOD below the maximum limit of 100%. DOD corresponds to the percentage of withdrawn capacity of stored energy from the battery, i.e., 0% DOD is a fully charged battery and 100% DOD is an empty battery. The battery capacity should not drop below 80% before recharging; otherwise, the number of cycles, or *cycle life*, of a battery will reduce. Figure 2.7a illustrates that increasing the depth of discharge, or the *withdrawn capacity*, considerably reduces the number of cycles of a battery. Charging/discharging rates also significantly influence battery life. Increasing the discharge rates reduces the capacity since the transformation of the active chemicals cannot happen at the same rate with the current drawn. Forcing the battery with too much current during either charging/discharging process may cause irreversible capacity loss. Working at high charging/discharging cycles will reduce the storage capacity of the battery, and the battery will reach the end of its useful life more quickly. Figure 2.7b compares the number of cycles and capacity of batteries that are working at different charging/discharging rates. Working temperature of the battery also has a significant effect on battery life. As illustrated in Fig. 2.7c, long battery life could be achieved if the operating temperature is maintained within a specific temperature range. Chemical reactions may occur faster as temperature increases. Increasing the rate of chemical reactions can improve the performance of a battery if the temperature of the battery remains within the safety zone. Increment of battery temperature beyond the critical limit will cause deterioration, and ultimately the battery will be out of use. To overcome such an overheating problem, special care should be taken for the thermal management of the batteries. In Chap. 4 we will discuss some case studies on passive thermal management with phase change materials (PCMs) in various applications. Such active or passive thermal management strategies can help maintain battery temperature within a prescribed range in which the battery can work with higher performance.

There are different types of batteries, and the most widely preferred ones are (*i*) *lead-acid*, (*ii*) *nickel-cadmium (Ni-Cd)*, (*iii*) *nickel-metal hydride (Ni-MH)*, (*iv*) *sodium-sulfur (NaS)*, (*v*) *sodium-nickel chloride (Na-NiCl$_2$)*, and (*vi*) *lithium-ion (Li-ion)*. In the following sections, the characteristics of each type of electrochemical storage technologies are discussed (Gallo et al. 2016).

***Lead-Acid***

The lead-acid battery is a mature technology and has the following technical characteristics:

- *Advantages:* Low cost and moderate efficiency (70–80%)
- *Disadvantages:* Poor cycle life (at 80% DOD cycle life is 400–1000), low specific energy (30–50 Wh/kg), requirement of temperature control, and toxicity

***Nickel-Cadmium***

Ni-Cd battery is also a mature technology and has the following pros and cons:

- *Advantages:* Longer cycle life (at 80% DOD cycle life is 1000–1500) and high specific energy (55–75 Wh/kg)
- *Disadvantages:* Toxicity

***Nickel-Metal Hydride***

Ni-MH battery is developed to improve Ni-Cd technology and to make the battery more environmentally friendly. Ni-MH possesses the following characteristics:

- *Advantages:* Improved energy density (over 80 Wh/kg)
- *Disadvantages:* Poor cycle life (at 80% DOD cycle life is 800–1200), higher self-discharge rate (0.4–1.2%/day)

***Sodium-Sulfur***

NaS batteries require high temperatures to keep the sulfur and sodium in a liquid state, that is, these types of batteries are also known as high-temperature batteries. NaS batteries have the following technical characteristics:

- *Advantages:* Good cycle life (at 80% DOD cycle life is 4500–5000), almost no self-discharge rate (0.05%/day), and high efficiency (75–90%).
- *Disadvantages:* Molten sodium is highly reactive, so it is hazardous.

***Sodium-Nickel Chloride***

Na-$NiCl_2$-type batteries also work at high temperatures with molten materials. The following technical specifications are listed for this type of batteries:

- *Advantages:* Good cycle life (at 80% DOD cycle life is 4000–4500), high efficiency (86–88%), and high energy density (94–120 Wh/kg).
- *Disadvantages:* Initial working temperature is low and requires long heating up periods (nearly 12 h).

***Lithium-Ion Batteries***

Due to high energy density and power density of lithium-ion batteries, they are one of the most widely used rechargeable batteries in portable devices (laptops or cell phones) and electrical or hybrid vehicles. Various positive and negative electrode materials are developed to improve performance and ensure safety. Some widely used positive electrodes are as follows: $LiTiS_2$ (lithium-titanium disulfide, LTS), $LiCoO_2$ (lithium cobalt oxide, LCO), $LiNiO_2$ (lithium nickel oxide, LNO), $LiNi_{1-x-y}Mn_xCo_yO_2$ (lithium nickel manganese cobalt, NMC), $LiNi_{0.8}Co_{0.15}Al_{0.05}O_2$ (nickel cobalt aluminum, NCA), $LiMn_2O_4$ (lithium manganese oxide, LMO), and $LiFePO_4$ (lithium iron phosphate, LFP). A wide range of materials are also developed as negative electrodes: $Li_4Ti_5O_{12}$ (lithium titanate, LTO) and $Li_xSi_y$ and $Li_xSn_y$ (silicon- or tin-based lithium metal alloys). Li-ion batteries possess the following characteristics:

- *Advantages:* High specific energy (150–250 Wh/kg), high specific power (500–1000 W/kg), high efficiency (85–90%), low self-discharge rate (0.1–0.4%/day), long cycle life (at 80% DOD cycle life is 3000–5000+)
- *Disadvantages:* Requires specific safety measures since it includes toxic, flammable, and highly reactive components

### *2.3.3 Electrical Energy Storage*

Capacitors, supercapacitors, and superconducting magnetic energy storage systems are used to store electrical energy. Conventional capacitors have low energy density (2–10 kWh/$m^3$), moderate efficiency (60–85%), and high self-discharge rate (over 40%/day). Some improvements are achieved by modifying conventional capacitors. Electrochemical double-layer capacitor (EDLC), pseudo-capacitors, and asymmetric (or hybrid) supercapacitors are the widely known supercapacitors, or so-called ultracapacitors. Supercapacitors have a superior advantage over *batteries* regarding speed of charging/discharging and cycle life. Supercapacitors have higher energy density (10–30 kW/m$^3$), enhanced efficiency (85–95%), and low self-discharge rate (20–40%/day) compared to the conventional ones. Superconducting magnetic energy storage (SMES), on the other hand, has limited fields of application since it requires high operating costs due to the requirements of cryogenic cooling systems (<100 K) and high magnetic fields. Despite higher costs, SMESs possess high efficiency (80–90%), high power density (1–4 MW/m$^3$), and fast response time with high discharge rates.

## 2.3.4 Mechanical Energy Storage

Electrical energy that is generated from a renewable resource can be converted to mechanical energy, either *potential* (gravitational) or *kinetic*, to store the excessive amount of power generation for further usage. In traditional mechanical storage techniques, one of the following methods is used: (*i*) *pumped hydro*, (*ii*) *compressed air*, and (*iii*) *flywheel.* Among other energy storage strategies, mechanical energy storage applications provide an opportunity to store a significant amount of power rates.

***Pumped Hydro Storage (PHS)***
Pumped hydro (so called: hydro pumping) can be incorporated to a renewable-based power plant, such as solar or wind, to overcome intermittency and provide continuous power generation or to produce electrical power when there is excessive demand while the source is not accessible. In the case of solar power plants, the excessive power generated could be used during daytime to pump water from a lower reservoir to one that is uphill. At nighttime, water is released from the upper to the lower reservoir to produce electricity. The concept of a conventional PHS technology is illustrated in Fig. 2.8. The PHS systems require a considerable amount of investment as it includes separated water reservoirs, pipelines, and turbomachinery units. Large amounts of energy can be stored in a water reservoir for long durations with small losses. However, due to the enormous volume of water reservoirs, the energy density of PHSs is quite small (0.5–1.5 Wh/L). Some mechanical losses may occur both inside the pipelines and turbomachinery unit (pump and turbine). The efficiencies of conventional PHSs stand in the range of 50–85%. The major benefits of a PHS are a long lifetime (30–60 years) and rapid response time.

PHS systems require two reservoirs of water which are at different elevations. That is, the installation of PHS is limited to the type of geographical location. A PHS also has a significant influence on the environment. Recently, there have been some improvements to provide flexible usage of the stored energy inside the reservoirs and also reduce the impact on the environment. The former is achieved by implementation of variable speed PHS so that there is optimal load control during charging

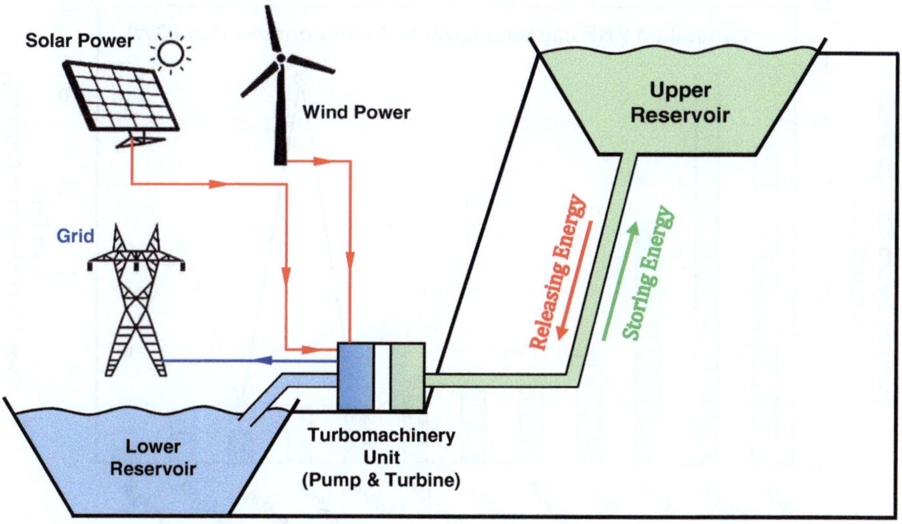


**Fig. 2.8** Schematic of a conventional PHS

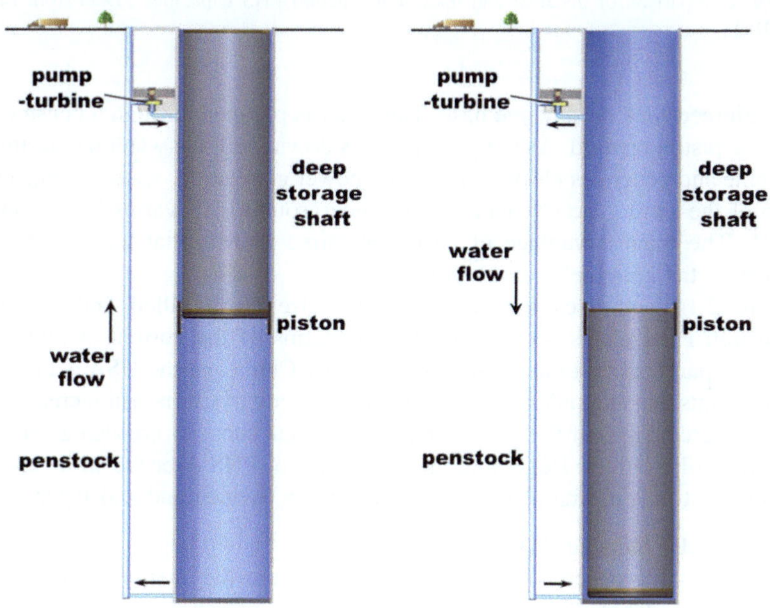


**Fig. 2.9** Schematic of the underground PHS (Adapted from Gravity Power 2017)

(pumping) and discharging (generation). The latter one is acquired by developing ocean-pumped storage and underground PHS. The conceptual schematic of the underground PHS is given in Fig. 2.9. The system consists of a deep water-filled shaft, turbomachinery unit, and penstock. The water-filled piston should be located

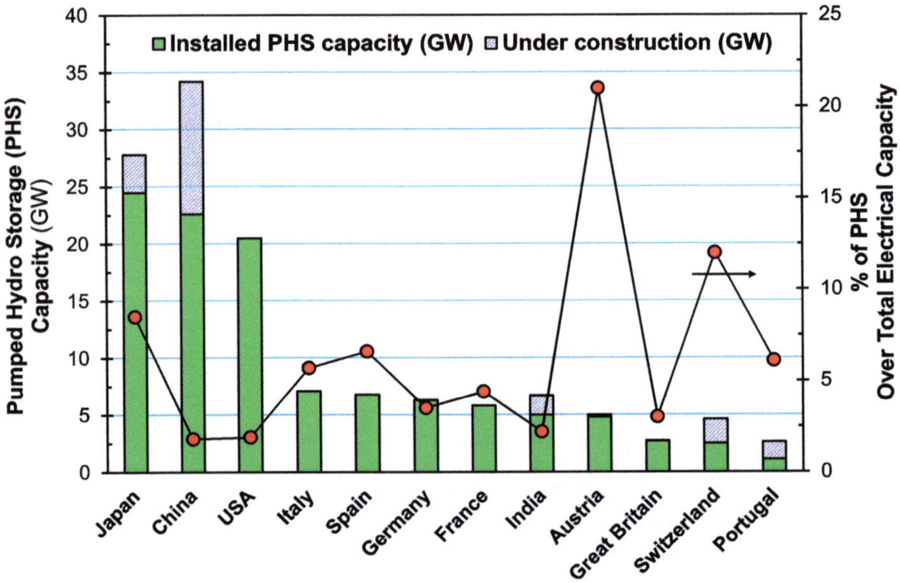


**Fig. 2.10** Comparison of installed and under-construction PHS capacities. (Data from Barbour et al. 2016)

on a reinforced rock or concrete base. The water is pumped through a penstock and moves the piston upward. As the piston moves downward, the water passes through the turbine and generates electricity. During power generation, water is reinjected at the top of the shaft. The efficiency of an underground PHS varies in the range of 75–85%. The main advantage of an underground PHS is that the system is not restricted by the geography of the site.

Figure 2.10 compares the capacities of currently installed and the under-construction PHS plants for 12 countries. In terms of the current available PHS capacity, Japan is in first place and is followed by China and the USA. Considering the PHS plants that are under construction, the capacity of China will increase nearly 50%. The percentages of PHS over the total electrical capacity are also given in this figure with solid circles. Regarding the percentage of PHS over the entire electrical capacity, Austria is in first place and is followed by Switzerland and Japan.

***Compressed Air Storage (CAS)***

In this type of ES systems, air is compressed to store the energy during off-peak hours in large underground reservoirs or aboveground pressure vessels. In traditional CAS units, the air is compressed with compressors and is heated up with an additional preheater to increase its temperature. During the discharge period, the air is expanded through a turbine and generates electricity. The conceptual schematic of a CAS unit is given in Fig. 2.11. CAS requires an enormous amount of power to drive the compressor during the charging period, making such systems

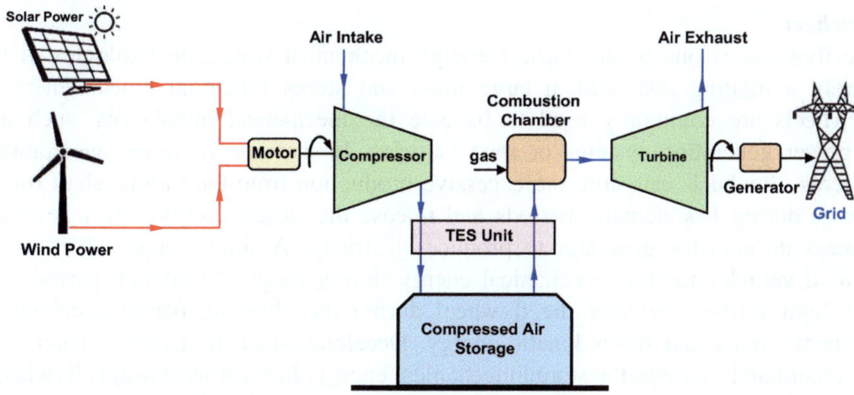


**Fig. 2.11** Schematic of a compressed air storage system

**Fig. 2.12** Underground compressed air storage system. (Source: Hydrostor 2017)

applicable for nuclear-based energy storage applications. Currently used diabatic-CAS systems have efficiencies in the range of 40–60%. To improve the diabatic-CAS systems, currently, there is ongoing research on the development of adiabatic-CAS, isothermal-CAS, and underwater-CAS systems. In adiabatic-CAS systems, an auxiliary thermal energy storage (TES) is utilized to store the generated heat during the compression of air. Injection of water ensures the isothermal process during the compression period. Warm water is then injected into the air during expansion to maintain the air at a constant temperature. In underwater-CAS systems, compressed air is stored inside submerged airbags (see Fig. 2.12). The hydrostatic pressure helps to maintain the isobaric state. During the discharge period, the hydrostatic pressure forces air through the turbine to produce electricity.

***Flywheel***

The flywheel is one of the earliest energy mechanical storage technologies. It is merely a rotating disk with a large mass and stores rotational kinetic energy. Flywheels are commonly used to balance the mechanical imbalances, such as in power generation systems or road vehicles. In the case of power generation systems, flywheels can store the excessive production from the turbine shaft (or a *motor*) during low demand periods and release the stored energy with a reverse process through the generator to produce electricity. A similar approach is used in road vehicles to store mechanical energy during frequent start/stop periods in rush-hour traffic. Spinning the flywheel during the charging period accelerates the heavy mass and stores kinetic energy. Deceleration of the flywheel rotor, on the other hand, releases the stored mechanical energy. In short-term usage, flywheel efficiency can reach up to 75–85% by using a low-friction environment around the rotating disk. A low-friction environment can be achieved by using low-friction gas, evacuation, and magnetic bearings. Flywheels are not suitable for long-term storage of mechanical energy as they have very high self-discharge rate (55–100%/day).

The mechanical energy (*rotational kinetic energy*) that is stored within a flywheel can be evaluated by using the following definition:

$$\mathrm{KE}_{rotational} = \frac{1}{2} I \omega^2 \quad (\text{in J}) \tag{2.3}$$

where $I$ (in kgm$^2$) is the moment of inertia and $\omega$ is the angular velocity of the disc (see Fig. 2.13). In the case of a cylindrical flywheel, the moment of inertia can be expressed as

$$I = kmr^2 \quad (\text{in kg.m}^2) \tag{2.4}$$

where $m$ (in kg) and $r$ (in m) are the mass and radius of the flywheel, respectively. $k$ is the inertial constant of the rotating body, and it is taken as unity for the hoop or thin cylindrical shells. For solid cylinders or disks, $k$ is defined as ½.

Not only the bearing, either *vacuum* or *magnetic*, but also the type of material that is used as flywheel is crucial. The appropriate type of material should be selected based on the energy density and specific strength. Table 2.2 compares commonly used flywheel materials. One more issue about the usage of the flywheel is its high level of noise. Some advanced methods are under development on alternative bearing designs to address this environmental issue.

**Fig. 2.13** Schematic of a flywheel energy storage system

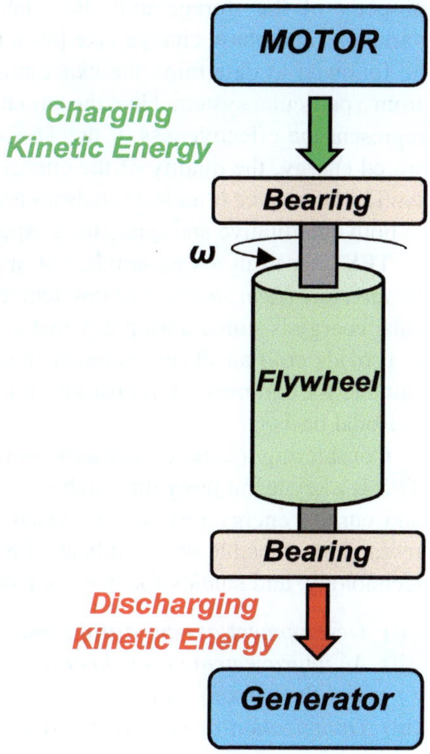


**Table 2.2** Different types of flywheel materials

| Material | Density (kg/m$^3$) | Strength (MN/m$^2$) | Specific strength (MNm/kg) |
|---|---|---|---|
| Steel (AISI 430) | 7800 | 1800 | 0.22 |
| Alloy (AlMnMg) | 2700 | 600 | 0.22 |
| Titanium ($TiA1_6Zr_5$) | 4500 | 1200 | 0.27 |
| GFRP (60 vol% E-glass) | 2000 | 1600 | 0.80 |
| CFRP (60 vol% HT C) | 1500 | 2400 | 1.60 |

GFRP: Glass fiber reinforced polymer; CFRP: Carbon fiber-reinforced polymer; HT C: Heat treated carbon

Adapted from Dincer and Rosen (2011)

## 2.3.5 *Thermal Energy Storage*

Thermal energy storage (TES) is achieved by temperature variation of a substance (*sensible heat*), the phase change of a substance (*latent heat*), and the thermochemical reaction of the storage medium. In the sensible heat TES, the amount of stored energy ($Q$ in J) is related to the mass ($m$), volumetric heat capacity ($C = \rho c$), and temperature variation ($\Delta T$) of the storage material. In latent heat TES, as well as the mass of the storage unit, the latent heat of fusion ($h_{sf}$ or $h_{fg}$) is the critical thermophysical property of the storage unit. In a latent heat TES, generally, both temperature variations and phase change take place in a storage unit, so a unified approach must be followed to determine the exact amount of stored (or *released*) thermal energy from a particular system. Here the amount of stored energy (*quantity*) does not simply represent the effectiveness of the TES unit itself. In addition to the quantity of the stored energy, the quality of the energy must be considered. That is, *energetic* (1st law) and *exergetic* (2nd law) analyses are conducted to evaluate system performances in both quantitative and qualitative aspects.

TES is used in a vast number of applications for short-term or long-term high-temperature (*heat storage*) or low-temperature (*cold storage*) purposes. For instance, solar energy is stored during daytime for night heating, or ice is stored during winter to provide cold air during summertime. Such short- or long-term TES systems are suitable for commercial, industrial, or residential applications on a daily, weekly, or seasonal basis.

Considering the increasing trend of the heating and cooling demand of humanity, TES is a key technology that enables sustainable usage of renewable energy systems and current energy production systems more efficiently with low capacities. The research on sensible or latent heat TES is mainly focused on improving the current technology, and studies focus on one of the following items:

(i) *Development of new storage materials*
(ii) *An improvement on the heat transfer mechanism between the storage material and the working fluid*
(iii) *Optimization of the overall system with a real-world implementation (industrial, commercial, or residential)*

In the following chapters, we provide in-depth knowledge of thermal energy storage and particularly *heat storage applications*:

- Chapter 3, Thermal Energy Storage Methods
- Chapter 4, Thermal Energy Storage Applications
- Chapter 5, System Modeling and Analysis
- Chapter 6, System Optimization
- Chapter 7, System Characterization and Case Studies

## 2.4 Comparison of Energy Storage Technologies

In this chapter, a brief overview of the energy storage techniques is given. Each form of energy storage technique has significant importance since ES

- Makes it possible to use intermittent renewable sources to meet fluctuating demands
- Maintains a continuous power supply
- Improves the overall system performance

**Table 2.3** Comparison of ES techniques

| Energy storage technique | | Power rate (MW) | Discharge time | Self-discharge rate (%/day) | Capital costs ($/kW) |
|---|---|---|---|---|---|
| Chemical | Hydrogen | 0.1–1000+ | 1–24 + h | Very small | 1900–6300 |
| | Methane | 0.1–1000+ | 1–24 + h | Very small | 3500–5000 |
| Electrochemical | Lead-acid | 0.001–50 | s–h | 0.033–0.3 | 300–600 |
| | Ni-Cd | 0.01–40 | s–h | 0.067–0.6 | 500–1500 |
| | Ni-MH | 0.01–1 | h | 0.4–1.2 | 600–1800 |
| | Li-ion | 0.1–50 | min–h | 0.1–0.3 | 1200–4000 |
| | NaS | 0.05–50 | s–h | 0.05–20 | 1000–3000 |
| | Na-$NiCl_2$ | 0.001–1 | min–h | 15 | ~400–1800 |
| Electrical | Supercapacitors | 0.001–10 | ms–h | 20–40 | 100–360 |
| Mechanical | Conventional PHS | 100–5000+ | 1–24 + h | 0.005–0.02 | 600–2000 |
| | UG-PHS | 200–1000 | 0.3–4 h | Very small | 1000 |
| | A-CAES | 0.1–100+ | 1–24 + h | 0.5–1 | 700–1000 |
| | I-CAES | 0.1–10 | 1–12 h | Very small | 500–1000 |
| | UW-CAES | 1–1000+ | 1–24 + h | Very small | 750–2000 |
| | Flywheel | 0.1–10 | s–min | 55–100 | 250–350 |
| **Thermal** | | **Power rate ($MW_{th}$)** | **Storage period** | **Efficiency (%)** | **Capital costs ($/kW)** |
| Sensible heat TES | | 0.001–10 | Day–month | 50–90 | 3400–4500 |
| Latent heat TES | | 0.001–1 | Hour–day | 75–90 | 6000–15,000 |
| Thermochemical | | 0.01–1 | Hour–month | 75–100 | 1000–3000 |

Adapted from Gallo et al. (2016)

The abovementioned advantages could be achieved when a suitable ES system is selected for a particular application. Table 2.3 briefly compares the popular ES techniques considering their power rates, discharge times, self-discharge rates, and capital costs. For the case of thermal energy storage, two merits revised as storage period and efficiency. Mechanical and chemical energy storage techniques have highest power rates. Electrochemical storage technology provides the highest discharging rates. Considering the self-discharge rate, the supercapacitors and flywheels have the highest rates, making them suitable for short-term storage. The maturity levels and investment costs of several types of energy storage technologies are given in Fig. 2.14. For each type of ES technique, one can infer that:

- Chemical ES techniques, especially power-to-gas with hydrogen and synthetic natural gas, are currently under research and development. The capital cost is at moderate levels, but the current situation is far from commercialization.

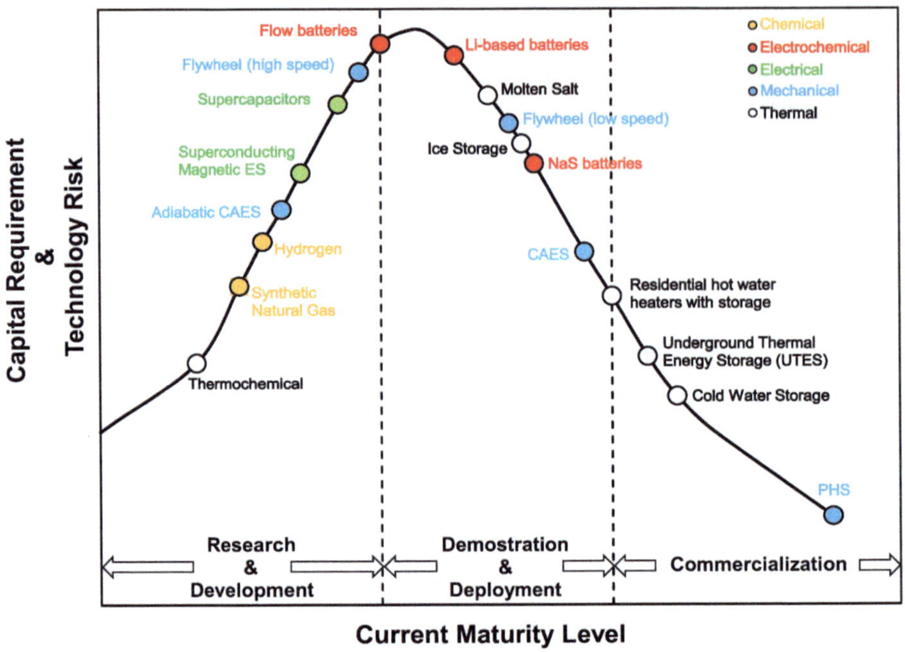


**Fig. 2.14** Maturity of ES technologies (Data from IEA 2014)

- Electrical ES techniques, supercapacitors, and SMES are also currently in research and development stage. The capital cost and technology risks are higher for this type of storage methods.
- Electrochemical ES techniques, or batteries, are close to commercialization, but the investment costs are high.
- Mechanical ES techniques appear in each range. Adiabatic compressed air storage (A-CAES) and high-speed flywheel technologies are under R&D stage. Low-speed flywheel and classical CAES techniques, on the other hand, are mature technologies. The PHS technique has the lowest risk with highest commercialization potential over all types of ES technologies.
- Thermal energy storage (TES) techniques also present different stages of technological maturity. Thermochemical TES is currently in the development stage. The most widely used energy storage techniques are cold water storage, underground TES, and domestic hot water storage. These types of TES systems have low risk and high level of maturity. Molten salt and ice storage methods of TES are close to commercialization.

## 2.5 Closing Remarks

This particular chapter concerns key energy storage methods and provides some important aspects and formulations of chemical, electrochemical, electrical, mechanical, and thermal energy storage techniques. Numerous examples and case studies are presented to highlight the importance of such methods and their deployment in various applications. Some recent advances are discussed, and alternative storage approaches are compared based on power rate, discharge time, and capital cost for various applications where energy storage technology is applicable from kW to GW scale to store energy in various forms, and where storage duration varies widely from seconds to months.

## Nomenclature

| | |
|---|---|
| $c$ | Specific heat, J/kgK |
| $C$ | Volumetric heat capacity, J/m$^3$K |
| $h$ | Specific enthalpy, J/kg |
| $I$ | Moment of inertia, kgm$^2$ |
| $k$ | Inertial constant |
| $m$ | Mass, kg |
| $Q$ | Energy, J |
| $r$ | Radial position, m |
| $T$ | Temperature, K |

## Greek letters

| | |
|---|---|
| $\omega$ | Angular velocity, (1/s) |
| $\rho$ | Density, kg/m$^3$ |

## Subscripts

| | |
|---|---|
| $fg$ | Liquid to gas |
| $sf$ | Solid to liquid |

## References


Barbour, E., Wilson, I. G., Radcliffe, J., Ding, Y., & Li, Y. (2016). A review of pumped hydro energy storage development in significant international electricity markets. *Renewable and Sustainable Energy Reviews, 61*, 421–432.

Choi, S. S., & Lim, H. S. (2002). Factors that affect cycle-life and possible degradation mechanisms of a Li-ion cell based on $LiCoO_2$. *Journal of Power Sources, 111*(1), 130–136.

Dincer, I. (1997). *Heat transfer in food cooling applications*. Washington, DC: Taylor & Francis.

Dincer, I., & Rosen, M. (2011). *Thermal energy storage: Systems and applications* (2nd ed.). Hoboken: Wiley.

DTU International Energy Report. (2013, November). Energy storage options for future sustainable energy systems. Edited by H. H. Larsen, & L. Sønderberg Petersen. DTU National Laboratory for Sustainable Energy.

EnergyStorageSense. http://energystoragesense.com/. Accessed 12 Nov 2017.

Gallo, A. B., Simões-Moreira, J. R., Costa, H. K. M., Santos, M. M., & dos Santos, E. M. (2016). Energy storage in the energy transition context: A technology review. *Renewable and Sustainable Energy Reviews, 65*, 800–822.

Gravity Power Grid-Scale Energy Storage. http://www.gravitypower.net/. Accessed 12 Nov 2017.

Guney, M. S., & Tepe, Y. (2017). Classification and assessment of energy storage systems. *Renewable and Sustainable Energy Reviews, 75*, 1187–1197.

Hydrostor. https://hydrostor.ca/. Accessed 12 Nov 2017.

International Energy Agency (IEA). (2014). *Technology roadmap energy storage.*

International Renewable Energy Agency (IRENA). (2016). *The power to change: Solar and wind cost reduction potential to 2025.*

Population Reference Bureau. (2012). *World population data sheet.*

Sakintuna, B., Lamari-Darkrim, F., & Hirscher, M. (2007). Metal hydride materials for solid hydrogen storage: A review. *International Journal of Hydrogen Energy, 32*(9), 1121–1140.

U.S. Energy Information Administration. (2012).

World Population Balance. http://www.worldpopulationbalance.org/population_energy. Accessed 12 Nov 2017.

# Chapter 3
# Thermal Energy Storage Methods

## 3.1 Introduction

It is important to note that energy demands in various sectors, including commercial, industrial, and utility sectors, vary on daily, weekly, and seasonal bases. Such demands can only be matched with the help of thermal energy storage (TES) systems that operate synergistically. The deployment of TES for thermal applications, namely, space and water heating, cooling, air-conditioning, etc., has recently received much attention. A variety of TES techniques have been developed during the past two decades. TES systems have an enormous potential in using thermal energy equipment in a more effective and efficient manner. Various types of energy storage systems play an important role in energy conservation methods. In Fig. 3.1, widely used energy storage techniques are briefly listed. Thermal energy storage (TES) is the storage of thermal energy at high or low temperatures for future use. This chapter focuses on the fundamental aspects of sensible, latent, and thermochemical TES techniques. TES systems for short-term or long-term heat or cool storage which are adequate for a variety of engineering applications are discussed.

## 3.2 Basics of Thermal Energy Storage

Sustainability of energy is one of the most critical problems in using renewable energy resources in engineering applications in either heating/cooling or power generation systems. Many researchers have investigated energy storage systems as a solution for the intermittent nature of renewable energy resources such as *solar* and *wind* power. In Chapter 2, we overviewed the commonly used energy storage strategies. In the current chapter, we will discuss the fundamentals of thermal energy storage (TES) methods. One of the earliest and well-known usages of the TES is storing solar energy for later use. Solar energy is stored during daytime for using at



I. Dincer, M. A. Ezan, *Heat Storage: A Unique Solution For Energy Systems*, Green Energy and Technology, https://doi.org/10.1007/978-3-319-91893-8_3

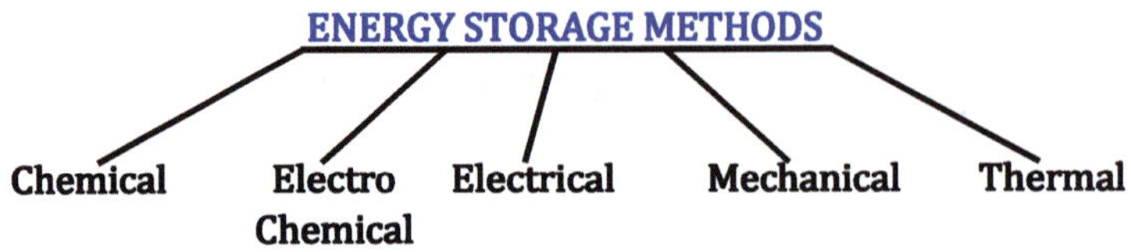


**Fig. 3.1** Classification of ES methods

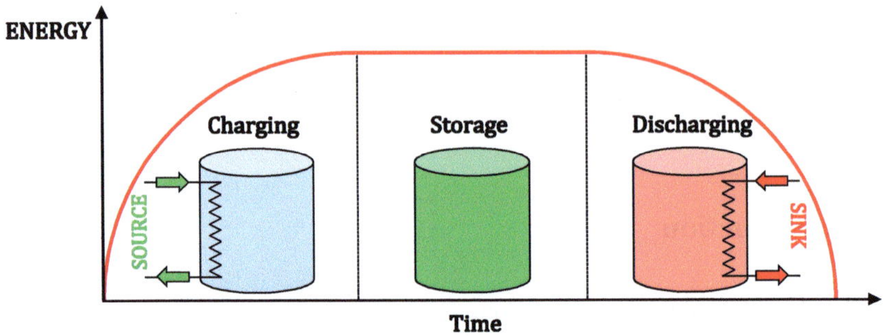


**Fig. 3.2** A typical cycle of a TES unit

nighttime. This procedure can allow continuous usage of solar energy throughout the day. A typical charging/discharging cycle of a TES unit is illustrated in Fig. 3.2. Thermal energy is charged while the energy source is widely available or is cheap. After a short- or long-term storage time, the stored energy is discharged when the source is not available or is expensive. The charging, storage, and discharging processes repeat consecutively in a cyclic manner.

Thermal energy storage is essential for using conventional energy systems in a manner that is sustainable, efficient, economical, and environmentally friendly. Thermal energy storage (TES) is a key technology in reducing the mismatch between energy supply and demand for thermal systems. TES methods are commonly used for residential or commercial heating/cooling applications or for providing continuous power generation in renewable-based power plants. TES systems not only provide a balance between supply and demand but also increase the performance and reliability of energy systems.

In TES systems, energy can be stored via changing the internal energy of the storage medium as:

1. *Sensible heat*
2. *Latent heat*
3. *Thermochemical heat*

Mature TES techniques that are preferred for heating or cooling applications are *sensible heat storage* and *latent heat storage*. *Thermochemical heat storage*, on the other hand, is a promising technique with a high energy storage density; however, currently, there are only lab-scale units, and research is ongoing to

**Table 3.1** Comparison of TES methods

| Thermal energy storage method | Capacity (kWh/t) | Efficiency (%) | Cost (€/kWh) |
|---|---|---|---|
| Sensible heat | 10–50 | 50–90 | 0.1–10 |
| Latent heat | 50–150 | 75–90 | 10–50 |
| Thermochemical heat | 120–250 | 75–100 | 8–100 |

Data from Safari et al. (2017)

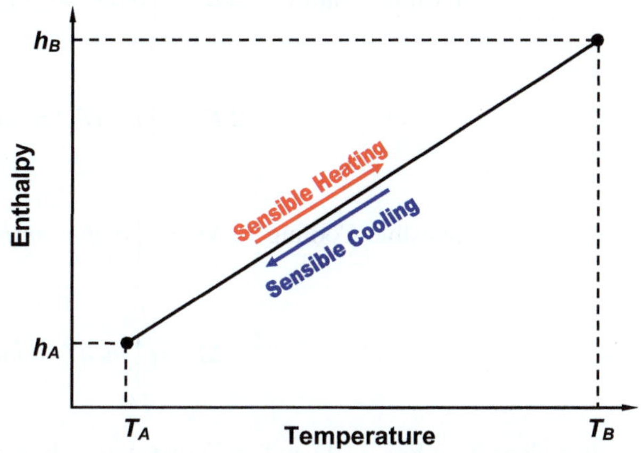


**Fig. 3.3** Enthalpy variation of a medium during sensible heating or cooling

reveal commercial systems. In *sensible heat storage* (SHS) systems, energy is stored in a storage medium using temperature difference. In *latent heat storage* (LHS) systems, energy is stored via changing the phase of the storage medium from one phase to another, by melting (*solid* to *liquid*), solidification (*liquid* to *solid*), evaporation (*liquid* to *gas*), or condensation (*gas* to *liquid*). *Thermochemical heat storage* is achieved by thermochemical reactions and sorption processes.

Table 3.1 compares the ranges of capacities, efficiencies, and invention costs of each TES technique. Sensible heat storage is the most mature TES technique with low cost. Latent heat and thermochemical heat, on the other hand, are promising techniques as they have higher energy storage density and efficiency. However, latent heat and thermochemical heat storage methods require high investment and maintenance costs. The basic concepts of each TES method are discussed in the following sections.

## 3.3 Sensible Heat Thermal Energy Storage

Sensible heat storage is achieved by increasing (*heating*) or decreasing (*cooling*) the temperature of the storage medium. A typical cycle of sensible heat thermal energy storage (SHTES) system involves *sensible heating* and *cooling* processes as given in Fig. 3.3. The heating (or cooling) process increases (or reduces) the enthalpy of the storage medium. For an incompressible storage medium, the amount of heat stored (*or* rejected) energy in (*or* from) the system depends on the specific heat ($c$ in J/kgK) of the medium, the temperature difference ($\Delta T$), and the mass of the storage medium ($m$ in kg). The total energy variation of an SHTES system can be defined as follows:

$$\text{Sensible Heating:} \quad \Delta E = \int_{h_A}^{h_B} m dh = m(h_B - h_A) \text{ or} \quad \Delta E = \int_{T_A}^{T_B} mc dT = mc(T_B - T_A) \tag{3.1}$$

$$\text{Sensible Cooling:} \quad \Delta E = \int_{h_B}^{h_A} m dh = m(h_A - h_B) \text{ or} \quad \Delta E = \int_{T_B}^{T_A} mc dT = mc(T_A - T_B) \tag{3.2}$$

It is clear from Eqs. (3.1) and (3.2) that the total amount of energy stored within an SHTES system is related to the temperature variation, volume of the storage medium, density, and specific heat of the material. The amount of energy that is charged (or *discharged*) inside the SHTES system is directly proportional to the temperature change of the medium. Improving temperature variation enhances the stored energy. However, increasing the temperature between the storage medium and the surroundings will lead heat losses (or gains). That is, insulation has a great importance for designing effective storage units for SHTES systems. In SHTES, it is desirable for the storage medium to have high volumetric heat capacity ($C = \rho c$ in J/m$^3$K), long-term stability under multiple thermal cycles, to be nontoxic, and most importantly low cost. Possessing high volumetric heat capacity will reduce the required storage volume and investment costs. Lowering the tank volume can also help prevent heat interactions between the storage medium and the surroundings.

The charging and discharging durations in SHTES systems are mostly limited, so the rate of heat transfer is also an important parameter to design an effective storage unit. Two facts may affect the rate of heat transfer within the storage unit: (*i*) heat interaction between the storage medium and the heat transfer fluid and (*ii*) thermal diffusivity of the storage medium. In SHTES systems, various types of liquids (i.e., molten salts, water, and oils) or solids (i.e., rocks or metals) are used as storage medium. Table 3.2 presents the thermophysical properties of commonly used storage mediums. In the following subsections, SHTES systems with *solid* or *liquid* storage mediums are introduced.

**Table 3.2** Thermophysical properties of commonly used SHTES mediums

| Material | Density (kg/m$^3$) | Specific heat (J/kgK) | Volumetric heat capacity (10$^6$J/m$^3$K) |
|---|---|---|---|
| Clay | 1458 | 879 | 1.28 |
| Brick | 1800 | 837 | 1.51 |
| Sandstone | 2200 | 712 | 1.57 |
| Wood | 700 | 2390 | 1.67 |
| Concrete | 2000 | 880 | 1.76 |
| Glass | 2710 | 837 | 2.27 |
| Aluminum | 2710 | 896 | 2.43 |
| Iron | 7900 | 452 | 3.57 |
| Steel | 7840 | 465 | 3.68 |
| Gravelly earth | 2050 | 1840 | 3.77 |
| Magnetite | 5177 | 752 | 3.89 |
| Water | 988 | 4182 | 4.17 |

Adapted from Dincer and Rosen (2011)

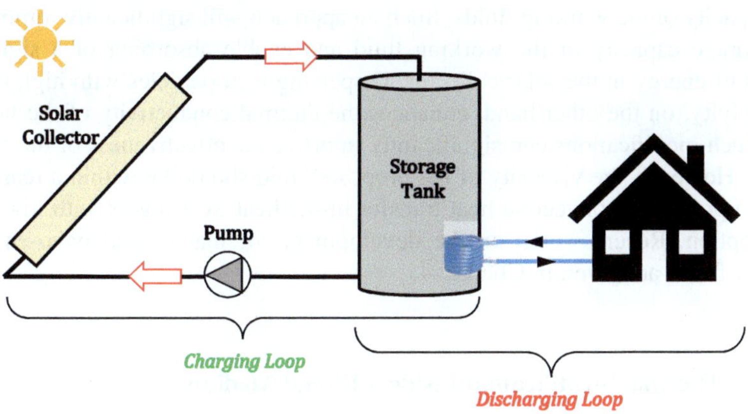


**Fig. 3.4** Simple SHTES system for solar-based water heating unit

### *3.3.1 Liquid Storage Medium*

There are numerous reviews in the literature that represent the thermophysical properties of commonly used solid and liquid materials for SHTES systems and compare the candidate material with various aspects such as thermal, chemical, economic, and environmental aspects. Water is one of the best suitable liquids for SHTES systems, possessing high specific heat and lower price for moderate-temperature applications. The significant adverse effect of water is its corrosiveness. Some chemical additives must be used in water storage tanks to prevent corrosion and freezing problems during subzero temperatures. One of the most widely used water-based SHTES applications is short-term solar thermal energy storage in residential applications. Such an application is schematically illustrated in Fig. 3.4. During daytime, a solar collector receives solar energy, and working fluid (either *water* or another *liquid*) transfers the thermal energy to water inside the storage tank. There are two possible ways of circulating the working fluid: (*i*) *direct* and (*ii*) *indirect* approaches. In the case of direct use, water (*or any other storage liquid*) is directly circulated from the collector through the tank without a heat exchanger. In the direct mode of sensible heat storage technique, forced or natural circulation alternatives are widely used.

In the forced convection approach, a pump is used to maintain a constant flow rate. In the second approach, temperature gradients within the storage tank provide a natural circulation through the solar collector. In the indirect design, on the other hand, there is a heat exchanger inside the tank, and working fluid does not mix with the water (*or any other storage liquid*). Different types of working fluids can be used to improve system performance. As a recent trend, researchers are trying to develop working fluids with enhanced thermal properties. Increasing the heat capacity or thermal conductivity of the working fluid can allow collecting higher amounts of thermal energy from the source with a lower surface area. For instance, dispersing microscale-encapsulated phase change materials within the base fluid increases the heat capacity of the working fluids. Such an approach will significantly improve the heat storage capacity of the working fluid and enable absorbing of a significant amount of energy in the solar collector. Dispersing nanoparticles with high thermal conductivity, on the other hand, enhances the thermal conductivity of the working fluid. Such modifications can significantly improve the effectiveness of the SHTES system. However, the viscosity of the proposed fluid should be within a reasonable range to provide a convective heat transfer inside heat exchangers with low power consumption. Recent works on the developments of macro- and nano-enhanced working fluids are given in Chapter 4.

#### 3.3.1.1 Thermal Stratification Inside a Liquid Medium

The variation in the density of storage liquids as a function of temperature causes thermal stratification within storage tanks. The mixing effect inside a tank reduces system efficiency. Stratification should be achieved by placing the inlet and outlet ports of a tank in such a way that the highest and lowest temperatures are at the top and bottom sections of the tank, respectively. Such a design minimizes the mixing effects of hot and cold water and ensures supplying of hot water to the outlet section. Dincer and Rosen (2011) list the following guidelines to design an effective SHTES tank:

- Inside the SHTES tank, the liquid medium should be stratified, and the volumes of liquid with warm and cold temperatures should be kept in separate regions of the tank. The mixing zones of warm and cold liquid should be as minimum as possible even during the charging and discharging periods. Figure 3.5 illustrates tanks with different degrees of thermal stratification. In Fig. 3.5a, b, the warm water is at the top, and the cold water is at the bottom. There is a mixing layer between the hot and cold regions. The mixing layer is also known as *thermocline*

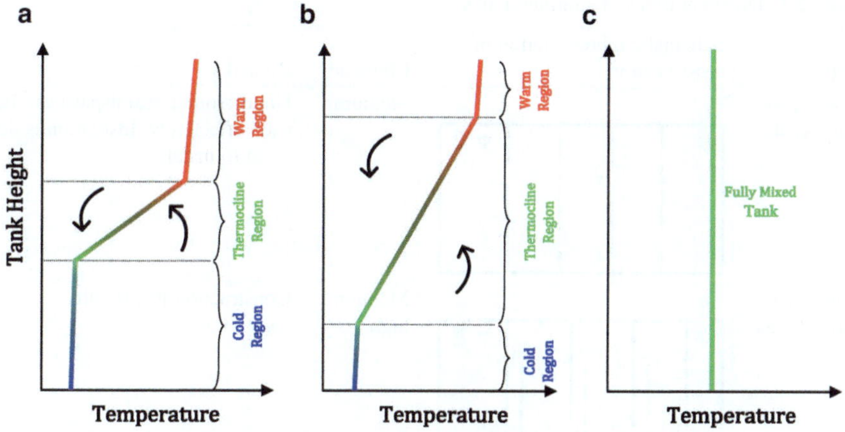


**Fig. 3.5** Degree of stratification inside SHTES tanks. (**a**) Highly stratified tank. (**b**) Moderately stratified tank. (**c**) Unstratified tank

**Fig. 3.6** Influence of inlet/outlet ports on the effective amount of storage liquid

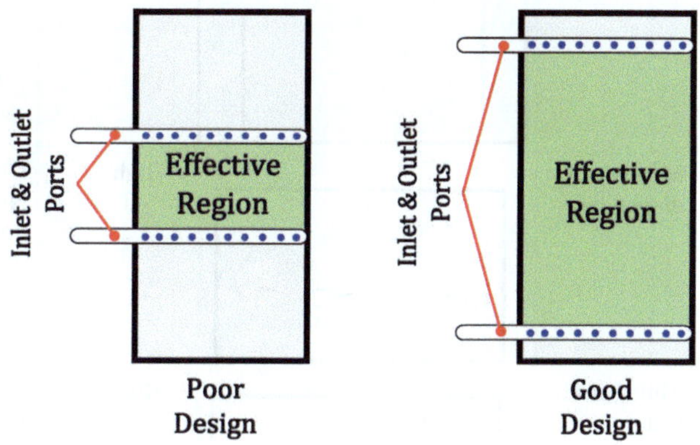


zone. The thickness of the thermocline zone should be narrow to maximize the amount of hot liquid within the tank. In Fig. 3.5a, the tank is highly stratified. The thermocline region is too narrow, that is, limited amount of cold and hot water mix with each other. In Fig. 3.5b, on the other hand, the thermocline region is extended. As the amount of hot water is reduced with increasing thermocline thickness, the tank efficiency reduces significantly. In Fig. 3.5c, an unstratified tank is illustrated. The cold and hot water thoroughly mix inside the tank, and there is a uniform temperature distribution from the top to bottom of the tank. Such a storage tank is inappropriate as the supplied temperature from the tank will be low.

- The inlet and outlet ports should be placed wisely to increase the active liquid zone as much as possible. A poor design and a good design based on effective amount of storage liquid are illustrated in Fig. 3.6.

**Table 3.3** Different types of stratified tanks

| Type | Schematic representation of cross section | Efficiency | Remarks |
|---|---|---|---|
| Continuous multi-tank | 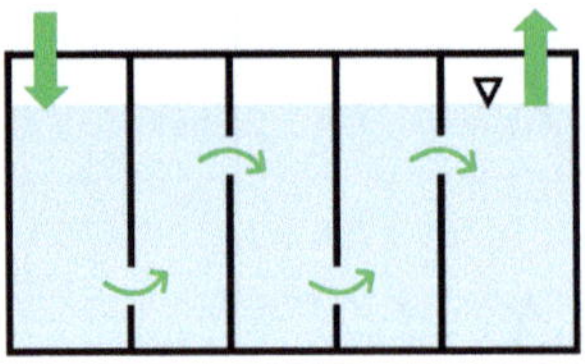 | Medium | Underground beam space can be used effectively. Insulation is difficult to install |
| Improved dipped weir | 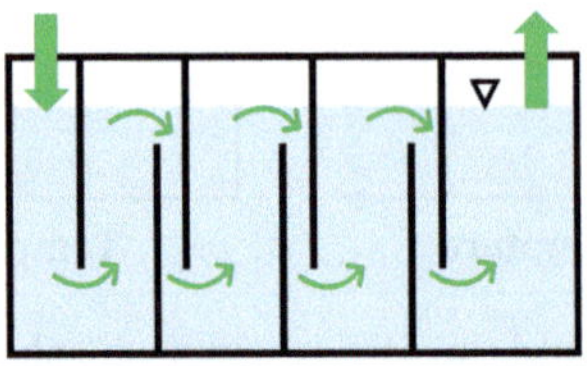 | Medium high | Construction is difficult |
| Thermally stratified | 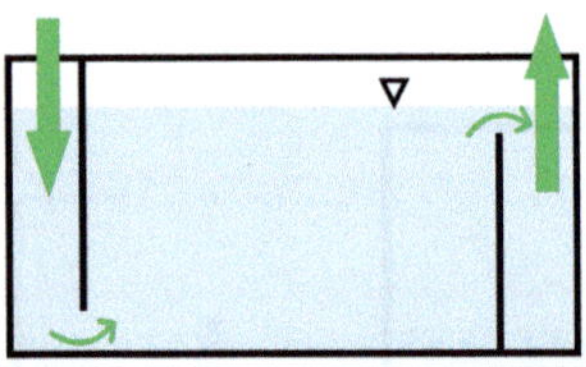 | High | Best suited for large-sized tank built aboveground |
| Movable diaphragm | 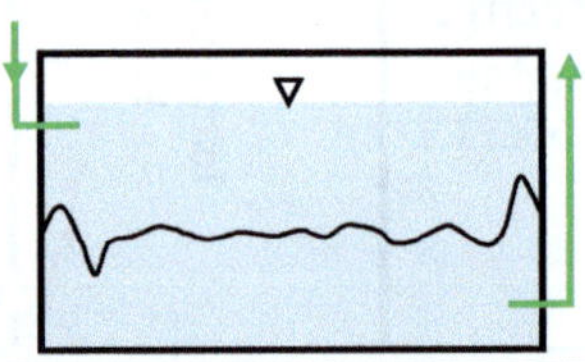 | High | Diaphragm material is problematic. Not easily adaptable to tanks with internal pillars and beams |
| Multi-tank water renewing | 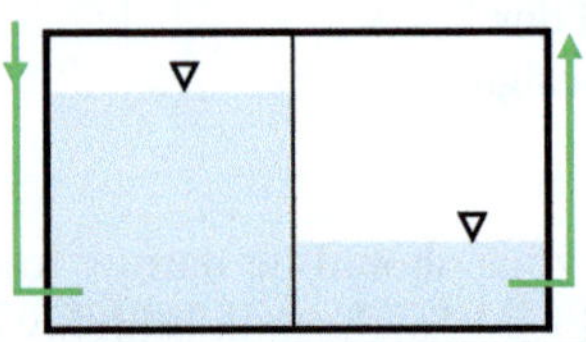 | High | Underground beam space can be utilized to some extent. Heat loss is high |

Adapted from Shimizu and Fujita (1985)

In addition to naturally stratified SHTES tanks, there are various approaches in the literature to maintain a desired temperature within a tank by using *baffles* or *diaphragms*. In Table 3.3, different types of stratified tank approaches and comments on each design are given. The stratification efficiency of a thoroughly mixed tank is 0%, and for the tank with perfect stratification, it is 100%. In the literature, there are various approaches to characterize the stratification performance of an SHTES tank by considering (i) the gradient of the thermocline, (ii) first law of thermodynamics, and (iii) second law of thermodynamics. Haller et al. (2009) prepared a comprehensive review to compare different methods that are proposed to characterize thermal

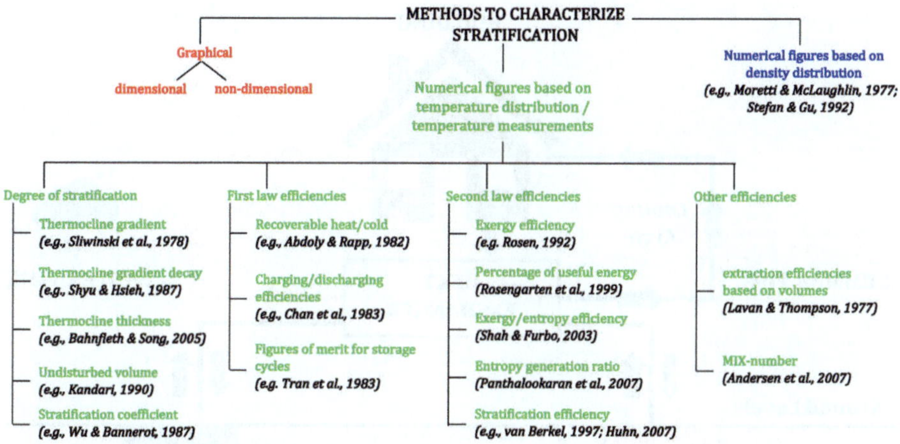


**Fig. 3.7** Typical methods to characterize the stratification within the storage tanks for SHTES (Data from Haller et al. 2009)

stratification in a water storage tank as illustrated in Fig. 3.7. There are two approaches in which graphical and numerical methods are followed. In numerical methods, either temperature or density distributions are considered to characterize the degree of stratification. In the temperature-distribution-based numerical approaches, the performance of the tank is evaluated from the stratification level or thermodynamic computations that are based on the first and second laws of thermodynamics.

### *3.3.2 Underground Thermal Energy Storage (Aquifer TES)*

For long-term thermal energy storage, one of the most common approaches is the usage of aquifer TES (ATES). Aquifers are freshwater sources and contain a large amount of water, often millions of cubic meters in volume. The huge amount of water volume inside aquifers increases the thermal energy storage capacity of the ATES. However, to achieve a good charging/discharging performance during the charging and discharging cycles, the following should be considered (Paksoy et al. 2009):

- Thermal gradient
- Aquifer depth and geometry
- Leakage factor of confining layers
- Direction of flow
- Water chemistry
- Static head
- Permeability
- Grain size distribution
- Natural groundwater flow

ATES basically stores thermal energy (either *warm* or *cold water*) within groundwater inside an aquifer for daily, weekly, or seasonal storage. A typical ATES

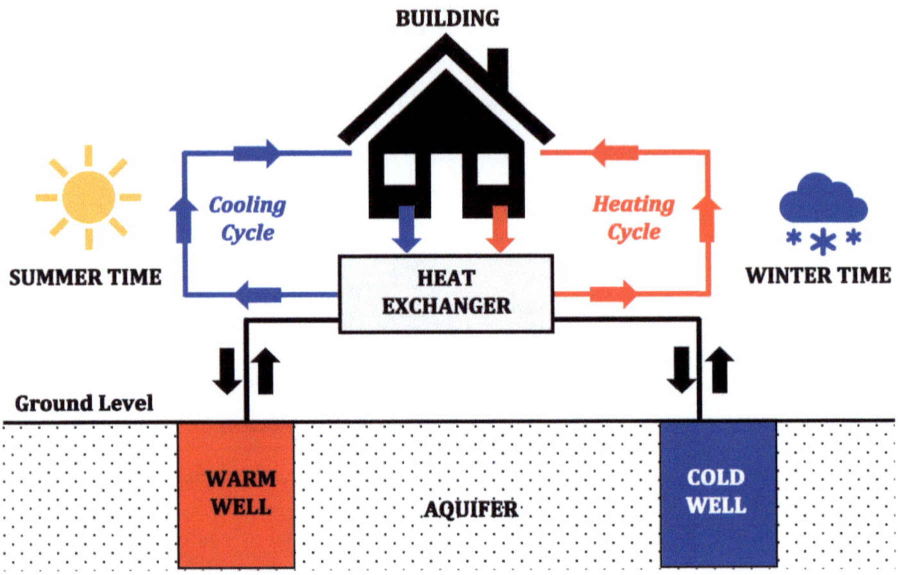


**Fig. 3.8** Schematic of aquifer thermal energy storage (ATES) system

system is schematically represented in Fig. 3.8 for long-term storage of thermal energy. The system merely consists of two wells (warm-water and cold-water wells) and water circulation loops to charge/discharge the thermal energy from the aquifers. During summertime, the excessive heat from the solar collector can be stored in the warm well to recover it during wintertime when there is demand for heat. Similarly, during wintertime, the cold return water from the building can be stored in the cold well and used for cooling purposes during summertime. Such systems are prevalent in Northern European countries and Canada. We will give some illustrative examples of the applications of ATES for long-term storing of thermal energy for heating and cooling purposes in Chapter 4.

ATES systems can be integrated into renewable energy systems to store excessive heat source during summertime directly. Solar- and geothermal-based renewables are possible heat sources for ATES systems. A large amount of stored heat can be used to supply district-heating networks or large buildings with high heating demand, such as airports, hotels, or hospitals. ATES systems are also integrated into industrial facilities to store excessive heat or waste heat from industrial processes. The ATES system charges the thermal energy during the low heating demand period, and the stored heat is unloaded from the ATES well to meet the high demand during peak heating periods. In Table 3.4, possible heat sources and users of ATES systems are listed.

Heat pump units can also be integrated into the ATES systems as an auxiliary heating/cooling equipment to meet dynamic heating or cooling demands of buildings. A schematic of the heat pump integrated ATES system is given in Fig. 3.9. During summertime, cold-water circulation is activated to extract heat from

**Table 3.4** Possible heat sources and users of ATES systems

| Heat sources | Heat users |
|---|---|
| *Renewable energy* | *Space heating* |
| Solar thermal | District heating |
| Geothermal | Large buildings |
| *Waste heat* | *Industrial heating* |
| Heat and power cogeneration | Batch or seasonal processes |
| Industrial process heat | Drying in the food industry |
| Waste incineration | Miscellaneous heating demands |
| *Load leveling* | *Agriculture* |
| In district-heating systems | Greenhouse heating |
| | Food drying |
| | Aquaculture |
| | *De-icing and snow melting* |
| | On roads, airports, sport centers, etc. |

Adapted from Dincer and Rosen (2011)

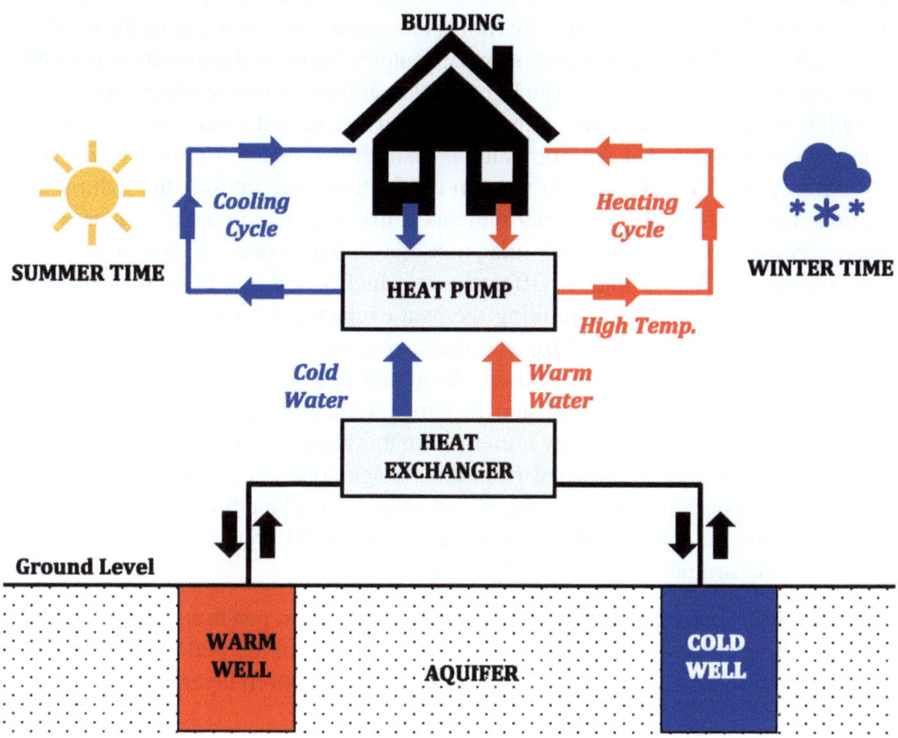


Fig. 3.9 Schematic of an ATES system with a heat pump

the building, and the relatively hot return water is injected into the warm well for use during wintertime. If further cooling is requested in the building, i.e., low-temperature or high cooling demand, the cold water is circulated through the heat pump unit to reduce the temperature of the water that is supplied to the building. A similar approach is followed during wintertime. If there is further heating demand or high-temperature request from the load side, i.e., building, the heat pump unit is activated. The warm water is circulated through the heat pump to achieve an increased temperature and enhanced heat transfer rates.

### 3.3.3 Solar Ponds

Inside a lake or pond, the incoming solar radiation increases the water temperature, and the hot water moves upward. The heat storage could not be established in an ordinary pond as there will be a heat loss through the cold surroundings from the hot water, which creates a layer at the upper regions of the pond. A thermal stratification inside the pond could be created by dissolving salt into the bottom layer of the pond. In this method, the salinity concentration is increased with the depth of the pond. As the density of water increases with salinity content, the natural convection is weaker at the bottom side of the pond. The lighter water at the top, on the other hand, moves more intensively. Between the intense upper convective and weak lower convective zones, there is a stagnant region, which is also known as the *non-convective zone*. The salinity concentration inside the non-convective zone inhibits the natural convection. The high-density hot water remains at the bottom, and the low-density cold water remains at the top of the non-convective zone. Consequently, the non-convective zone is like a buffer region which keeps the bottom of the solar pond at high temperature by limiting the heat exchange through the surface of the pond. The thermal regions and the variations in temperature and salinity concentration through the depth of the pond are illustrated in Fig. 3.10. The solar energy is trapped at the bottom of the pond, and the temperature of water increases in the lower convection zone. Thermal energy is charged in this region, so it is also called *storage zone*. The useful heat is extracted from the storage zone via a heat exchanger. The usage of solar ponds is attractive if there is a natural open land for water storage, high incoming solar radiation, and low-cost salt. The main differences between the solar ponds and the ordinary natural ponds are as follows:

- Clear water is used in solar ponds to ensure maximum penetration of sunlight.
- Salt is dissolved in water to limit natural convection at the bottom of the pond.
- Darkened bottom enhances the absorptivity of the bottom layer.

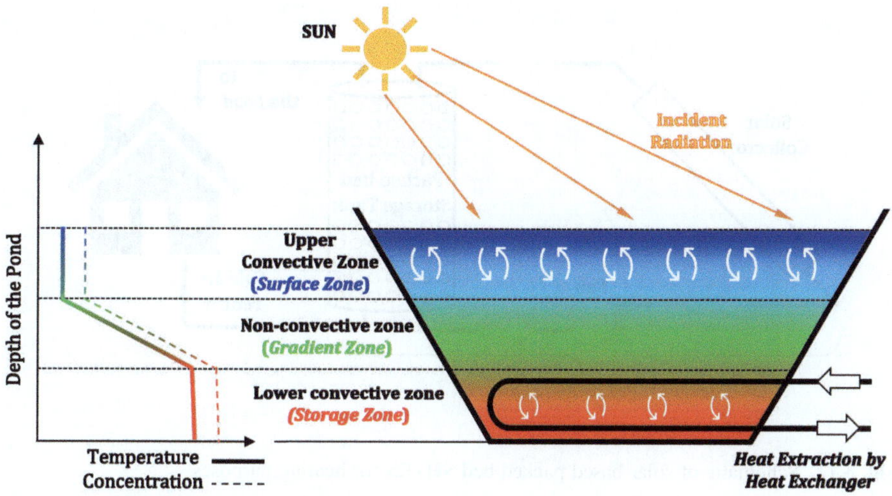


**Fig. 3.10** Layers of a typical solar pond with salinity gradient

### *3.3.4 Solid Storage Medium*

Rock beds, ceramic bricks, sand, and soil are the solid storage mediums mostly used in SHTES systems. Even though the volumetric heat capacities of the solids are lower than water (see Table 3.2), the temperature of the solid storage medium can reach up to 800 °C. That is, solid medium has the advantage of high-temperature variations. The solid medium has an additional advantage of possessing a natural thermal stratification. Unlike the liquid medium, there is no mixing effect inside the bed during the charging or discharging periods. Depending on the phase of the heat transfer fluid, there will be a weak natural convection current within the tank during the idle (or *storage*) period, but it does not cause a mixing zone as in the liquid storage mediums. The solid medium SHTES tanks are used in various applications such as residential or greenhouse heating and power generation systems. A packed bed storage tank system with a solar collector for residential heating is illustrated in Fig. 3.11. In this system, the hot air from the collector passes through the narrow gaps between the solid particles (e.g., rock or sand), and the temperature of the solid medium increases during daytime. The stored sensible energy inside the tank is retrieved by circulating the discharge loop during nighttime. In the case of an integrated SHTES system with a power generation plant, the thermal energy is retrieved either from solar or waste heat recovery unit for further usage. Various parameters that affect the charging or discharging performance of the solid storage SHTES system are as follows:

- The porosity of the tank
- The flow rate of the working fluid
- The type of the working fluid (e.g., water or air)

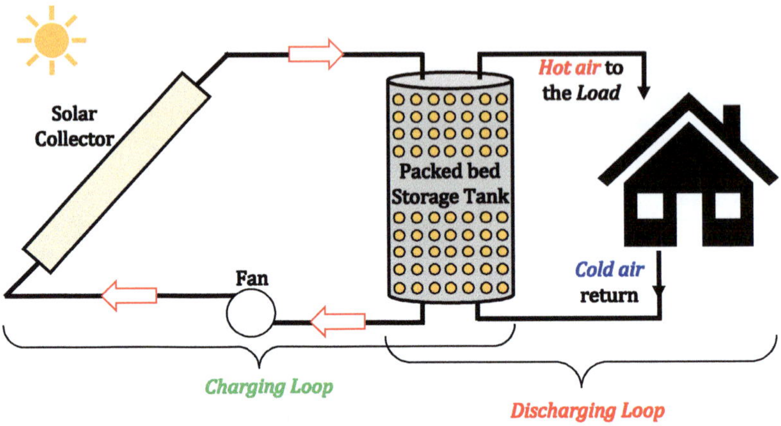


**Fig. 3.11** Schematic of solar-based packed bed SHTES for heating purposes

**Table 3.5** Thermal properties of candidate rocks for SHTES systems

| Material | Specific heat (J/kgK) | Density (kg/m$^3$) | Thermal conductivity (W/mK) |
|---|---|---|---|
| Quartzite | 626 ± 12 | 2615 ± 1.68 | 5.39 ± 0.07 |
| Calcareous sandstone | 652 ± 31 | 2661 ± 1.71 | 4.36 ± 0.15 |
| Helvetic siliceous limestone | 669 ± 15 | 2776 ± 2.95 | 3.60 ± 0.21 |
| Limestone | 683 ± 15 | 2697 ± 1.66 | 2.82 ± 0.06 |
| Gabbro | 643 ± 24 | 2911 ± 2.69 | 2.05 ± 0.08 |

Adapted from Jemmal et al. (2015)

- The type of solid medium
- The arrangement of the solid materials
- The shape of the tank
- The integration of the storage tank with the source and the demand

Rocks are widely used in high-temperature solid medium SHTES systems. Jemmal et al. (2015) recently characterized the thermal properties of different types of rocks (see Table 3.5) and listed the following criteria for selecting a suitable solid medium for an SHTES unit:

- High volumetric heat capacity allows storing a large amount of thermal energy in a volume.
- Possessing high thermal conductivity enhances dynamic response of the system.
- Low porosity increases the density and strength of the material.
- Low water content: possible water loss causes reduction in rock hardness and consequently physical damage.
- High strength increases the mechanical stability of the system undergoing long-term usage with thermal expansion/compression cycles.

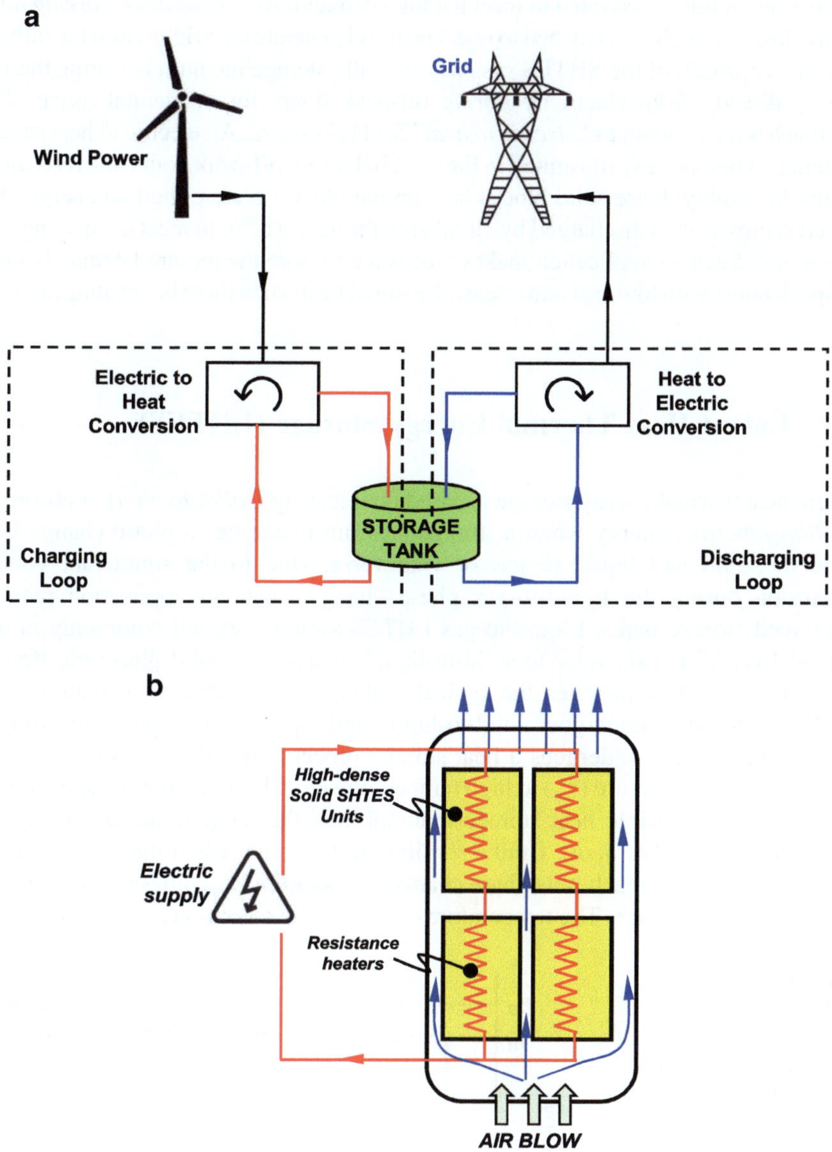


**Fig. 3.12** Sensible heat thermal energy storage from electricity. (**a**) Wind to thermal and thermal to electric SHTES system. (**b**) Electric to thermal SHTES system

Note that SHTES systems with a solid storage medium are also integrated into wind farms to store excess power production as thermal energy. In Fig. 3.12a, a conceptual design of a "*wind to heat*" SHTES system is illustrated. During off-peak hours, charging mode is activated, and the excessive electricity production is used to heat up the rock that is filled inside a well-insulated storage tank. During peak hours, the discharging mode is activated to meet the high demand. Air (or *another working fluid*) is circulated through the rock bed to reject heat and generate electricity from the turbine. Another approach of the SHTES system with solid storage medium is storing thermal energy directly from electricity during off-peak hours for residential usage. This approach is also known as "*electric to heat*" SHTES system. An electric to heat storage system is schematically illustrated in Fig. 3.12b. During off-peak hours, the resistances within the highly dense solid blocks heat up the blocks to store thermal energy. The stored energy is then discharged by circulating the air via a fan to meet the heating load of a house. Such an application makes it possible to store the required demand during off-peak hours with low cost and release the stored heat when there is a heating demand.

## 3.4 Latent Heat Thermal Energy Storage (LHTES)

Latent heat thermal energy storage is based on releasing (*solidification*) or absorbing (*melting*) thermal energy when a storage medium undergoes a phase change from solid to liquid and liquid to gas or vice versa. Due to the significant volume expansion during the liquid-to-gas phase change, such an application requires reinforced storage tanks. Liquid-to-gas LHTES systems are not commonly in use, that is, LHTES mostly refer to solid-to-liquid or liquid-to-solid phase transformations. Figure 3.13 illustrates the typical enthalpy – temperature variation of an LHTES unit which undergoes solid-to-liquid and liquid-to-solid phase transformations. If the system undergoes a heat storage process, one should follow the path from the low temperature ($T_A$) to high temperature ($T_B$). Initially, the storage medium behaves like a sensible heat storage material, and the temperature of the storage medium rises (*absorb heat*). Unlike sensible heat storage, when the temperature of the storage medium reaches its phase change temperature ($T_m$), phase transformation (or *melting*) takes place. The temperature of the storage medium remains at the phase

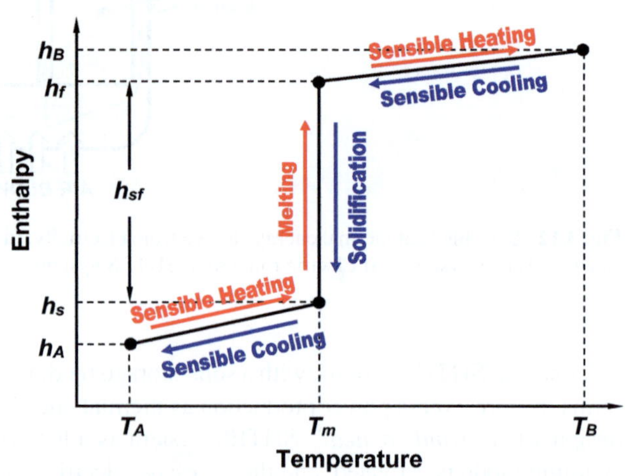


**Fig. 3.13** Enthalpy variation of a medium during charging–discharging periods of an LHTES unit

change temperature until the medium absorbs a specific amount of thermal energy. Complete phase change is achieved when the internal energy of the medium is increased by the value of the latent heat of fusion ($h_{sf}$). Further heat supply to the storage medium increases its temperature, and the internal energy of the system varies in the sensible heat form. In LHTES units, during heating or cooling processes, the storage medium undergoes a phase change (*solidification* or *melting*), that is, the storage medium of an LHTES unit is also known as a *phase change material* (PCM). The energy variation of an LHTES system during the heating or cooling processes is determined as follows:

$$\text{Heating Process:}\quad \Delta E = \int_{h_A}^{h_B} m\,dh = m(h_B - h_A) \text{ or}$$

$$\Delta E = \underbrace{\int_{T_A}^{T_m} (mc)_S dT}_{\textit{Sensible Heating}} + \underbrace{mh_{sf}}_{\textit{Melting}} + \underbrace{\int_{T_m}^{T_B} (mc)_L dT}_{\textit{Sensible Heating}} \tag{3.3}$$

$$\text{Cooling Process:}\quad \Delta E = \int_{h_B}^{h_A} m\,dh = m(h_A - h_B) \text{ or}$$

$$\Delta E = \underbrace{\int_{T_B}^{T_m} (mc)_L dT}_{\textit{Sensible Cooling}} + \underbrace{mh_{sf}}_{\textit{Solidification}} + \underbrace{\int_{T_m}^{T_A} (mc)_S dT}_{\textit{Sensible Cooling}} \tag{3.4}$$

where $h_{sf}$ (in J/kg) is the latent heat of fusion of the PCM and the subscripts $S$ and $L$ correspond to the solid and liquid phases of the PCM. Total energy variation in an LHTES unit involves two forms of internal energy: *sensible* heat and *latent* heat. Sensible heating or cooling is related to the specific heat of the storage medium and the temperature variation. The latent heat part of the energy variation is usually much higher than the sensible heating or cooling component. For instance, in the case of water, the latent heat of fusion (*solid-to-liquid* phase change) is $h_{sf} = 333.4$ kJ/kg. The total energy required to turn 1 kg of ice into water is 80 times greater than the energy that is desired to raise the temperature of 1 kg of water by 1 °C. It means that the latent heat storage has higher energy storage density in a smaller volume, or less material usage, in comparison to sensible heat storage. The latent heat of evaporation of water has the order of 2 MJ/kg which is almost ten times that of the solid-to-liquid transformation. However, the high-volume change during the liquid-to-vapor phase change is not useful in practical applications. Solid-to-liquid phase change is useful because a relatively large amount of heat can be stored in PCMs at a narrow temperature range without a corresponding high-volume change.

LHTES systems include the following primary components:

- *Phase change material* (PCM): Heat storage medium that undergoes a phase change. The phase transition temperature of the PCM should be selected considering the operating temperature range of the thermal application.
- *Heat transfer fluid* (HTF): The secondary material that transports the heat between the source (or *load*) and the PCM. Working conditions, phase, and the path of the HTF significantly influence the system efficiency.
- *PCM container*: The PCMs are encapsulated within rigid or elastic containers to prevent possible leakages of liquid PCM. PCM containers should be designed adequately to provide improved heat transfer between the PCM and the HTF.

### 3.4.1 Phase Change Material

PCM is one of the most critical parts of an LHTES system, and it should possess some characteristics to improve energy storage density and provide thermal stability under long-term cycling. Nomura et al. (2010) briefly listed the main characteristics of PCMs as given in Table 3.6. Thermal, physical, and chemical properties and economic factors should be taken into account to select a proper PCM in an LHTES system. There is a strong relationship between system performance and PCM characteristics. Zhang et al. (2009) prepared a simple chart to illustrate this relationship (see Fig. 3.14). PCM is the key component of an LHTES unit and should be selected wisely.

There are various types of materials that are used as PCM in different LHTES applications. Figure 3.15 presents a classification of PCMs that are used for solid-to-liquid phase transformation. There are three main groups: (*i*) *organics* (e.g., paraffin, fatty acids, glycols), (*ii*) *inorganics* (e.g., salt hydrates and metallic), and (*iii*) *eutectics* (e.g., a mixture of organic and/or inorganic materials). In Tables 3.7 and

**Table 3.6** Characteristics of PCMs

| Thermal properties | Physical properties | Chemical properties | Economic factors |
|---|---|---|---|
| 1. Suitable phase change temperature range | 5. Large density | 8. Long-term chemical stability | 12. Cheap |
| | 6. Small volume change during phase change | | 13. Abundant |
| 2. Large heat of fusion | | 9. Compatibility with construction and container materials | |
| 3. Large specific heat | 7. No sub-cooling | | |
| 4. Large thermal conductivity in both solid and liquid phases | | 10. Completely reversible phase change | |
| | | 11. Nontoxic, nonflammable | |

Adapted from Nomura et al. (2010)

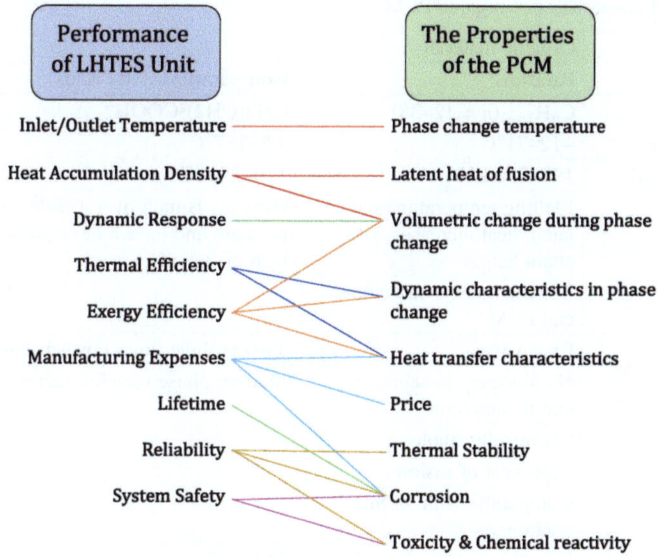


**Fig. 3.14** Relationship between system performance and PCM properties (Data from Zhang et al. 2009)

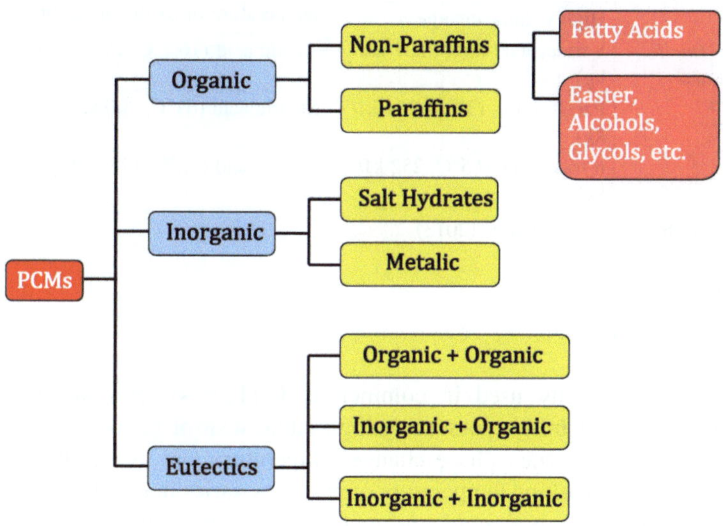


**Fig. 3.15** Main groups of phase change materials (PCMs)

3.8, the main characteristics of each material group are compared considering the thermophysical properties, cost, advantages, and disadvantages of system integration.

**Table 3.7** Characteristics of organic PCMs

| | Organic material | |
|---|---|---|
| Property | Paraffins | Non-paraffins (fatty acids) |
| Formula | $C_nH_{2n+2}$($n$ = 12–38) | CH3(CH2)•COOH |
| Melting point | −12–71°C | 7.8–187°C |
| Heat of fusion | 190–260 kJ/kg | 130–250 kJ/kg |
| Features | Melting temperature and latent heat increases with chain length | Fatty acids melt over a wide range of temperature and have a large variation in latent heat of fusion |
| | It is the most used commercial PCM | |
| Cost | Expensive | Two to three times costlier than paraffins |
| Advantages | No tendency to segregate and to supercool | Sharper phase transformation |
| | Chemically stable | |
| | High heat of fusion | |
| | Compatible with all metal containers | |
| Disadvantages | Low thermal conductivity | Mildly corrosive |
| | Do not have a sharp well-defined melting point | Highly flammable |
| | Flammable | Should not be exposed to excessively high temperature or oxidizing agents |
| | High volume change | |
| Examples *(melting point and latent heat)* | *n*-tridecane (4.5°C, 231 kJ/kg) | Acetic acid (16.7°C, 187 kJ/kg) |
| | Paraffin wax (32°C, 251 kJ/kg) | Stearic acid (61°C, 200 kJ/kg) |
| | n-tricontane (65°C, 252 kJ/kg) | Lauric acid (42°C, 178 kJ/kg) |

Adapted from Rathod and Banerjee (2013)

#### 3.4.1.1 Paraffins

Paraffins are commonly used in commercial LHTES systems since they have excellent chemical stability, possessing high latent heat of fusion and maintaining the thermal properties (i.e., phase change temperature and latent heat) after long thermal cycles. The melting temperature of paraffins varies in a wide range, and they are good candidates for both *cold storage* (*subzero temperatures*) and *heat storage* (*high temperature*) applications. However, paraffins possess low thermal conductivity (~0.2 W/mK). This is the primary drawback of paraffins. For systems with limited durations for charging and discharging, paraffins with low thermal conductivity will reduce system performance. There are various techniques to improve the heat transfer between the paraffin and the working fluid. The usage of extended surfaces (*fins*) on the inner or outer walls of the containers, dispersing micro-/nanoparticles with high thermal conductivity inside the paraffin, and developing

**Table 3.8** Characteristics of inorganic and eutectic PCMs

| | Inorganic material | |
|---|---|---|
| Property | Salt hydrates | Eutectics |
| Formula | AB • $nH_2O$ | – |
| Melting point | 11–120 °C | 4–93 °C |
| Heat of fusion | 100–200 kJ/kg | 100–230 kJ/kg |
| Features | Oldest and most studied | Composition of two or more components |
| | Its alloys of inorganic salts and water | Melts and freeze without segregation |
| Cost | Low cost | Costly |
| Advantages | Easy availability | High heat of fusion per unit volume |
| | Sharp melting point | High conductivity |
| | High thermal conductivity and density | |
| | Low volume change | |
| Disadvantages | Supercooling | Low heat of fusion per unit weight |
| | Corrosion on metal container | |
| | Due to higher density, salts settle at bottom and reduce active volume | |
| Examples *(melting point and latent heat)* | $CaCl_2$ • $6H_2O$ (30 °C 170–192 kJ/kg) | $Na_2SO_4$ + NaCl + KCl + $H_2O$ (4 °C, 234 kJ/kg) |
| | $Na_2SO_4$ • $10H_2O$ (32 °C, 251 kJ/kg) | $NH_2CONH_2$ + $NH_4NO_3$ (46 °C, 95 kJ/kg) |
| | $MgSO_4$ • $7H_2O$ (48.4 °C, 200 kJ/kg) | |

Adapted from Rathod and Banerjee (2013)

paraffin-embedded metal matrix composites are commonly followed approaches to improve heat penetration within paraffins. Some illustrative examples of heat transfer enhancements for PCMs are given in Fig. 3.16. In Fig. 3.16a, fins are placed inside the PCM cavity to improve heat transfer and reduce the time for complete melting/solidification (Pakrouh et al. 2015). Figure 3.16b, on the other hand, represents a novel composite PCM composed of paraffin, expanded graphite, and epoxy (Wang et al. 2017). The proposed shape-stable composite PCM has improved thermal conductivity (~2.0 W/mK) with high mechanical strength.

#### 3.4.1.2 Salt Hydrates

In comparison to paraffins, salt hydrates have higher thermal conductivity (~0.5 W/mK) with low cost. However, there are two critical issues for salt hydrates: (*i*) *supercooling* and (*ii*) *phase segregation*. These problems are encountered for the

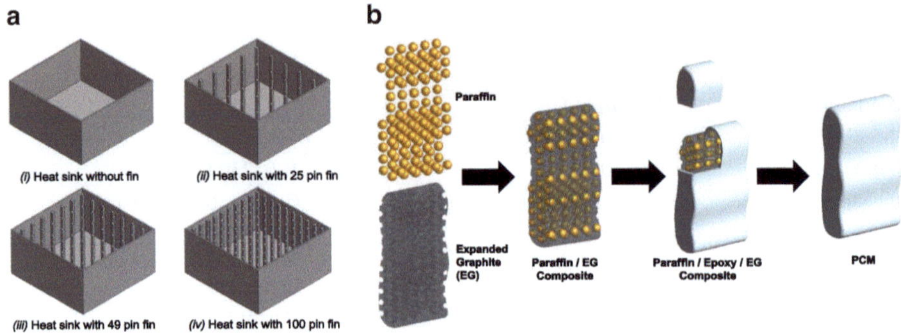


**Fig. 3.16** Some techniques to improve heat transfer rate of paraffins. (**a**) Extended surfaces inside PCM cavity (Pakrouh et al. 2015). (**b**) Paraffin with expanded graphite matrix (Wang et al. 2017)

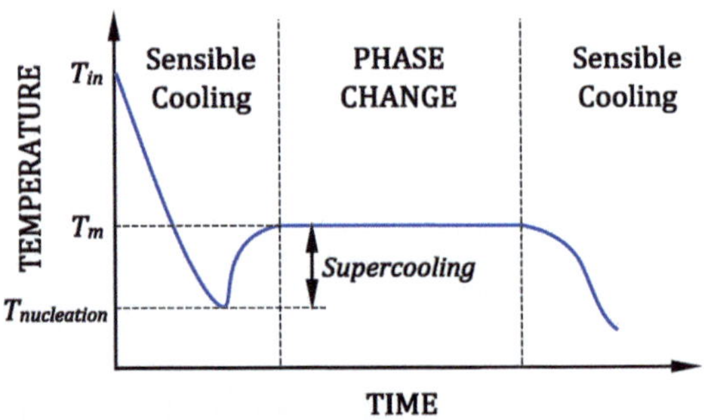


**Fig. 3.17** The supercooling of a PCM

salt hydrates depending on the production process, the integrants, or the amount of water. Supercooling is a crucial problem that causes poor nucleation of the substance. The supercooling phenomenon is illustrated in Fig. 3.17. During the cooling of a substance, the temperature drops below the melting point without initiating solidification. The nucleation starts at a temperature that is far below the melting point, and the temperature of the substance suddenly increases to the melting temperature. The degree of supercooling is defined as the difference between the melting temperature and the nucleation temperature ($T_m - T_{nucleation}$). The degree of supercooling should be kept small to maintain reversible charging/discharging cycles within an acceptable temperature range. Some nucleation agents are dispersed into the PCM to reduce the supercooling effect. Another approach to resolve the supercooling is the incomplete melting of the PCM during the charging process. The solid PCM particles that remain inside the liquid phase at the end of charging ensure initiation of solidification during the cooling mode. Intentionally placing weak insulations at some parts of PCM containers, so-called cold fingers, also helps to form solid particles within the tank. Putting some impurities inside the pure PCMs is a simple and common approach to catalyze the nucleation. Several advanced methods are also under development to trigger nucleation, such as the application of vibration, ultrasound, magnetic, and electrical fields.

Fig. 3.18 The phase segregation of a PCM

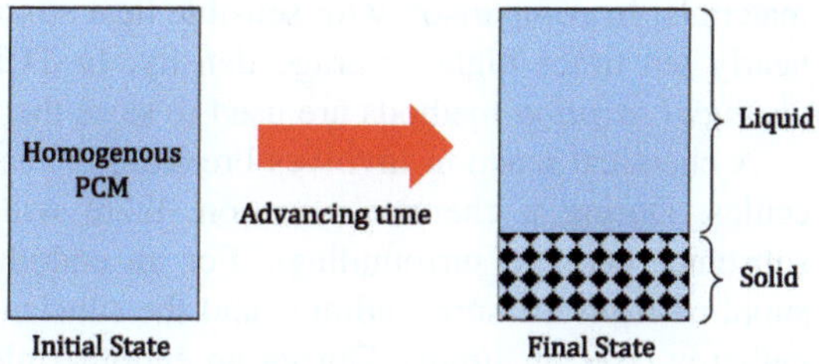


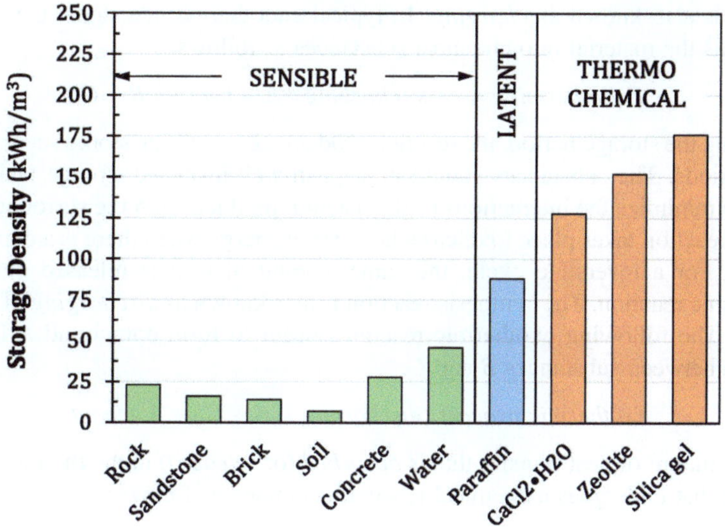


**Fig. 3.19** Storage densities of commonly used TES materials (Data from Aydin et al. 2015)

Phase segregation is another crucial problem of salt hydrates. For an improperly produced salt hydrate, the dense salt can split from the water and settle at the bottom of the container (Fig. 3.18). It is an undesirable situation since the segregated PCM does not retain the desired thermal properties (*melting temperature* or *latent heat*), and it is not possible to use this multiphase material in an LHTES. The PCM should retain its thermal, chemical, and physical properties during long thermal melting/freezing cycles without phase separation or supercooling. Thickening agents are mostly used to prevent phase segregation.

## 3.5 Thermochemical Energy Storage

Thermochemical energy storage (TCES) is a promising technique with great energy storage density and high cyclic efficiency. Figure 3.19 compares the storage densities of commonly used sensible heat TES, latent heat TES, and thermochemical TES materials. In comparison with sensible heat storage, the TCES method can provide nearly ten times higher storage density. In TCES systems, *chemical reaction* or *chemical sorption* methods are used to store thermal energy.

A chemical reaction involves breaking or restoring the chemical bonds of molecules. During a chemical reaction, there will be heat interaction between the substance and the surroundings. For an endothermic reaction, thermal energy is supplied from the surroundings, and the substance absorbs the heat as the reaction enthalpy ($\Delta H_r$ in J/mol). During an endothermic reaction, a significant amount of thermal energy is stored within the chemical bonds of the substance, and hence this process is also known as *charging*. In typical endothermic reactions, the supplied heat splits the material into chemical substances as follows:

$$\textit{Endothermic process (charging): } A + heat \rightarrow B + C$$

During the storage period, the reaction products, *B* and *C*, are stored separately for long periods. The substances should be kept in well-insulated storage tanks separately to minimize the interactions of the reaction products with the surroundings. A reverse reaction takes place to release the thermal energy when there is a demand for heating. For a reversible cycle, the same amount of heat is released during the exothermic reaction. The synthesis reaction is also known as *discharging* of thermal energy. The following exothermic reactions occur to form compound *A* from the reaction between substances *B* and *C*:

$$\textit{Exothermic process (discharging): } B + C \rightarrow A + heat$$

The amount of heat transfer that is *absorbed* (or *released*) in the thermochemical material that undergoes a chemical reaction is defined as follows:

$$Q_{TCES} = n_A \Delta H_r \tag{3.5}$$

where $n_A$ and $\Delta H_r$ correspond to mol number and the reaction enthalpy of compound *A*. A reversible TCES charging and discharging cycle is schematically illustrated in Fig. 3.20. Endothermic reaction absorbs the thermal energy that is provided by a renewable source or waste heat of an industrial process. The reaction splits compound *A* into substances *B* and *C*. The reaction products are then stored for further usage. After a long-term storage period, the stored energy is released to meet demand by an exothermic reaction. Compound *A* is formed by the reaction between substances *B* and *C*.

Heat storage with chemical reaction is achieved by one of the following interactions: *gas-gas*, *liquid-gas*, and *solid-gas*. Representative chemical reactions are listed in Table 3.9 for each type of phase interactions.

Adsorption (*solid* materials) or absorption (*liquid* medium) processes are also used to store/release a significant amount of thermal energy for a long or short term (Hauer 2007). A schematic of the adsorption (*exothermic*)/desorption (*endothermic*) cycle is given in Fig. 3.21. Water molecules (either in *liquid* or *gas* phase) are bonded on the inner surface of adsorbents during the adsorption process. The reverse process is called desorption. During the desorption (*charging*) process, thermal

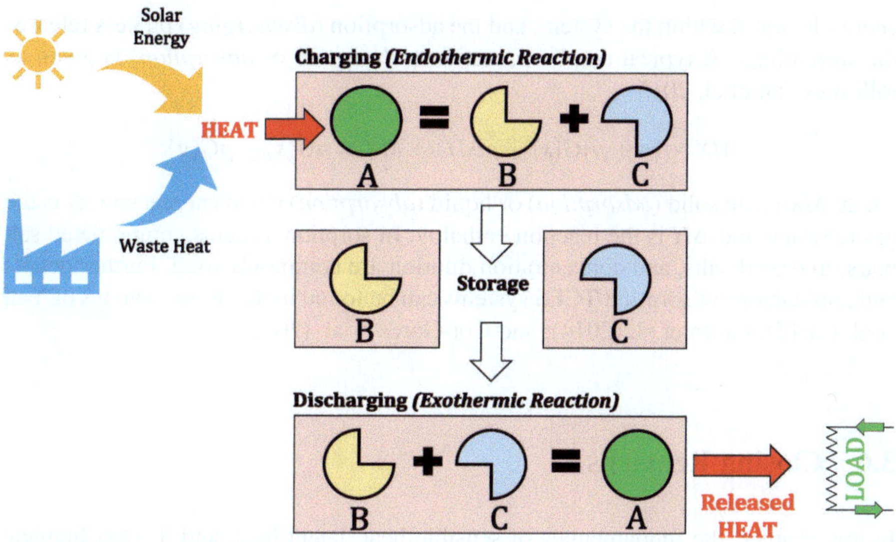


**Fig. 3.20** Schematic of the thermochemical energy storage cycle

**Table 3.9** Representative reactions for *gas–gas*, *liquid–gas*, and *solid–gas* TCES

| Gas–gas reaction | Liquid–gas reaction | Solid–gas reaction |
|---|---|---|
| $2NH_3(g) \leftrightarrow N_2(g) + 3H_2(g)$ | $(CH_3)_2CHOH(l) \rightarrow (CH_3)_2CO(g) + H_2(g)$ | $MO_{2x+1}$ + heat $\rightarrow MO + xO_2$ |
| $2SO_3(g) \leftrightarrow 2SO_2(g) + O_2(g)$ | $NH_4HSO_4(l) \rightarrow NH_3(g) + H_2O(g) + SO_3(g)$ | $BaO_2 \rightarrow BaO + 1/2O_2$ |
| $CH_3OH(g) \leftrightarrow 2H_2(g) + CO(g)$ | | $ZnO \rightarrow Zn + 1/2O_2$ |
| $C_6H_{12}(g) \leftrightarrow C_6H_6(g) + 3H_2(g)$ | | $PbCO_3(g) \rightarrow PbO(s) + CO_2(g)$ |
| $CH_4(g) + H_2O(g) \rightarrow CO(g) + 3H_2(g)$ | | $Ca(OH)_2(s) \rightarrow CaO(s) + H_2O(g)$ |
| $CH_4(g) + CO_2(g) \rightarrow 2CO(g) + 2H_2(g)$ | | $Mg(OH)_2(s) \rightarrow MgO(s) + H_2O(g)$ |

**Fig. 3.21** Charging and discharging of thermochemical heat by sorption reactions

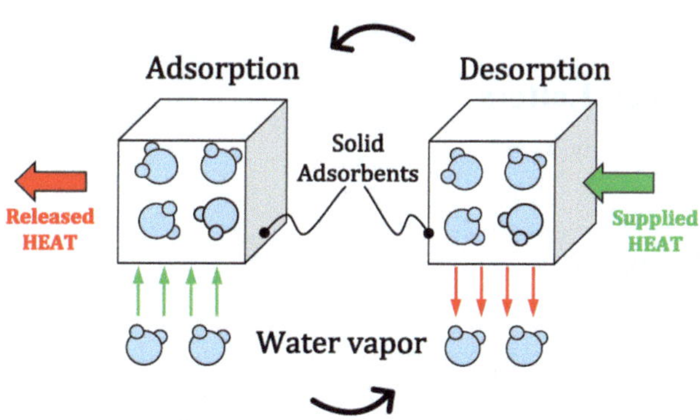

energy is stored within the system, and the adsorption (*discharging*) process releases the stored heat. A typical sorption reaction (*adsorption* or *absorption*) is given as follows (Yan et al. 2015):

$$MX \times (n+p)G(s) + p\Delta H \Leftrightarrow MX \times nG(s) + pG(g)$$

where *MX* is the solid (*adsorption*) or liquid (*absorption*) chemical reactant. *G* is the gas reactant, and $\Delta H$ is the reaction enthalpy. In sorption systems, ammoniated salt pairs, hydrated salts, and concentration dilution are commonly used. Further details and applications of sorption TCES systems can be found in the recent reviews of Yan et al. (2015), Pardo et al. (2014), and Cot-Gores et al. (2012).

## 3.6 Closing Remarks

In this chapter, the fundamentals of sensible heat, latent heat, and thermochemical TES techniques are discussed. The details of liquid and solid medium sensible heat TES methods are introduced to outline the main concerns about each method. The advantages and disadvantages of the commonly used phase change materials (PCMs) are also presented in detail. Some novel approaches to overcome the drawbacks of PCMs are also revealed. The ongoing research in the field of TES focuses on developing systems with high performance and low cost to widen TES applications.

## Nomenclature

| | |
|---|---|
| $c$ | Specific heat, J/kgK |
| $C$ | Volumetric heat capacity, J/m$^3$K |
| $E$ | Energy, J |
| $\Delta H$ | Reaction enthalpy, J/mol |
| $h$ | Specific enthalpy, J/kg |
| $m$ | Mass, kg |
| $n$ | Mole number |
| $T$ | Temperature, K or °C |

## Greek Letters

| | |
|---|---|
| $\rho$ | Density, kg/m$^3$ |

## Subscripts

| | |
|---|---|
| $L$ | Liquid |
| $S$ | Solid |
| $sf$ | Solid to liquid |

## References


Abdoly, M. A., & Rapp, D. (1982). Theoretical and experimental studies of stratified thermocline storage of hot water. *Energy Conversion and Management, 22*, 275–285.

Andersen, E., Furbo, S., & Fan, J. (2007). Multilayer fabric stratification pipes for solar tanks. *Solar Energy, 81*, 1219–1226.

Aydin, D., Casey, S. P., & Riffat, S. (2015). The latest advancements on thermochemical heat storage systems. *Renewable and Sustainable Energy Reviews, 41*, 356–367.

Bahnfleth, W. P., & Song, J. (2005). Constant flow rate charging characteristics of a full-scale stratified chilled water storage tank with double ring slotted pipe diffusers. *Applied Thermal Engineering, 25*, 3067–3082.

Chan, A. M. C., Smereka, P. S., & Giusti, D. (1983). A numerical study of transient mixed convection flows in a thermal storage tank. *Transactions of the ASME. Journal of Solar Energy Engineering, 105*, 246–253.

Cot-Gores, J., Castell, A., & Cabeza, L. F. (2012). Thermochemical energy storage and conversion: A-state-of-the-art review of the experimental research under practical conditions. *Renewable and Sustainable Energy Reviews, 16*(7), 5207–5224.

Dincer, I., & Rosen, M. (2011). *Thermal energy storage: Systems and applications* (2nd ed.). Hoboken: Wiley.

Haller, M. Y., Cruickshank, C. A., Streicher, W., Harrison, S. J., Andersen, E., & Furbo, S. (2009). Methods to determine stratification efficiency of thermal energy storage processes–review and theoretical comparison. *Solar Energy, 83*(10), 1847–1860.

Hasnain, S. M. (1998). Review on sustainable thermal energy storage technologies, part I: Heat storage materials and techniques. *Energy Conversation and Management, 39*, 1127–1138.

Hauer, A. (2007). *Thermal energy storage for sustainable energy consumption. Sorption theory for thermal energy storage* (pp. 393–408). Dordrecht: Springer.

Huhn, R. (2007). *Beitrag zur thermodynamischen Analyse und Bewertung von Wasserwarmespeichern in Energieumwandlungsketten*. PhD thesis, Technische Universitat Dresden, Germany, p. 21.

Jemmal, Y., Zari, N., & Maaroufi, M. (2015). *Study of rock suitability for high temperature thermal energy storage in concentrated solar tower power plants*. In Renewable and Sustainable Energy Conference (IRSEC), 2015 3rd International (pp. 1–6). IEEE.

Kandari, A. M. (1990). Thermal stratification in hot storage-tanks. *Applied Energy, 35*, 299–315.

Lavan, Z., & Thompson, J. (1977). Experimental study of thermally stratified hot water storage tanks. *Solar Energy, 19*, 519–524.

Moretti, P. M., & McLaughlin, D. K. (1977). Hydraulic modeling of mixing in stratified lakes. *Journal of the Hydraulics Division, 103*, 367–380.

Nomura, T., Okinaka, N., & Akiyama, T. (2010). Technology of latent heat storage for high temperature application: A review. *ISIJ International, 50*, 1229–1239.

Pakrouh, R., Hosseini, M. J., Ranjbar, A. A., & Bahrampoury, R. (2015). A numerical method for PCM-based pin fin heat sinks optimization. *Energy Conversion and Management, 103*, 542–552.

Paksoy, H., Snijders, A., & Stiles, L. (2009). *State-of-the-art review of aquifer thermal energy storage systems for heating and cooling buildings*. In Proceedings of the EFFSTOCK conference, Stockholm.

Panthalookaran, V., Heidemann, W., & Müller-Steinhagen, H. (2007). A new method of characterization for stratified thermal energy stores. *Solar Energy, 81*, 1043–1054.

Pardo, P., Deydier, A., Anxionnaz-Minvielle, Z., Rougé, S., Cabassud, M., & Cognet, P. (2014). A review on high temperature thermochemical heat energy storage. *Renewable and Sustainable Energy Reviews, 32*, 591–610.

Rathod, M. K., & Banerjee, J. (2013). Thermal stability of phase change materials used in latent heat energy storage systems: A review. *Renewable and Sustainable Energy Reviews, 18*, 246–258.

Rosen, M. A. (1992). Appropriate thermodynamic performance measures for closed systems for thermal energy storage. *Transactions of the ASME. Journal of Solar Energy Engineering, 114*, 100–105.

Rosengarten, G. (1999). A second law approach to characterising thermally stratified hot water storage with application to solar water heaters. *Transactions of the ASME. Journal of Solar Energy Engineering, 121*, 194–200.

Safari, A., Saidur, R., Sulaiman, F. A., Xu, Y., & Dong, J. (2017). A review on supercooling of Phase Change Materials in thermal energy storage systems. *Renewable and Sustainable Energy Reviews, 70*, 905–919.

Shah, L. J., & Furbo, S. (2003). Entrance effects in solar storage tanks. *Solar Energy, 75*, 337–348.

Sharma, A., Tyagi, V. V., Chen, C. R., & Buddhi, D. (2009). Review on thermal energy storage with phase change materials and applications. *Renewable and Sustainable Energy Reviews, 13*, 318–345.

Shimizu, M., & Fujita, K. (1985). Actual efficiencies of thermally-stratified thermal storage tanks. *IEA Heat Pump Center Newsletter, 3*(1/2), 20–25.

Shyu, R. J., & Hsieh, C. K. (1987). Unsteady natural convection in enclosure with stratified medium. *Transactions of the ASME. Journal of Solar Energy Engineering, 109*, 127–133.

Sliwinski, B. J., Mech, A. R., & Shih, T. S. (1978). Stratification in thermal storage during charging. *Proceedings of the 6th International Heat Transfer Conference, Toronto, 4*, 149–154.

Stefan, H. G., & Gu, R. (1992). Efficiency of jet-mixing of temperature stratified water. *Journal of Environmental Engineering, ASCE, 118*, 363–379.

Tran, N., Kreider, J. F., & Brothers, P. (1989). Field measurements of chilled water storage thermal performance. *ASHRAE Transactions, 95*, 1106–1112.

van Berkel, J., Rindt, C. C. M., & van Steenhoven, A. (1999). Modelling of two-layer stratified stores. *Solar Energy, 67*, 65–78.

Wang, Z., Situ, W., Li, X., Zhang, G., Huang, Z., Yuan, W., et al. (2017). Novel shape stabilized phase change material based on epoxy matrix with ultrahigh cycle life for thermal energy storage. *Applied Thermal Engineering 123*, 1006–1012.

Wu, L., & Bannerot, R. B. (1987). Experimental study of the effect of water extraction on thermal stratification in storage. *Proceedings of the 1987 ASME-JSME-JSES Solar Energy Conference, Honolulu, 1*, 445–451.

Yan, T., Wang, R. Z., Li, T. X., Wang, L. W., & Fred, I. T. (2015). A review of promising candidate reactions for chemical heat storage. *Renewable and Sustainable Energy Reviews, 43*, 13–31.

Zhang, R. Y., et al. (2009). *Phase change material and phase change energy storage technology* (pp. 130–181). Beijing: Science Press.

# Chapter 4
# Thermal Energy Storage Applications

## 4.1 Introduction

Thermal energy storage (TES) methods are integrated into a variety of thermal applications, such as in *buildings* (for *hot water, heating, and cooling purposes*), *solar power generation systems*, and *greenhouses* (for *heating or cooling purposes*) to achieve one or more of the following advantages:

- Remove mismatch between supply and demand
- Improve system efficiency
- Balance fluctuations in temperature or thermal load
- Provide economic advantages
- Reduce investment, working, and maintenance costs

In this chapter, we focus on some common applications of TES processes and systems. For each TES method, i.e., sensible or latent, illustrative applications are represented by providing alternative system designs, comparative results, recent advances, and developments.

## 4.2 Building Applications with TES

Energy consumption of residential buildings may vary depending on the economic level of the household, the location of the building, and the climatic conditions. Figure 4.1 shows the results of a survey that was conducted by the US Energy Information Administration on residential energy consumption in 2009. According to Fig. 4.1a, the total energy consumption related to space heating, water heating, and air conditioning is 66% of the overall energy demand of a home in the USA. Figure 4.1b, on the other hand, compares the share of total energy consumption in 1980 and 2009. Even though the population in the USA and number of homes



I. Dincer, M. A. Ezan, *Heat Storage: A Unique Solution For Energy Systems*, Green Energy and Technology, https://doi.org/10.1007/978-3-319-91893-8_4

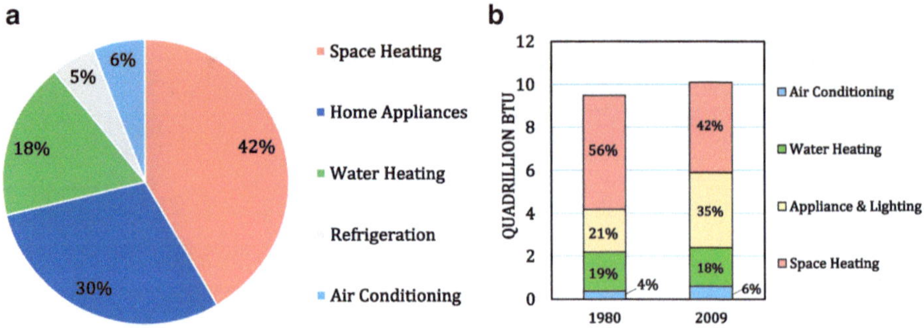


**Fig. 4.1** Statistics on energy consumption in the USA. (**a**) The share of residential energy consumption in the USA in 2009, (**b**) comparison between 1980 and 2009. (Data from US Energy Information Administration, 2018)

increased by 30% and 40%, respectively, there is a slight increment in the overall energy consumption. Two approaches helped slow down the increment in the overall energy consumption of residential buildings: (*i*) improvements in building insulation significantly reduced heating/cooling loads, and (*ii*) enhanced efficiencies of heating/cooling devices and water heaters decreased electricity consumption. Further improvements in building envelope or in heating/cooling equipment could be achieved by integration of TES units in buildings.

Figure 4.2 outlines the TES techniques that are used to reduce heating/cooling demand of buildings. TES systems provide not only improved efficiency of the current systems but also the integration of renewables to heating/cooling systems throughout the day or year and regulate temperature fluctuations. In *passive* systems, the thermal mass of the building envelope is increased by using materials with higher thermal energy storage capacity, either sensible or latent. The improved thermal mass maintains the indoor temperature within a comfort range with less effort. In *active* systems, high-temperature (*heat* storage) or low-temperature (*cold* storage) thermal energy can be stored within dedicated tanks or inside the channels of the air-conditioning system to future use. There are various applications for long-term or short-term heat/cold storage in buildings. For instance, in Northern Europe, snow is stored within large snow pools in wintertime for long-term storage. The stored energy is retrieved to cool buildings during the summertime. As a short-term application, the air-conditioning unit makes ice within large insulated tanks during nighttime, when the purchased electricity is cheap and the ambient temperature is lower. The stored *cold* energy is then discharged from the tank during daytime to maintain building temperature within the comfort range. Passive and active TES techniques in buildings are explained in the following sections.

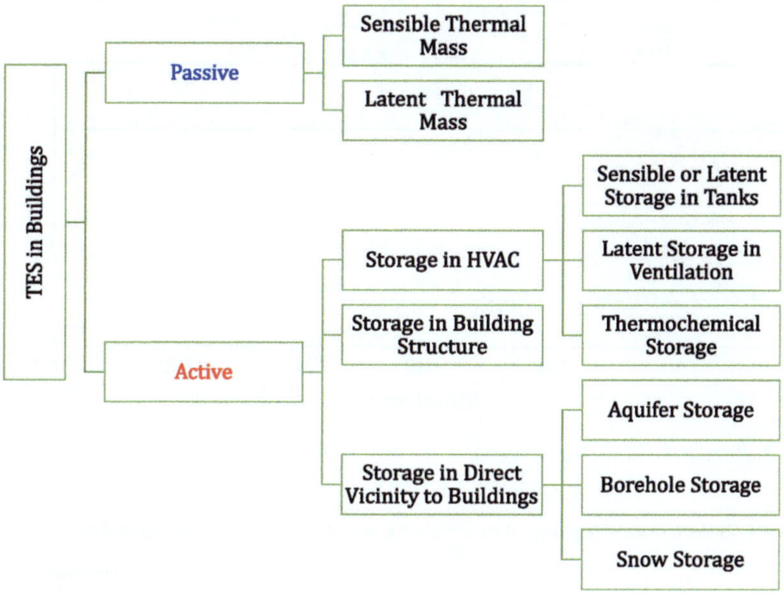


**Fig. 4.2** TES techniques for buildings. (Reproduced from Heier et al. 2015)

### *4.2.1 Increasing the Thermal Mass of Building Envelope (Passive TES)*

Various facts affect the heating and cooling demand of a building throughout the year such as environmental conditions (outdoor temperature, wind speed, and solar radiation), the thermal and structural properties of the building envelope, and indoor thermal inertia. Figure 4.3 illustrates the heating and cooling loads of a building throughout the year. Heat transfer through the wall is reduced by increasing the thermal resistance of the envelope. However, the fluctuations (*or* time lag) within the space are related to the thermal energy storage capacity of the building envelope. The timewise variation of the indoor temperature is regulated by increasing the thermal mass of the building by either sensible or latent heat techniques. Such a design reduces the energy consumption of the building and provides better thermal comfort. Table 4.1 lists the commonly used solid and liquid materials for sensible heat thermal energy storage applications in buildings to improve the thermal mass.

Al-Sanea et al. (2012) developed a one-dimensional transient model to simulate the influence of the amount of thermal mass (*heavyweight concrete* with the volumetric heat capacity of 1.884 kJ/m$^3$K) on the heating and cooling loads of a building. They have studied two different designs in which the thermal mass is placed adjacent to indoor (W1) and outdoor (W2). In either case, they have varied the thickness of the thermal mass layer in the range of $0 \leq L_{mass} \leq 50$ cm by keeping

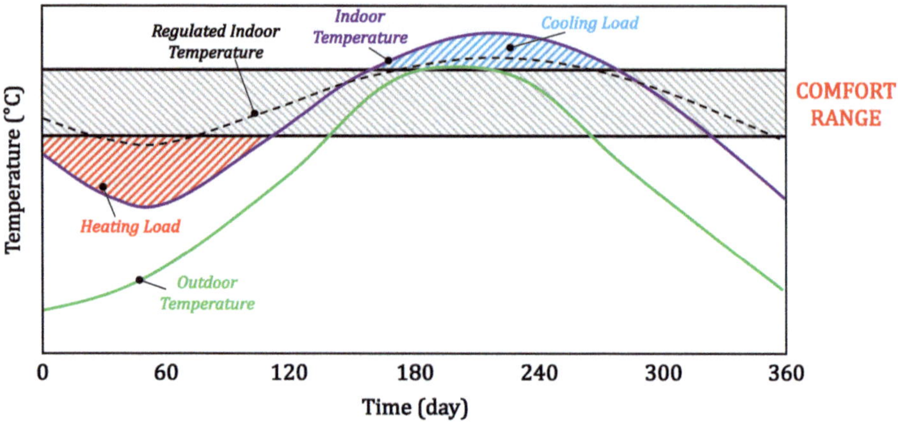


**Fig. 4.3** Regulation of indoor temperature fluctuation via TES

**Table 4.1** Selected solid and liquid materials for sensible energy storage in buildings

| Material | Type | Temperature range (°C) | Density (kg/m$^3$) | Specific heat (kJ/kgK) | Thermal conductivity (W/mK) |
|---|---|---|---|---|---|
| Water | Liquid | 0–100 | 1000 | 4.19 | 0.58 |
| Caloriea HT43 | Liquid | 12–260 | 867 | 2.2 | – |
| Ethanol | Liquid | Up to 78 | 790 | 2.4 | 0.171 |
| Propanol | Liquid | Up to 97 | 800 | 2.5 | 0.161 |
| Butanol | Liquid | Up to 118 | 809 | 2.4 | 0.167 |
| Isobutanol | Liquid | Up to 100 | 808 | 3 | 0.133 |
| Isopentanol | Liquid | Up to 148 | 831 | 2.2 | 0.141 |
| Octane | Liquid | Up to 126 | 704 | 2.4 | 0.134 |
| Engine oil | Liquid | Up to 160 | 888 | 1.88 | – |
| Brick | Solid | 20–70 | 1600 | 0.84 | 1.20 |
| Concrete | Solid | 20–70 | 2240 | 1.13 | 0.9–1.3 |
| Cement sheet | Solid | 20–70 | 700 | 1.050 | 0.36 |
| Gypsum plastering | Solid | – | 1200 | 0.837 | 0.42 |
| Granite | Solid | 20–70 | 2650 | 0.900 | 2.90 |
| Marble | Solid | 20–70 | 2500 | 0.880 | 2.00 |
| Sandstone | Solid | 20–70 | 2200 | 0.712 | 1.83 |
| Clay sheet | Solid | – | 1900 | 0.837 | 0.85 |
| Asphalt sheet | Solid | – | 2300 | 1.700 | 1.20 |
| Steel slab | Solid | 20–70 | 7800 | 0.502 | 50 |
| Corkboard | Solid | – | 160 | 1.888 | 0.04 |
| Wood | Solid | – | 800 | 2.093 | 0.16 |
| Plastic board | Solid | – | 1050 | 0.837 | 0.50 |
| Rubber board | Solid | – | 1600 | 0.200 | 0.30 |
| Stone, granite | Solid | Up to 160 | 2640 | 0.82 | 1.7–3.98 |

(continued)

**Table 4.1** (continued)

| Material | Type | Temperature range (°C) | Density (kg/m$^3$) | Specific heat (kJ/kgK) | Thermal conductivity (W/mK) |
|---|---|---|---|---|---|
| Stone, sandstone | Solid | Up to 160 | 2200 | 0.71 | 1.83 |
| PVC board | Solid | – | 1379 | 1.004 | 0.16 |
| Asbestos board | Solid | – | 2500 | 1.050 | 0.16 |
| Formaldehyde board | Solid | – | 30 | 1.674 | 0.03 |
| Thermalite board | Solid | – | 753 | 0.837 | 0.19 |
| Fiberboard | Solid | – | 300 | 1 | 0.06 |
| Siporex board | Solid | – | 550 | 1.004 | 0.12 |
| Polyurethane board | Solid | – | 30 | 0.837 | 0.03 |
| Light plaster | Solid | – | 600 | 1 | 0.16 |
| Dense plaster | Solid | – | 1300 | 1 | 0.50 |
| Aluminum | Solid | Up to 160 | 2707 | 0.896 | 204 |
| Aluminum oxide | Solid | Up to 160 | 3900 | 0.84 | 30 |
| Aluminum sulfate | Solid | Up to 160 | 2710 | 0.75 | – |
| Cast iron | Solid | Up to 160 | 7900 | 0.837 | 29.3 |
| Pure iron | Solid | Up to 160 | 7897 | 0.452 | 73 |
| Calcium chloride | Solid | Up to 160 | 2510 | 0.67 | – |
| Copper | Solid | Up to 160 | 8954 | 0.383 | 385 |

Data from Tatsidjodoung et al. (2013)

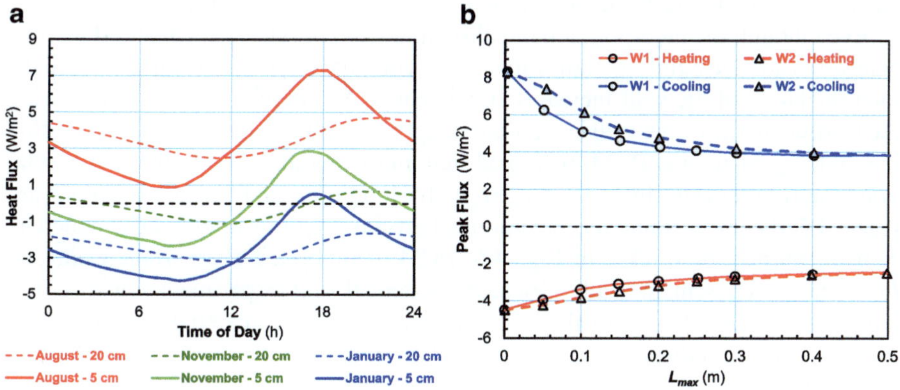


**Fig. 4.4** Regulation of indoor heat load via thermal mass. (**a**) Transient variation of thermal load, (**b**) influence of thermal mass on peak thermal load. (Adapted from Al-Sanea et al. 2012)

the thermal resistance of the wall the same. In Fig. 4.4a, the daily variations of thermal loads are given for three selected months with two different thermal masses. Results reveal that increasing the thickness of the thermal mass within the wall damps the fluctuations in thermal loads of the buildings both in summer and winter

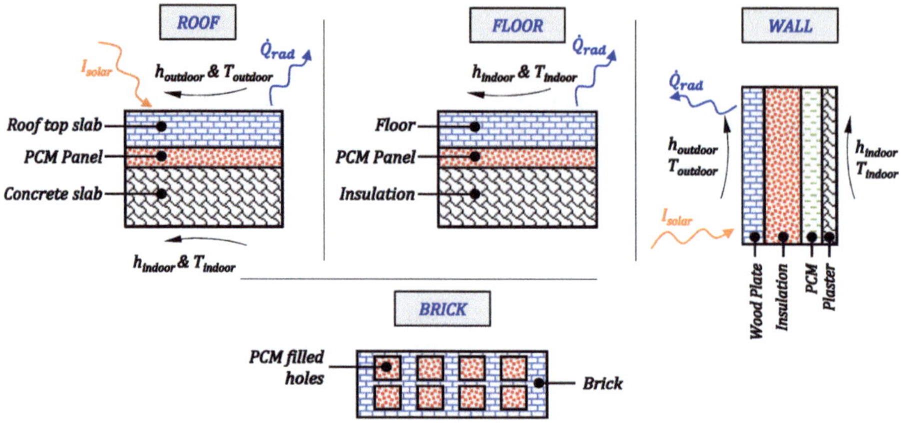


**Fig. 4.5** PCM-incorporated building materials

months. It is also reported that the daily total cooling and heating of thermal loads are significantly decreased by improving the thermal mass. For instance, in November, the daily total heating load is reduced from 2.126 to 0.986 kWh/m$^2$day by increasing the thickness of thermal mass from 5 cm to 20 cm for W2 design. The influence of increasing thermal mass on peak thermal load (heating/cooling) is represented in Fig. 4.4b. The peak load is reduced almost by half with the incorporation of 40 cm of heavy concrete. Beyond 40 cm, no significant change is observed in thermal loads. It is also interesting to note that the location of the thermal mass (adjacent to *inside* or *outside*) affects the peak load when the thickness is less than 20 cm.

Even though the incorporation of substantial mass provides an enhanced thermal comfort inside the building, it may not be an economic or practical solution since the total thickness of the wall increases significantly. An alternative way of increasing the thermal inertia of the wall is incorporating PCMs (phase change materials) within the building materials. PCM-integrated building materials provide thermal regulation requiring less spacing so that it has both economical and constructional advantages. PCMs are commonly incorporated inside gypsum wallboards, concrete, cement, plaster, or bricks. In Fig. 4.5, PCM-incorporated *floor*, *roof*, *wall*, and *brick* are given schematically. Various numerical and experimental works deal with the thermal behavior of the PCM-embedded building materials under various environmental conditions. Iten et al. (2016) recently reviewed the studies on PCM applications in buildings for heating and cooling purposes. There are several parameters that affect the performance of PCM-embedded building designs: thermophysical properties (i.e., melting point, latent heat of fusion, and thermal conductivity), thermal stability under long-term cycles, the type of encapsulation, and the position of the PCM. PCMs have to have high latent heat and thermal conductivity to procure high thermal inertia and have quick thermal response, respectively.

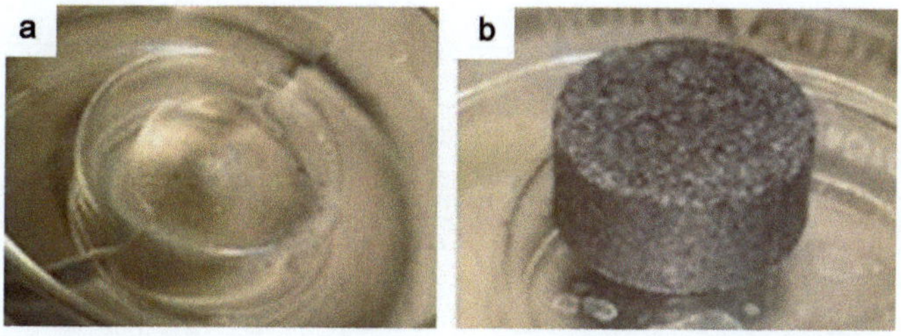


**Fig. 4.6** The physical states of (**a**) the reference PCM and (**b**) shape-stable nano-PCM (Sayyar et al. 2014)

One of the most challenging tasks is incorporating PCMs into building materials. The macro- or microencapsulation method is used to prepare PCM-embedded materials. Sealed cavities should be used to prevent leakage of liquid PCM. In recent works, microencapsulated PCM powders are implemented into building materials, such as concrete or gypsum. PCMs may cause some adverse effects on the structural or mechanical properties of the concrete. In addition to the capsule material, the fraction and type of PCM should be selected properly. It has been noted that decreasing the size of microcapsules causes a subcooling effect (Yamagishi et al. 1996). Nucleation agents are used to triggering the solidification of PCM within microcapsules. Preparation of shape-stabilized PCMs is another option to prevent liquid leakage within building materials. High-density polyethylene (HDPE) is mostly preferred as a supporting material while preparing shape-stable PCMs. However, due to the low thermal conductivity of HDPE, carbon-based nano-additives, such as graphene or graphite, are used to improve the transient response of shape-stable PCMs. Sayyar et al. (2014) developed a novel shape-stable nano-PCM and conducted experimental and numerical studies to evaluate its performance on the thermal behavior of a building wall. Nanocomposite PCM (nano-PCM) is prepared by using a eutectic mixture of capric acid (CA)/palmitic acid (PA) and graphite nanosheets. Figure 4.6 shows the physical states of CA-PA mixture and the nano-PCM under the same environment which is 10 °C higher than the melting temperature of the PCM overnight. One can deduce that the nano-PCM preserves its shape without any leakage. Figure 4.7 compares the variations of heat loads of different wall designs. Here the control wall corresponds to the conventional building envelope without PCM. In the nano-PCM design, on the other hand, the gypsum board is divided into three layers: 5 mm of nano-PCM, 5 mm of nano-plaster, and 3 mm of plaster. The differences between with and without air-conditioning modes are shown as colored areas in both control wall and nano-PCM wall. The incorporation of nano-PCM into the building wall considerably reduces energy consumption for heating/cooling when compared with the control wall. It is reported that in the heating and cooling modes, a total of 75% and 83% reductions are achieved, respectively.

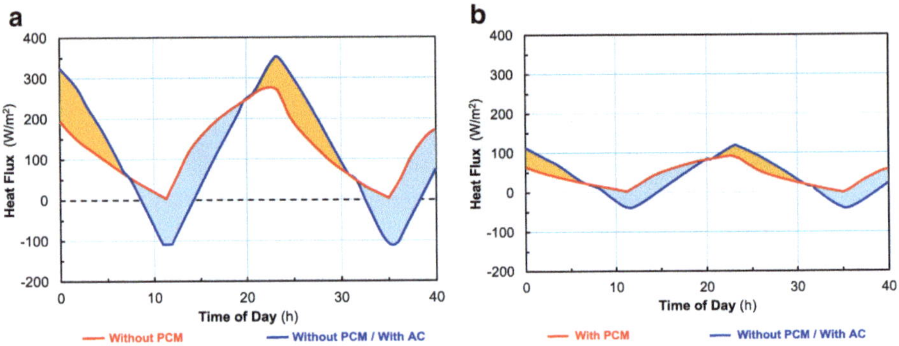


**Fig. 4.7** Heating/cooling load variations for two different envelope designs. (**a**) Control wall, (**b**) nano-PCM wall. (Adapted from Sayyar et al. 2014)

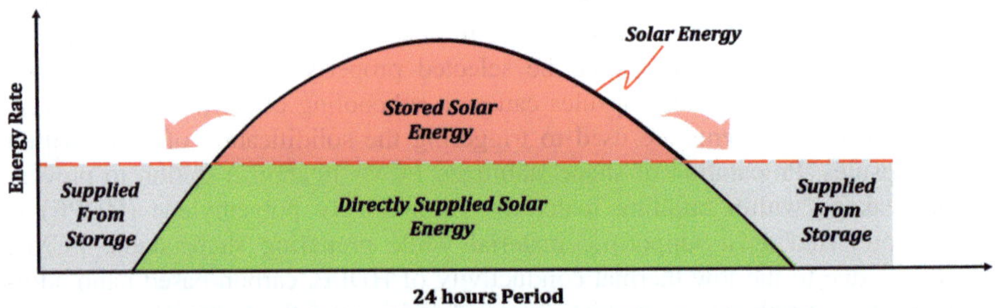


**Fig. 4.8** Variation in the solar energy throughout the day

## 4.2.2 *TES-Embedded Thermal Facilities in Buildings (Active TES)*

One of the most commonly used TES methods in buildings is to supply domestic hot water (DHW) from solar energy. There are various tank designs (either *sensible* or *latent*) to store thermal energy during off-peak hours to reduce the installed heating power. The variation in the solar energy throughout the day is illustrated in Fig. 4.8. Depending on the climatic conditions and the season, the available solar energy may exceed the thermal demand during daytime. Excessive solar energy could be stored in a storage tank for future use. Integration of TES units to the heating or cooling systems improves indoor comfort conditions, reduces the size of the conventional mechanical HVAC systems, and provides economic advantages by reducing power consumption during peak hours.

Active heating/cooling methods with TES can be divided into (*i*) *direct* and (*ii*) *indirect* methods. In the direct methods, the outdoor air is directly circulated through a TES unit (either *sensible* or *latent*) to store thermal energy for future use. Direct methods do not require additional heating/cooling equipment, instead, they include simple channel networks with flow controllers. Besides, in indirect methods, the thermal energy that is obtained from an external source, such as a solar collector or a chiller, is stored within the TES unit for future use. In the following, some illustrative applications are given for TES-integrated DHW units and active heating/cooling systems with TES.

#### 4.2.2.1 Solar Domestic Hot Water

Conventional solar domestic hot water (SDHW) units include a water tank and an auxiliary heater to produce hot water for residential, commercial, or industrial buildings. The high-temperature water that is supplied to the buildings is at around 50–60 °C (Sharif et al., 2015). In Fig. 4.9a, a simple SDHW system with a TES unit is presented schematically. When solar energy is accessible, it is stored in the tank, in either sensible or latent form, and then released from the tank when there is demand. The design of the tank, such as positions of the inlet/outlet sections and flow paths, has crucial importance to achieve high-temperature thermal energy from the tank. Rosen (2001) reported that the exergy efficiency of the stratified tank is higher than that of the thoroughly mixed tanks. Figure 4.9b illustrates the variation in temperature within the storage tank along the vertical direction. Hot water is in the upper region of the tank. Thanks to the thermal stratification effect, the low-temperature water at the bottom does not mix with the hot water.

The volume of hot water region should be increased to enhance the useful high-temperature thermal energy within stratified tanks. In several studies, PCMs were integrated inside stratified sensible storage tanks to improve thermal energy storage density (Cabeza et al. 2002, 2006; Mehling et al. 2003). As shown in Fig. 4.10a, PCM capsules are mostly placed in the upper regions of the storage tanks. The total

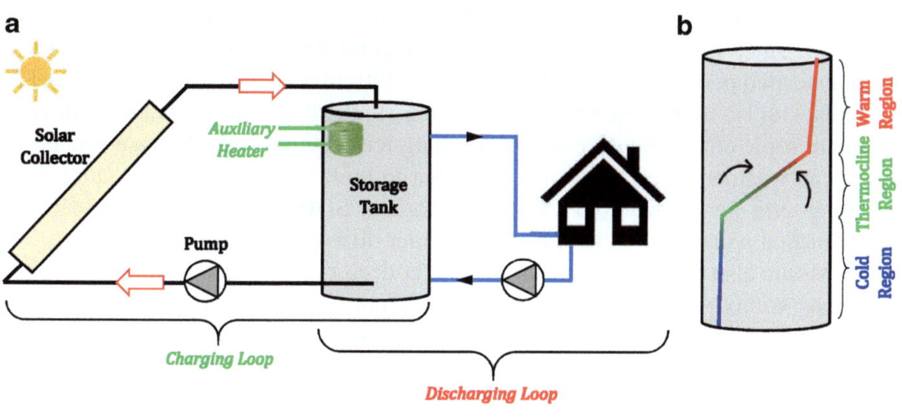


**Fig. 4.9** Solar domestic hot water storage system with stratified storage tank. (**a**) SDHW system, (**b**) thermal stratification

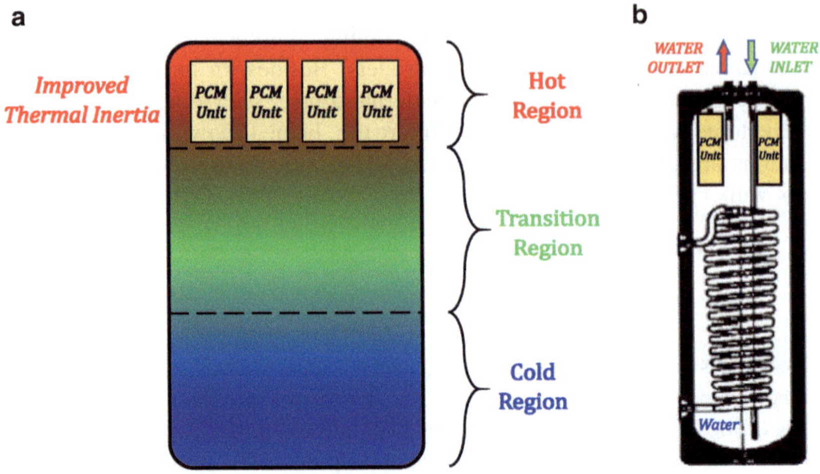


**Fig. 4.10** The enhanced thermal inertia of SDHW with PCM. (**a**) PCM-embedded SDHW tank, (**b**) SDHW tank design of Cabeza et al. (2006)

amount of PCM and its thermophysical properties, i.e., melting temperature and latent heat, are critical parameters that affect the performance of PCM-incorporated SDHW tanks. The storage tank that is designed by Cabeza et al. (2006) is given in Fig. 4.10b. The cold water enters the tank from the top and is then directed through the bottom of the tank with a dip tube. Unlike the conventional SDHW storage tanks, there are PCM modules at the top region of the tank to provide hot water in more extended periods with the same tank volume. Different types of PCMs are tested in the experimental unit such as paraffin, sodium acetate trihydrate, and fatty acids. As the primary purpose of this system is to evaluate hot water for domestic usage, the melting temperature of the PCMs is around 60 °C. The experimental and numerical works of team Cabeza reveal that the incorporation of PCM modules is a promising technique since it helps to maintain water temperature at a higher value and for a more extended period. Recently, Seddegh et al. (2015) reviewed the work on SDHW systems with latent heat thermal energy storage units. It was noted that even though there are some comprehensive studies on long-term simulations, they are limited to European weather. They have suggested that to build up a better understanding of the long-term performance of PCM-incorporated SDHW systems, numerical and experimental works should be conducted under different climatic conditions.

PCMs are also directly integrated to solar collectors to obtain hot water for longer durations without using an additional storage unit. In such a design, the PCM is placed around the collector pipes or underneath the absorber plate to improve the thermal inertia of the collector. In Fig. 4.11, three different collector designs with PCMs are presented. Rabin et al. (1995) proposed an integrated collector storage concept with a eutectic PCM. At a solid-liquid phase transition temperature of 27–29 °C 13.4 kg of PCM is used, and the latent heat of the eutectic PCM is

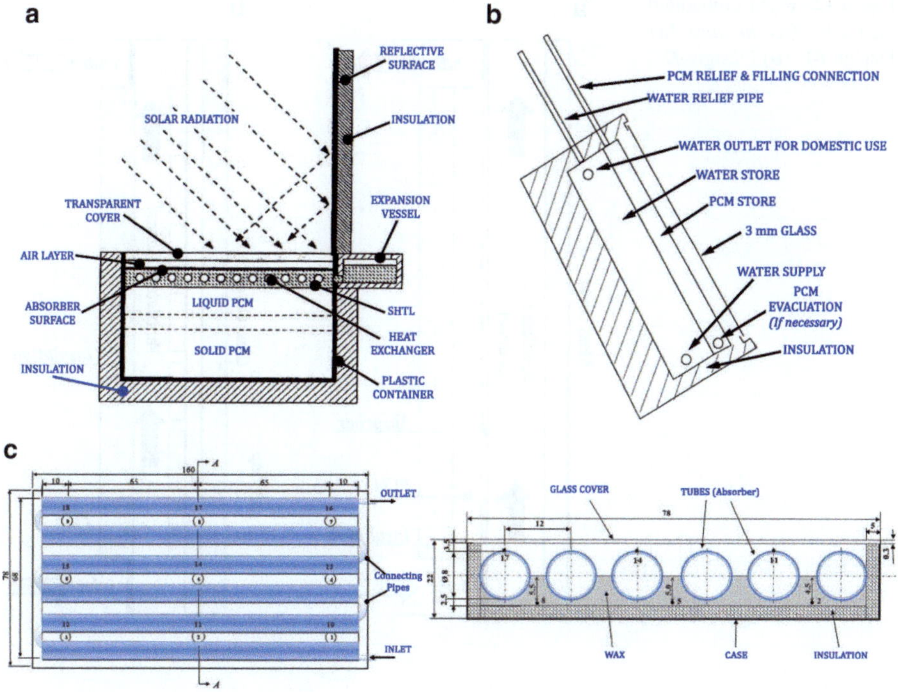


**Fig. 4.11** PCM-embedded collector designs for SDHW. (**a**) Rabin et al. (1995), (**b**) Kürklü et al. (2002), (**c**) Khalifa et al. (2013)

164.5 kJ/kg. Experimental results reveal that two-thirds of the total daily solar radiation falling on the solar collector is stored within the PCM-based collector system. Kürklü et al. (2002), on the other hand, proposed a solar collector with a hybrid sensible and latent heat storage partitions. Solar collector consists of parts that are filled with water and PCM. Paraffin wax of 50 kg, with a melting range of 45–50 °C, is used in the collector. It is reported that the PCM-embedded solar collector maintains supply of hot water above 36 °C during nighttime. Khalifa et al. (2013) placed a paraffin wax underneath the collector tubes. In such a design, the paraffin changes to liquid state and stores thermal energy in latent form. Improved thermal mass within the collector acts as a heat source after sunset and makes it possible to receive hot water over a more extended period.

#### 4.2.2.2 Heating/Cooling Buildings with TES-Embedded Units

Heating and cooling loads of the buildings are regulated by integrating the TES units into HVAC systems. There are various applications in which high-temperature (*heat* storage) or low-temperature (*cold* storage) thermal energy is stored for further usage.

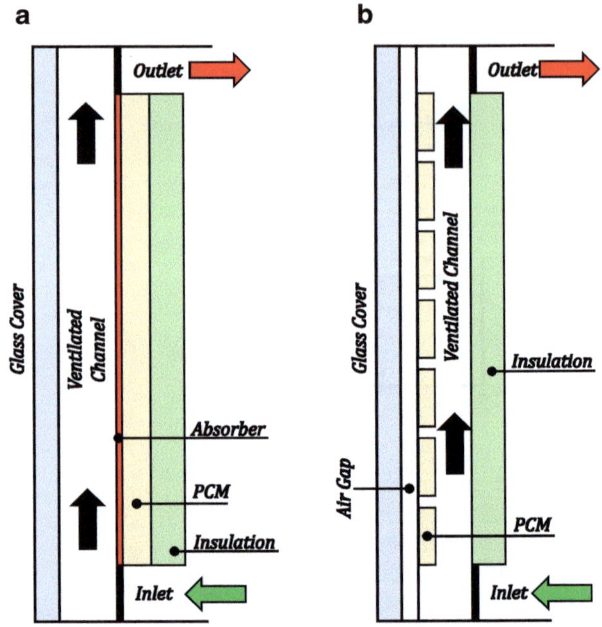


**Fig. 4.12** PCM-embedded façade for *free heating*. (**a**) Design #1, (**b**) Design #2

TES-embedded active heating/cooling units are categorized into two: (*i*) *direct* approaches and (*ii*) *indirect* approaches.

#### Free Heating and Cooling (Direct Approach)

The direct approach is also called *free heating* or *free cooling*. In Fig. 4.12, two illustrative designs are given for active TES for free heating in buildings. In these systems, solar energy is collected within PCM slabs during daytime. In each configuration, the air gap between the PCM and the glass cover reduces the convective heat losses from the storage medium through the cold environment. The PCM changes to liquid state throughout daytime and stores thermal energy in latent heat form. The inlet/outlet ports of the façade are kept closed during the storage mode, and there will be a weak circulation that is driven by buoyancy forces. After sunset, or when there is heating demand, the indoor air is circulated inside the ventilated channel to reject the thermal energy that is stored in the PCM. Free cooling is widely used to reduce cooling loads of buildings during summer. In Fig. 4.13, a simple design for free cooling is shown. The outside air is circulated through the storage unit at nighttime to store low-temperature thermal energy. The stored energy is then released from the TES unit during daytime by circulating the indoor air through the TES unit. Such an application helps to maintain the indoor temperature within the comfort range for a prolonged duration without utilizing the HVAC unit.

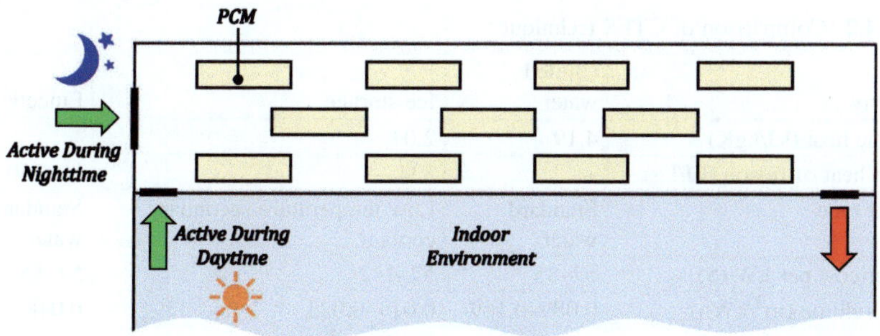


**Fig. 4.13** PCM-embedded ventilation unit for *free cooling*

## Cold and Heat TES (Indirect Approach)

TES units are integrated into the heating or cooling systems to provide improved thermal efficiency and enhanced thermal comfort. TES integration into a conventional HVAC unit also reduces power consumption and investment cost by shaving the peak load. The heating or cooling load of a building can be met entirely or partially by the storage tank. In the following, the basic concepts of high-temperature and low-temperature TES applications that are used for heating and cooling buildings are discussed.

### Cold TES

Cold TES (CTES) is storing thermal energy at low temperatures. This is done by one of the following methods: (*i*) *chilled water storage*, (*ii*) *ice storage*, and (*iii*) *eutectic salt storage*. Table 4.2 compares each of the CTES alternatives considering the economic and thermal aspects. In Fig. 4.14, commonly used CTES techniques are schematically given.

In the stratified chilled water tanks, the primary concern is to restrain the mixing of the chilled cold water with the return warm water. There are various tank designs with different inlet/outlet piping configurations to obtain thermally stratified temperature distribution without mixing. In Chap. 3, sensible heat storage tanks with baffles and membranes are discussed in detail. CTES with chilled water storage is easy to use as it requires cold flow at around 4–6 °C. Compared to other CTES techniques, chilled water has the lowest investment cost. The chiller charging efficiency (COP) reaches as much as six for CTES systems with chilled water. Besides, chilled water requires large tanks to store the same amount of cold energy when compared with ice or eutectic salt storage techniques.

In the ice slurry, ice-on-coil, or encapsulated PCM (water or eutectic salt) techniques, the storage medium undergoes a phase change (either *partially* or *entirely*) during the charging and discharging processes so that such systems make it possible to improve storage density and reduce the tank volumes. The schematic of the ice slurry generation facility is shown in Fig. 4.15. The weak glycol/water solution flows through the evaporator of the chiller to produce ice crystals. The

**Table 4.2** Comparison of CTES techniques

| Property | Chilled water | Ice storage | Eutectic salt |
|---|---|---|---|
| Specific heat (kJ/kgK) | 4.19 | 2.04 | – |
| Latent heat of fusion (kJ/kg) | – | 334 | 80–250 |
| Chiller type | Standard water | Low-temperature secondary coolant | Standard water |
| Chiller cost per kW ($) | 57–85 | 57–142 | 57–85 |
| Tank volume ($m^3$/kWh) | 0.089–0.169 | 0.019–0.023 | 0.048 |
| Storage installed cost per kWh ($) | 8.5–28 | 14–20 | 28–43 |
| Charging temperature (°C) | 4–6 | −6 to −3 | 4–6 |
| Chiller charging efficiency (COP) | 5.0–6.0 | 2.7–4.0 | 5.0–6.0 |
| Discharge temperature (°C) | 1–4 | 1–3 | 9–10 |
| Discharge fluid | Water | Secondary coolant | Water |
| Tank interface | Open tank | Closed system | Open tank |
| Maintenance | High | Medium | Medium |

Adapted from Hasnain (1998)

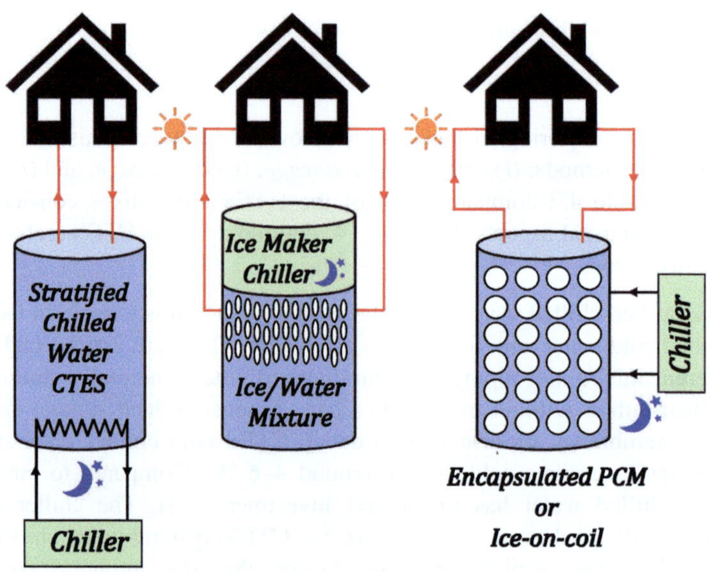


**Fig. 4.14** CTES storage tank in building application

small ice crystals (*slush*) are then either pumped or dropped into the storage tank for further usage to meet the cooling demand of the building. Increasing the fraction of ice improves the thermal energy capacity of the tank. However, increasing the fraction of ice also significantly increases the pressure drop and restricts flow

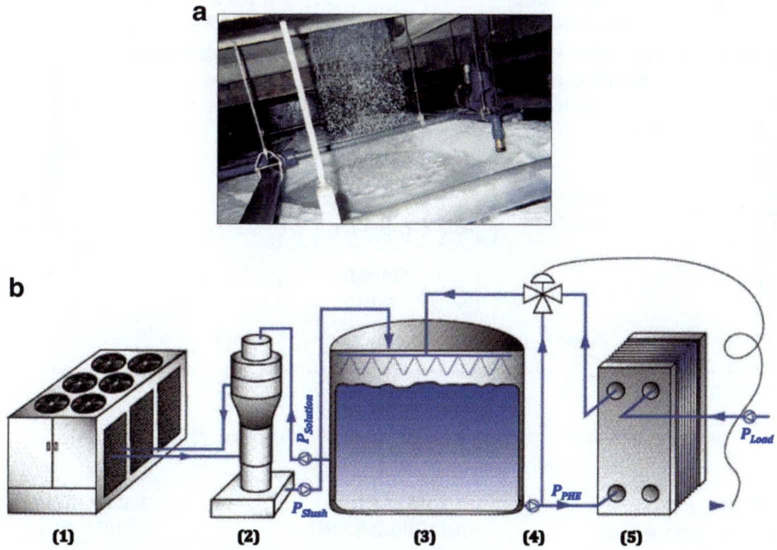


**Fig. 4.15** Components of an ice slurry generator system (Ice Synergy, 2018). (**a**) Ice slurry is falling into a tank, (**b**) (1) compressor and condenser of the chiller, (2) ice slurry generator (evaporator), (3) ice slurry storage tank, (4) controllers and valves, (5) plate heat exchanger and load (Adapted from Dincer and Rosen, 2011)

through the storage tank. It is suggested to limit the fraction of ice below 15–20% (Ice Synergy 2018).

In encapsulated PCM CTES systems, water or eutectic salts with low melting temperatures are filled within rectangular, cylindrical, or spherical capsules. Capsules are placed in a storage tank, and a heat transfer fluid (HTF, or secondary fluid) flows around the capsules during the charging and discharging periods. Figure 4.16 presents a schematic diagram of the charging and discharging periods of a CTES system with encapsulated PCM. The tank shape, the capsule material, and the arrangement of the capsules are critical design parameters that are widely studied to develop storage tanks with high performance. Inward solidification and melting take place during the charging and discharging periods, respectively. The heat transfer between the capsule and the HTF must be improved to build an efficient storage tank. Ethylene glycol aqueous solutions are commonly used as HTFs. The flow rate and inlet temperature of the HTF significantly affect the phase change process within the capsule. The pressure drop across the tank should also be considered to achieve the optimum working condition.

The heat transfer within the capsule is quite complicated. During the inward melting and solidification process, the interface of the solid/liquid moves, and the shape of the interface is affected by the local convective heat transfer coefficient around the capsule and the natural convection currents within the liquid phase of the PCM within the capsule. Figure 4.17 illustrates the distorted liquid-solid interfaces during inward melting and solidification inside the spherical capsule.

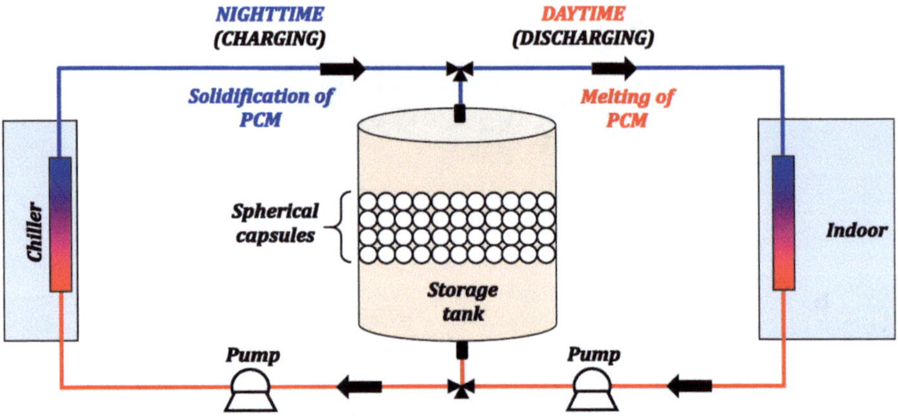


**Fig. 4.16** Charging and discharging of a CTES system with spherical capsules

**Fig. 4.17** Inward solidification and melting within the spherical capsule

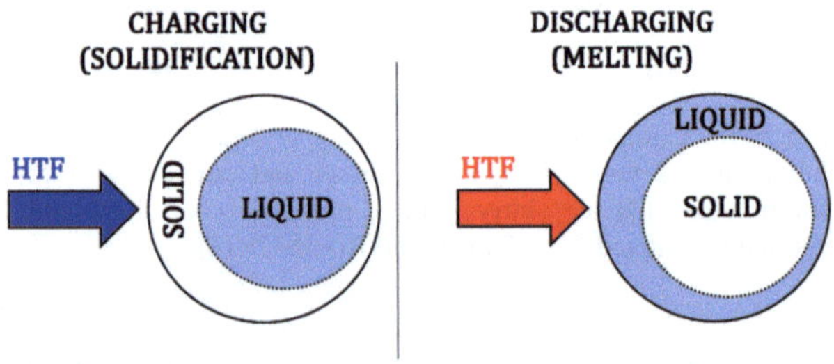


There are various alternative materials (pure substances, eutectics, or mixtures) that can be used as PCM in low-temperature CTES applications. In the recent review of Oró et al. (2012), the thermophysical properties of PCMs (both noncommercial and commercial) that are suitable for CTES applications are compiled for a wide temperature range of $-86\ °C \leq T_m \leq 20\ °C$. In Table 4.3, noncommercial PCMs with melting temperatures in the range of $-33.5\ °C \leq T_m \leq 10\ °C$ are listed. In Table 4.4, on the other hand, commercial PCMs are given for the same temperature range. The melting temperature, latent heat of fusion, and the thermal conductivity of a PCM determine the performance of a latent heat CTES unit. Melting temperature of a PCM should be within the working range of the system to completely change to solid and liquid phases throughout the charging/discharging cycles. The latent heat of fusion determines the energy storage capacity of the system and the tank volume. The rate of heat transfer within a PCM capsule is related to the thermal conductivity of the material. The time for complete melting or solidification is prolonged if the thermal conductivity of the PCM is low. Various methods have been experimented to improve heat transfer within a PCM by inserting nano-, micro-, or macro-sized materials within it. Dispersion of nanoparticles within a base PCM significantly improves its thermal conductivity; however, for large-scale CTES applications, it may not be a feasible solution. The heat transfer speed could be improved by using

**Table 4.3** Noncommercial PCMs for CTES applications

| Composition | Type | Melting temperature (°C) | Heat of fusion (kJ/kg) | Thermal conductivity (W/mK) | Density (kg/m$^3$) |
|---|---|---|---|---|---|
| 21.01 *wt.%* $MgCl_2$ | Eutectic water-salt solution | −33.5 | −36.30 kJ/mol | – | – |
| $Al(NO_3)_3$ (30.5 wt.%) + $H_2O$ | Eutectic water-salt solution | −30.6 | 131 | – | 1283 L 1251 s |
| 27.9 *wt%* $Li_2SO_4$ | Eutectic water-salt solution | −23 | −26.10 kJ/mol | – | – |
| NaCl (22.4 *wt.%*) + $H_2O$ | Eutectic water-salt solution | −21.2 | 222 | – | 1165 L 1108 s |
| 23.3 *wt.%* NaCl | Eutectic water-salt solution | −21.2 | 233 (5.0) kJ/mol | – | – |
| Diethylene glycol | Eutectic | −10 | 247 | – | 1200 L |
| 6 *wt.%* KCl + $H_2O$ | Inorganic | −10 | – | – | – |
| Dodecane | Organic | −9.6 | 216 | 2.21 L | – |
| 22.1 *wt.%* $BaCl_2$ | Eutectic water-salt solution | −7.7 | −10.2 kJ/mol | – | – |
| Triethylene glycol | Organic | −7 | 247 | – | 1200 L |
| 16.5 *wt.%* $KHCO_3$ + $H_2O$ | Inorganic | −6 | – | – | – |
| 18.63 *wt.%* $MgSO_4$ | Eutectic water-salt solution | −4.8 | −84.96 kJ/mol | – | – |
| Tetradecane + octadecane | Eutectic organic | −4.02 | 227.52 | – | – |
| 20.5 *wt.%* $NaCO_3$ + $H_2O$ | Inorganic | −3 | – | – | – |
| 6.49 *wt.%* $KsSO_4$ | Eutectic water-salt solution | −1.55 | 26.88 kJ/mol | – | – |
| 4.03 *wt.%* $Na_2SO_4$ | Eutectic water-salt solution | −1.2 | −1.07 kJ/mol | – | – |
| $H_2O$ | | 0 | 333 | 0.6 L 2.2 s | 998 L 917 s |
| $H_2O$ + polyacrylamide | Compound | 0 | 295 | 0.486 | 1047 L |
| 91.67%tetradecane+8.33% hexadecane | Eutectic organic | 1.7 | 156.2 | – | – |
| Tetradecane + docosane | Eutectic organic | 1.5–5.6 | 234.33 | – | – |
| Tetradecane + heneicosane | Eutectic organic | 3.54–5.56 | 200.28 | – | – |

(continued)

**Table 4.3** (continued)

| Composition | Type | Melting temperature (°C) | Heat of fusion (kJ/kg) | Thermal conductivity (W/mK) | Density (kg/m$^3$) |
|---|---|---|---|---|---|
| $K_2HPO_4 \cdot 6H_2O$ | Salt hydrates | 4 | 109 | – | – |
| $Na_2SO_4$ (31 *wt.%*) | Inorganic eutectic compounds | 4 | 234 | – | – |
| KF $4H_2O$ | Paraffin | 4.5 | 165 | – | – |
| Tetrahydrofurano (THF) | Eutectic organic | 5 | 280 | – | 970 s |
| Microencapsulated 94% tetradecane + 6% tetradecanol | Organic | 5.1 | 202.1 | – | – |
| Microencapsulated 100% tetradecane | Organic | 5.2 | 215 | – | – |
| Microencapsulated 96% tetradecane + 4% tetradecanol | Organic | 5.2 | 206.4 | – | – |
| Bulk 100% tetradecane | Organic | 5.5 | 215 | – | – |
| Bulk 96% tetradecane + 4% tetradecanol | Organic | 5.5 | 206.4 | – | – |
| Bulk 94% tetradecane + 6% tetradecanol | Organic | 5.5 | 202.1 | – | – |
| *n*-Tetradecane | Organic | 6 | 230 | – | – |
| Pentadecane + heneicosane | Eutectic organic | 6.23–7.21 | 128.25 | – | – |
| Formic acid | Fatty acid | 7.8 | 247 | – | – |
| Polyglycol E400 | Organic | 8 | 99.6 | 0.187 L | 1125 L |
| Paraffin $C_{15}$–$C_{16}$ | Organic | 8 | 153 | – | – |
| KF 4H2O | Paraffin | 8 | 153 | | |
| $LiClO_3 \cdot 3H_2O$ | Inorganic | 8.1 | 253 | – | – |
| Pentadecane + octadecane | Eutectic organic | 8.5–9.0 | 271.93 | – | – |
| Pentadecane + docosane | Eutectic organic | 7.6–8.99 | 214.83 | – | – |
| $ZnCl_2 \cdot 3H_2O$ | Inorganic (hydrated salt) | 10 | – | – | – |
| Paraffin $C_{15}$ | Organic | 10 | 205 | | |

Adapted from Oró et al. (2012)

**Table 4.4** Commercial PCMs for CTES applications

| Name | Type | Melting temperature (°C) | Heat of fusion (kJ/kg) | Producer |
|---|---|---|---|---|
| SN 33 | Salt solution | −33 | 245 | Cristopia |
| TH 31 | Salt hydrate | −31 | 131 | TEAP |
| MPCM (−30) | Paraffin | −30 | 140–150 | Microtek Laboratories, Inc. |
| SN 29 | Salt solution | −29 | 233 | Cristopia |
| SN 26 | Salt solution | −26 | 168 | Cristopia |
| – | Salt hydrate | −23 | 230 | TEAP |
| TH 21 | Salt hydrate | −21 | 222 | TEAP |
| SN 21 | Salt solution | −21 | 240 | Cristopia |
| STL 21 | Salt solution | −21 | 240 | Mitsubishi Chemical |
| ClimSel C-18 | Salt solution | −18 | 306 | Climator |
| SN 18 | Salt solution | −18 | 268 | Cristopia |
| TH 16 | Salt solution | −16 | 289 | TEAP |
| STL 16 | Salt solution | −16 | – | Mitsubishi Chemical |
| AN 15 | Salt solution | −15 | 311 | Cristopia |
| AN 12 | Salt solution | −12 | 306 | Cristopia |
| STLN 10 | Salt solution | −11 | 271 | Mitsubishi Chemical |
| AN 10 | Salt solution | −11 | 310 | Cristopia |
| TH 10 | Salt solution | −10 | 283 | TEAP |
| MPCM (−10) | Paraffin | −9.5 | 150–160 | Microtek Laboratories, Inc. |
| STL 6 | Salt solution | −6 | 284 | Mitsubishi Chemical |
| AN 06 | Salt solution | −6 | 284 | Cristopia |
| RT-4 | Paraffin | −4 | 179 | Rubitherm GmbH |
| TH 4 | Salt solution | −4 | 386 | TEAP |
| SLT 3 | Salt solution | −3 | 328 | Mitsubishi Chemical |
| AN 03 | Salt solution | −3 | 328 | Cristopia |
| – | Salt solution | 0 | 335 | TEAP |
| RT 3 | Paraffin | 4 | 198 | Rubitherm GmbH |
| RT 4 | Paraffin | 4 | 182 | Rubitherm GmbH |
| – | Salt solution | 4 | 105 | TEAP |
| RT5 | Paraffin | 5 | 198 | Rubitherm GmbH |
| RT6 | Paraffin | 6 | 175 | Rubitherm GmbH |
| MPCM (6) | Paraffin | 6 | 157–167 | Microtek Laboratories, Inc. |
| – | Salt solution | 7 | 300 | TEAP |
| ClimSel C 7 | Salt solution | 7 | 130 | Climator |
| | | 7 | 158.3 | |
| RT5 | Paraffin | 9 | 205 | Rubitherm GmbH |
| RT20 | Paraffin | 8 | 140 | Rubitherm GmbH |
| – | Salt solution | 10 | 170 | TEAP |

Adapted from Oró et al. (2012)

extended surfaces within the PCM domain. Such an application is given in Fig. 4.18. Shinde et al. (2015) compared the inward solidification of water inside rectangular cavities with and without fins. Fins are placed on the side walls of the cavity to increase the rate of freezing (see Fig. 4.18a). The volumes of liquid and solid phases are compared at selected instants in Fig. 4.18b. Consequently, it is reported that the

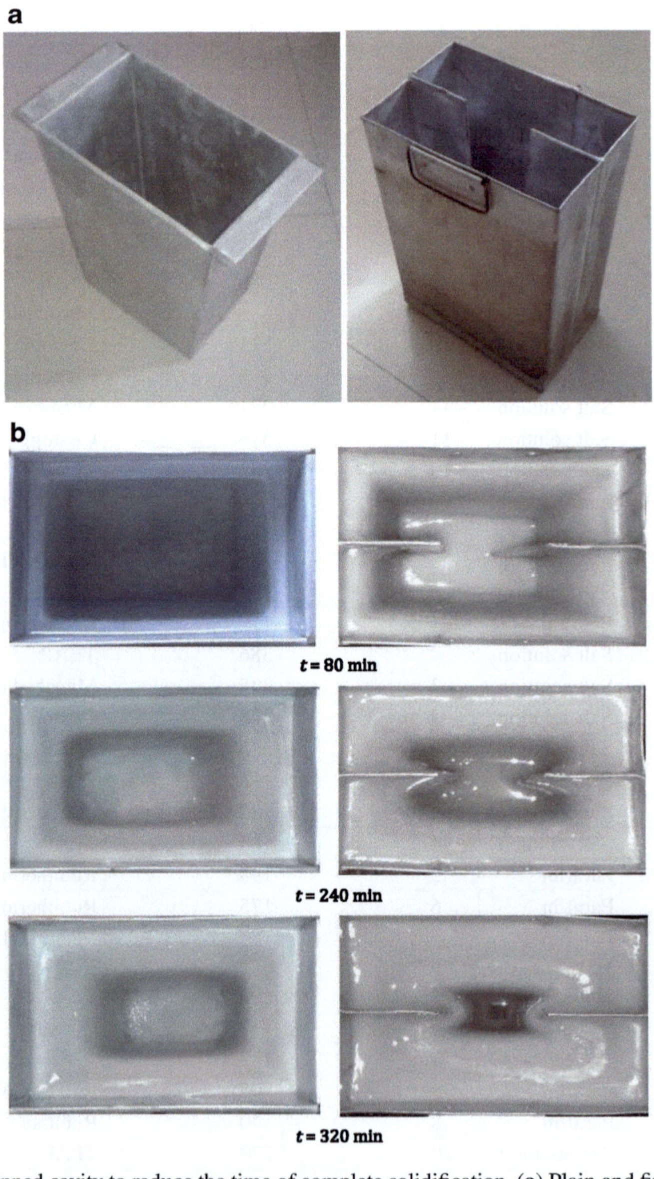


**Fig. 4.18** Finned cavity to reduce the time of complete solidification. (**a**) Plain and finned cavities, (**b**) inward solidification. (Shinde et al. 2015)

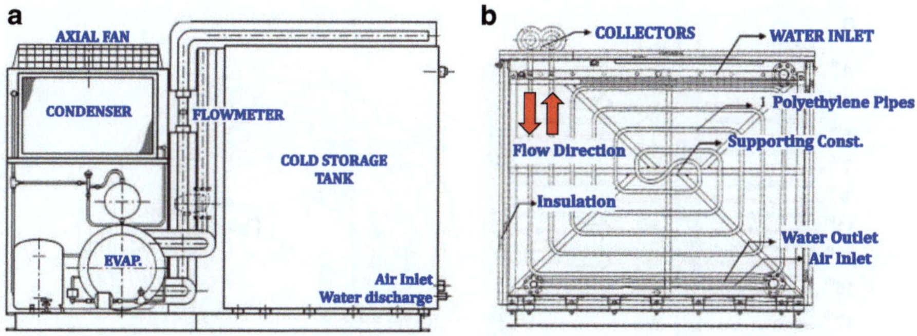


**Fig. 4.19** Ice-on-coil CTES with a chiller. (**a**) Chiller integrated to an ice-on-coil CTES, (**b**) flow path of the ice-on-coil CTES. (Ezan 2006)

finned cavity requires 22.72% less time for complete solidification when it is compared with the plain rectangular capsule.

Ice-on-coil CTES systems consist of a chiller, a storage tank, and a coil inside the tank so that it is easy to integrate into conventional cooling systems. In the charging mode, the heat transfer fluid (HTF) flows inside the coil, and water changes to the solid state. There are various designs to optimize the flow path of ice-on-coil tanks to achieve homogenous ice production throughout a tank. Figure 4.19 shows a storage tank with vertically aligned spiral-shaped coils. The HTF follows the path that is shown in Fig. 4.19b to provide a uniform temperature within the water and produce uniform ice around the coil.

An alternative tank design is given in Fig. 4.20. Here the coil is designed like the Archimedean spiral with a staggered pipe arrangement with counterflow to achieve a homogenous solidification/melting process. Figure 4.21 compares aligned and staggered pipe arrangements. As seen in this figure, the volume of the tank can be used more efficiently for a phase change process using staggered pipe arrangement instead of aligned pipe arrangement.

In both solidification and internal melting modes, the interface varies along the flow direction due to the varying wall temperature of the pipe. In Fig. 4.22, the variations of ice radius along the flow direction are illustrated for parallel and counterflow cases. Ice overlaps at the inlet section, and there are wasted spaces at the outlet of the pipes in the parallel flow design. The counterflow design, on the other hand, is a promising solution to reduce the wasted spaces and use the tank's volume more efficiently.

In CTES systems, the charging and discharging periods should be controlled either by a temperature sensor or an interface measurement device. Ezan (2011) developed a novel interface measurement method to monitor the solid-liquid interface variations inside an ice-on-coil storage tank. The proposed measurement method simply monitors the variations in the electrical conductivity of water at particular points to determine the time of phase change. Detailed information on the interface measurement method is given in Ezan et al. (2011). Measurement cards are

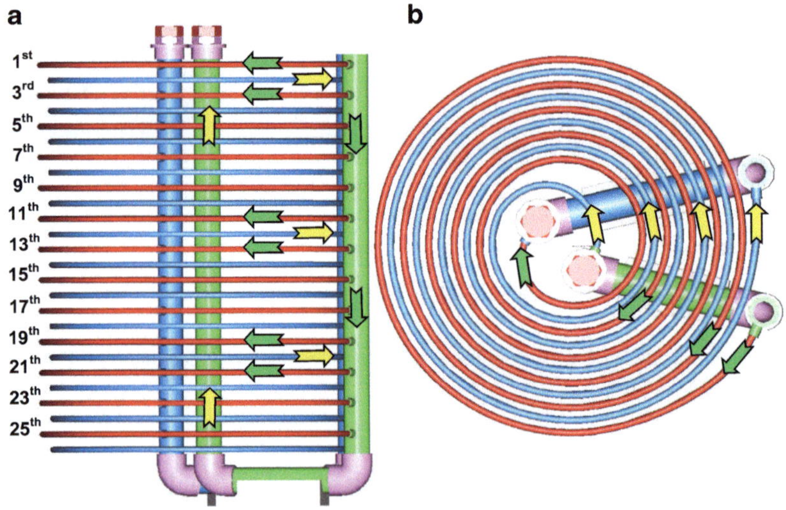


Fig. 4.20 Ice-on-coil CTES with Archimedean spiral. (a) Front view, (b) top view. (Ezan 2011)

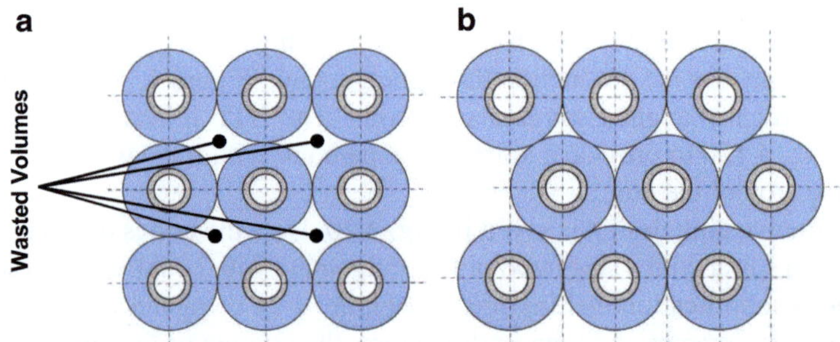


Fig. 4.21 Waste volumes in aligned and staggered arrangements. (a) Aligned, (b) staggered

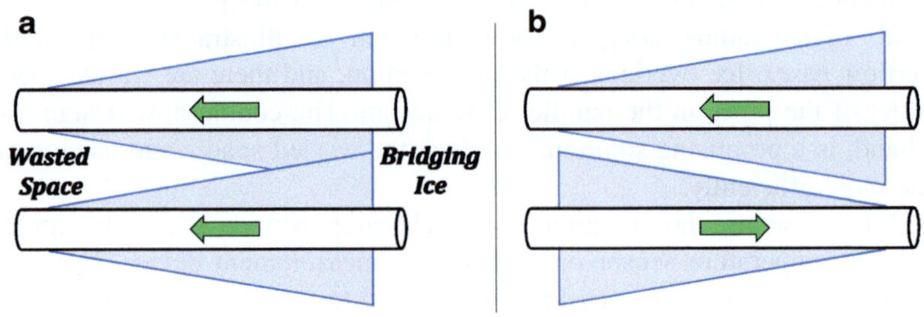


Fig. 4.22 Counter and parallel flow arrangements. (a) Parallel flow, (b) counterflow

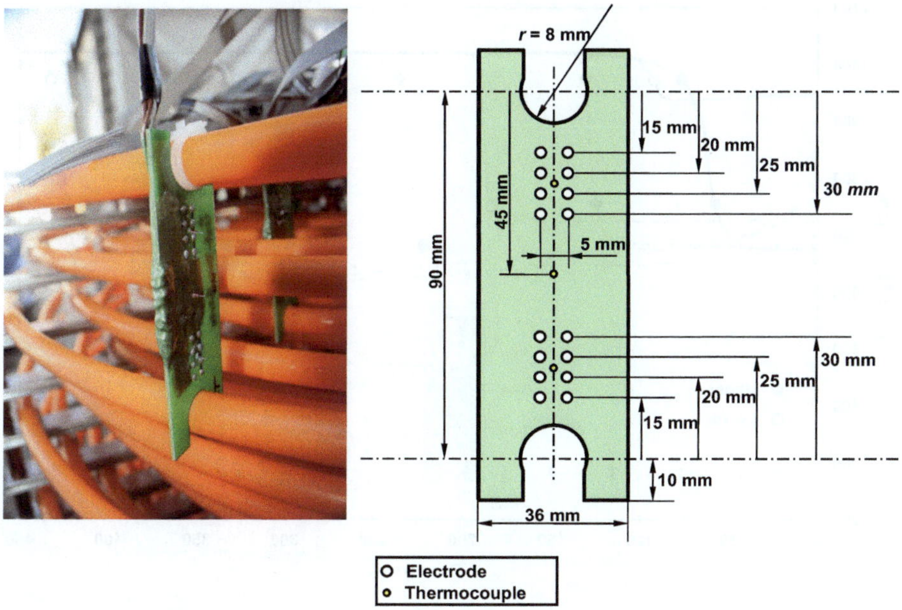


**Fig. 4.23** Solid-liquid interface measurement card. (Ezan 2011)

designed to be placed between two tubes in the vertical direction. In Fig. 4.23, the geometric parameters of the measurement cards are given together with an illustrative picture for the location of the card on the coil. Each card has eight pairs of electrodes and three thermocouples. In the figure, electrode and thermocouple positions are indicated with hollow and solid circles, respectively. Cards are designed to monitor the variations of solid-liquid interfaces for both upper and lower tubes. For this reason, four electrodes are placed near the upper tube, and the other four electrodes are settled near the lower tube. The interface measurement method is based on observation of variations in the electrical conductivity of the phase change medium. The signal variations are monitored during the experiments. Since the electrical conductivity of water changes dramatically while it freezes, when ice covers the pair of electrodes, the signal changes suddenly. The altering time of the signal designates the solidification time for each electrode position. Figure 4.24 illustrates the signal variations of a measurement card for eight electrode pairs, which are settled near the upper and lower tubes.

CTES systems are readily adaptable for residential or commercial buildings or even for district cooling systems. CTES is a mature thermal energy storage technology, and various types of cold storage units are commercially available. There are several manufacturers of CTES systems worldwide, and a few of them are listed in Table 4.5.

The cooling load variation of a conventional air-conditioning system throughout a day is illustrated in Fig. 4.25a. The cooling load of a building increases depending

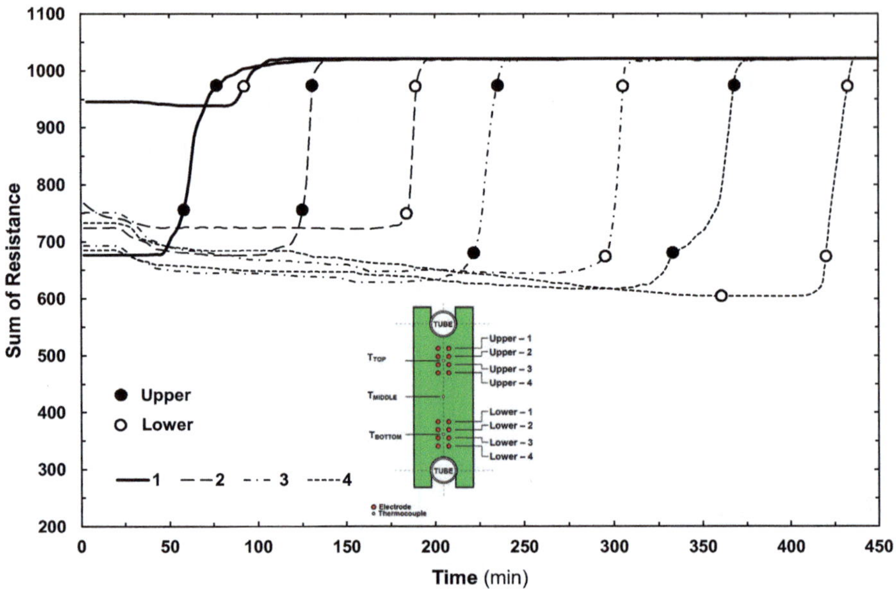


**Fig. 4.24** Timewise variations in the electrical resistance values. (Ezan 2011)

**Table 4.5** Commercial CTES system manufacturers

| Company | Web site |
|---|---|
| CalMac | http://www.calmac.com |
| Climator | http://climator.com |
| Cristopia – Thermofin | http://www.thermofin.net |
| ENGIE Axima | http://www.aximaref.com |
| Environmental Process Systems – EPS | http://www.epsltd.co.uk |
| Evapco | http://www.evapco.com |
| Ice Energy | http://www.ice-energy.com |
| Promethean Power Systems | http://coolectrica.com/ |
| Viking Cold | https://www.vikingcold.com |

on the high demand for energy in specific periods of the day, especially at midday. In a TES-integrated HVAC unit, a storage tank is attached to a chiller to store *cold* energy at nighttime and use this stored energy during daytime. At nighttime, chiller utilizes to store cold energy either in *sensible* (via temperature variation) or *latent* (via phase change) form. For instance, in the case of ice storage, large amounts of ice are produced within the storage tank during nighttime. During daytime, the cooling load of the air-conditioning unit is met by the cold storage tank and the indoor air temperature is maintained within the comfort range without utilizing the chiller during peak hours. This method is named *full storage strategy* and will balance the load distribution throughout the day, as shown in Fig. 4.25b.

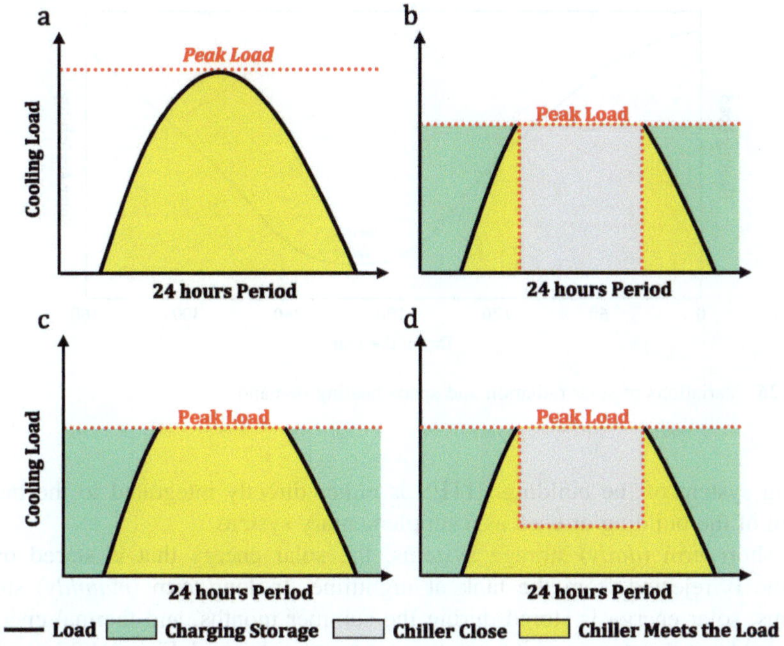


**Fig. 4.25** Daytime load distribution of an air-conditioning unit with and without cold TES. (**a**) Without cold TES, (**b**) cold TES with full storage, (**c**) cold TES with load leveling, (**d**) cold TES with demand leveling

Load leveling and demand leveling strategies are also given in Fig. 4.25. Various economic and environmental advantages can be achieved by shifting the cooling load from limited and expensive daytime period to a relatively broader and cheaper nighttime period. In comparison to the daytime, energy costs and the condensation temperature of cooling systems are relatively lower at nighttime; hence, cooling systems can be operated more efficiently and economically.

**Heat TES (Space Heating with Auxiliary Source)**

Solar thermal energy offers great potential to reduce the heating costs of buildings and provides energy savings. Previously, direct solar air-heating application was discussed as *free heating*. In this part, an overview of heat TES (HTES) (or high-temperature TES) is given. Figure 4.26 depicts the variations of the solar power and space heating requirements of a building throughout a year. The daily variations in the thermal load of the building and the available solar energy are given in Fig. 4.8. It is clear that, either on a daily or yearly basis, there is a mismatch between the available solar radiation and the heating demand. In HTES systems, thermal energy is stored within rock (*sensible*) or PCM-filled capsules (*latent*) when the solar energy is available. When the heating load is requested, the secondary fluid (or heat transfer fluid) is circulated within the storage tank to transfer the thermal energy to the

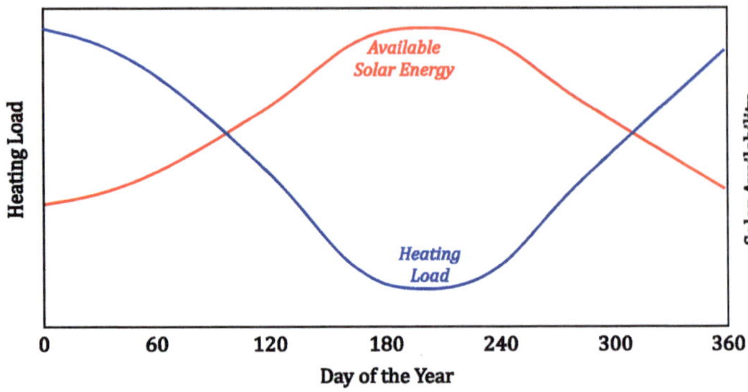


**Fig. 4.26** Variations of solar radiation and space heating demand

heating system of the building. HTES is either directly integrated to the heating system of the building or used as a supplementary system.

In short-term (*daily*) storage systems, the solar energy that is stored during daytime is rejected from the tank at nighttime. In long-term (*monthly*) storage systems, solar energy is stored during the summer months, and thermal energy is extracted from the storage unit when there is heating demand. Figure 4.27 illustrates solar-aided heating systems with TES, which are directly integrated to the conventional heating system of the building. In Fig. 4.27a, the energy that is stored in the storage tank during daytime is extracted from the tank during nighttime to meet the heating demand of the building. Such a system is an attractive solution when solar radiation is high, and the heating demand is small enough during daytime. Figure 4.27b, on the other hand, illustrates an application for long-term thermal energy storage for space heating. During summertime, the useful heat that is extracted from the solar collector is transferred to an underground TES unit. BTES (borehole TES) and ATES (aquifer TES) systems are widely in use for long-term thermal energy storage. BTES and ATES can easily be integrated to a solar-based energy source or a waste heat recovery unit to harvest thermal energy when accessible for future use.

A heat pump is an attractive heating utility for residential or commercial applications. The economic advantage of a heat pump is affected by the evaporation temperature. The performance of the heat pump enhances as the temperature difference between the heated space (condenser) and the low-temperature heat source (evaporator) is small. The performance of air source heat pumps (COP) significantly drops at the low outdoor temperatures. To maintain the evaporation temperature of the heat pump within a reasonable temperature range, one of the most widely followed approaches is to implement a ground source heat pump. In this way, the evaporator is immersed underground, and the refrigerant circulates in a closed-loop system to pump heat from ground to heated space. Another option to regulate the temperature fluctuations of a heat pump is the incorporation of a TES unit. A TES unit can be easily integrated into an air source heat pump to provide high evaporation

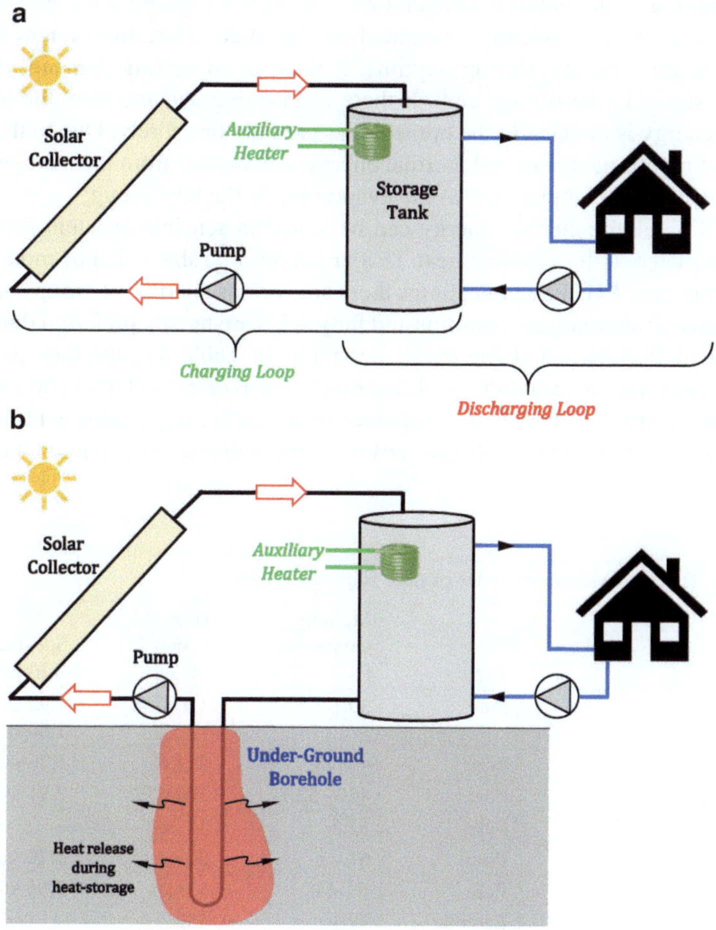


**Fig. 4.27** TES for space heating. (**a**) Short-term storage for heating, (**b**) long-term storage for heating

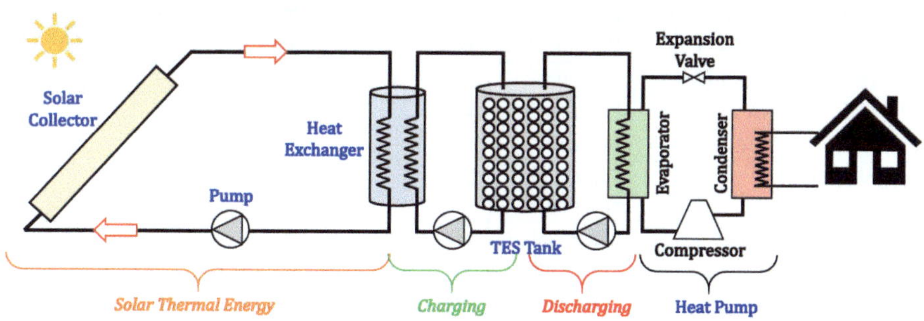


**Fig. 4.28** TES for space heating

temperatures at cold outdoor temperatures. A TES-integrated solar-assisted heat pump system is schematically illustrated in Fig. 4.28. Here the system has two different heating modes. During daytime, if there is no heating demand, the solar energy is stored in the storage tank. If there is a heating demand from the building, the solar energy is directed to the home either partially or entirely. Due to the excess demand at nighttime, the stored thermal energy is extracted from the storage tank to maintain a constant temperature at the evaporator of the heat pump.

In TES units, the thermal energy can be stored in sensible or latent forms. The candidate materials for sensible heat TES are given in Table 4.1. For moderate- or high-temperature LHTES applications, there are various alternative groups of PCMs. Among several alternatives, paraffins and fatty acids are mostly preferred due to their high heat of fusions and stable cyclic behavior. In Table 4.6, the thermophysical properties of some selected fatty acids are given. Two (or *more*) fatty acids are mixed to develop a eutectic PCM that is suitable for a specific application with a unique melting temperature. Table 4.6 also covers some eutectic fatty acids. One should

**Table 4.6** Thermophysical properties of fatty acids for HTES

| Name | Type | Melting temperature (°C) | Heat of fusion (kJ/kg) | Specific heat (kJ/kgK) |
|---|---|---|---|---|
| Capric acid (CA) | Pure | 30.1 | 150–158 | 1.95 s/ 1.60–1.72 L |
| Lauric acid (LA) | Pure | 41–43 | 211.6 | 1.76 s/2.27 L |
| Myristic acid (MA) | Pure | 49.7–52.7 | 190–201 | 2.91 s/3.67 L |
| Pentadecane acid (PDA) | Pure | 52.5 | 158.6 | – |
| Palmitic acid (PA) | Pure | 61.0 | 203.4 | 2.20 s/2.48 L |
| Stearic acid (SA) | Pure | 60–61 | 186.5 | 2.83 s/2.38 L |
| CA (61.5%) + LA (38.5%) | Eutectic | 19.1 | 132 | – |
| CA (73.5%) + MA (26.5%) | Eutectic | 21.4 | 152 | – |
| CA (75.2%) + PA (24.8%) | Eutectic | 22.1 | 153 | – |
| CA (86.6%) + SA (13.4%) | Eutectic | 26.8 | 160 | – |
| LA (62.6%) + MA (37.4%) | Eutectic | 32.6 | 156 | – |
| LA (64.0%) + PA (36.0%) | Eutectic | 32.8 | 165 | – |
| LA (75.5%) + SA (24.5%) | Eutectic | 37 | 182.7 | 1.92 s/2.10 L |
| MA (51.0%) + PA (49.0%) | Eutectic | 39.8 | 174 | – |
| MA (65.7%) + SA (34.7%) | Eutectic | 44 | 181 | – |
| PA (64.9%) + SA (35.1%) | Eutectic | 50.4 | 181 | |
| Methyl stearate | Chemical compound | 29.0 | 169 | – |
| Methyl palmitate | Chemical compound | 38.0 | 180 | – |
| Ethyl palmitate | Chemical compound | 23.0 | 122 | – |

Data from Kenisarin and Mahkamov (2007)

**Table 4.7** Thermophysical properties of some selected PCMs for HTES with 19 °C ≤ $T_m$ ≤ 28 °C

| Material | Type | Melting temperature (°C) | Heat of fusion (kJ/kg) | Thermal conductivity (W/mK) | Density (kg/m$^3$) |
|---|---|---|---|---|---|
| *n*-Heptadecane | Organic | 19 | 240 | 0.21 | 760 L |
| Paraffin $C_{16}$–$C_{18}$ | Organic | 20–22 | 152 | – | – |
| $FeBr_3 \cdot 6H_2O$ | Inorganic (salt hydrate) | 21 | 105 | – | – |
| Paraffin $C_{17}$ | Organic | 21.7 | 213 | – | – |
| Polyglycol E600 | Organic | 22 | 127.2 | 0.1897 L | 1126 L |
| Paraffin $C_{13}$–$C_{24}$ | Organic | 22–24 | 189 | 0.21 L | 0.760 L |
| 55–65% $LiNO_3 \cdot 3H_2O$ + 35–45% $Ni(NO_3)_2$ | Inorganic mixture | 24.2 | 230 | – | – |
| 45% $Ca(NO_3)_2 \cdot 6H_2O$ + 55% $Zn(NO_3)2 \cdot 6H_2O$ | Inorganic mixture | 25 | 130 | – | 1930 |
| 66.6% $CaCl_2 \cdot 6H_2O$ + 33.3% $MgCl_2 \cdot 6H_2O$ | Inorganic mixture | 25 | 127 | – | 1590 |
| 50% $CaCl_2$ + 50% $MgCl_2$ + $6H_2O$ | Inorganic eutectic | 25 | 95 | – | – |
| Octadecane + docosane | Organic eutectic | 25.5–27.0 | 203.8 | – | – |
| $Mn(NO_3) \cdot 6H_2O$ | Inorganic (salt hydrate) | 25.5 | 125.9 | – | 1738 L |
| Octadecane + heneicosane | Eutectic organic | 25.8–26.0 | 173.93 | – | – |
| 1-Dodecanol | Organic | 26 | 200 | – | – |
| 48% $CaCl_2$ + 4.3% NaCl +0.4% KCl + 47.3% $H_2O$ | Inorganic mixture | 26.8 | 188 | – | 1640 |
| 50% $CH_3CONH_2$ + 50% $NH_2CONH_2$ | Organic eutectic | 27 | 163 | | |
| 4.3% NaCl +0.4% KCl + 48% $CaCl_2$ + 47.3% $H_2O$ | Inorganic mixture | 27 | 188 | – | 1530 L |
| Paraffin $C_{18}$ | Organic | 28 | 244 | 0.148 L | 0.774 L |
| *n*-Octadecane | Organic | 28 | 200 | 0.148 L | 774 L |

Data from Cabeza et al. (2011)

note that the eutectic mixtures may have lower latent heat of fusion when compared with their constituents. The cyclic behavior of the mixture of the proposed PCM must be investigated. In Tables 4.7 and 4.8, the thermophysical properties of some selected PCMs that are suitable for HTES applications are summarized.

**Table 4.8** Thermophysical properties of some selected PCMs for HTES with 29 °C ≤ $T_m$ ≤ 60 °C

| Material | Type | Melting temperature (°C) | Heat of fusion (kJ/kg) | Thermal conductivity (W/mK) | Density (kg/m$^3$) |
|---|---|---|---|---|---|
| $CaCl_2 \cdot 6H_2O$ | Inorganic (salt hydrate) | 29 | 190.8 | 0.549 L | 1562 L |
| 62.5% Trimethylolethane + 37% water | Organic mixture | 29.8 | 218 | 0.65 | 1120 |
| $CaCl_2 \cdot 12H_2O$ | Inorganic (salt hydrate) | 29.8 | 174 | – | – |
| Ga | Inorganic eutectic | 30 | 80.9 | – | – |
| $LiNO_3 \cdot 3H_2O$ | Inorganic (salt hydrate) | 30 | 296 | – | – |
| $LiNO_3 \cdot 2H_2O$ | Inorganic (salt hydrate) | 30 | 296 | – | – |
| 67% $Ca(NO_3)_2$ + 33% $Mg(NO_3)_2$ | Inorganic mixture | 30 | 136 | – | 1670 |
| 47% $Ca(NO_3)_2 \cdot 4H_2O$ + 53% $Mg(NO_3)_2 \cdot 6H_2O$ | Inorganic eutectic | 30 | 136 | – | – |
| 60% $Na(CH_3COO) \cdot 3H_2O$ + 40% $CO(NH_2)_2$ | Inorganic eutectic | 31.5 | 226 | – | – |
| $Na_2SO_4 \cdot 3H_2O$ | Inorganic (salt hydrate) | 32 | 251 | – | – |
| $Na_2CO_3 \cdot 10H_2O$ | Inorganic (salt hydrate) | 32–36 | 246.5 | – | 1442 |
| Paraffin $C_{19}$ | Organic | 32 | 222 | – | – |
| $CaBr_2 \cdot 6H_2O$ | Inorganic (salt hydrate) | 34 | 115.5 | – | 1956 L |
| $LiBr_2 \cdot 2H_2O$ | Inorganic (salt hydrate) | 34 | 124 | – | – |
| $Na_2HPO_4 \cdot 12H_2O$ | Inorganic (salt hydrate) | 35–44 | 280 | 0.514 s | 1522 s |
| $Zn(NO_3)_2 \cdot 6H_2O$ | Inorganic (salt hydrate) | 36 | 146.9 | 0.464 L | 1828 L |
| Paraffin $C_{20}$ | Organic | 36.7 | 246 | – | – |
| $FeCl_3 \cdot 6H_2O$ | Inorganic (salt hydrate) | 37 | 223 | – | – |
| $Mn(NO_3) \cdot 6H_2O$ | Inorganic (salt hydrate) | 37.1 | 115 | – | 1738 L |
| 1-Tetradecanol | Organic | 38 | 205 | – | – |
| Heneicosane | Organic | 40–40.2 | 155.5–213 | – | 778 s |
| Paraffin $C_{21}$ | Organic | 40.2 | 200 | – | – |
| 50% $Na(CH_3COO) \cdot 3H_2O$ + 50% $HCONH_2$ | Organic eutectic | 40.5 | 255 | – | – |

(continued)

**Table 4.8** (continued)

| Material | Type | Melting temperature (°C) | Heat of fusion (kJ/kg) | Thermal conductivity (W/mK) | Density ($kg/m^3$) |
|---|---|---|---|---|---|
| $CoSO_4 \cdot 7H_2O$ | Inorganic (salt hydrate) | 40.7 | 170 | – | – |
| Phenol | Organic | 41 | 120 | – | – |
| Heptadecanone | Organic | 41 | 201 | – | – |
| 1-Cydodecyloctadecane | Organic | 41 | 218 | – | – |
| 4-Heptadecanone | Organic | 41 | 197 | – | – |
| $MgI_2 \cdot 8H_2O$ | Inorganic (salt hydrate) | 42 | 133 | – | – |
| $CaI_2 \cdot 6H_2O$ | Inorganic (salt hydrate) | 42 | 162 | – | – |
| $CaI_2 \cdot 6H_20$ | Inorganic (salt hydrate) | 42 | 162 | – | – |
| Paraffin $C_{16}$–$C_{28}$ | Organic | 42–44 | 189 | 0.21 | 0.765 L |
| Paraffin $C_{22}$ | Organic | 44 | 249 | – | – |
| Docosane | Organic | 44 | 196.5–252 | – | – |
| $K_2HPO_4 \cdot 7H_2O$ | Inorganic (salt hydrate) | 45 | 145 | – | – |
| 53% $NH_2CONH_2$ + 47% $NH_4NO_3$ | Inorganic eutectic | 46 | 95 | – | – |
| $Mg(NO_3) \cdot 2H_2O$ | Inorganic (salt hydrate) | 47 | 142 | – | – |
| $Fe(NO_3)_2 \cdot 9H_2O$ | Inorganic (salt hydrate) | 47 | 155 | – | – |
| Paraffin $C_{23}$ | Organic | 47.5 | 232 | – | – |
| $Na_2SiO_3 \cdot 4H_2O$ | Inorganic (salt hydrate) | 48 | 168 | – | – |
| $K_2HPO_4 \cdot 3H_2O$ | Inorganic (salt hydrate) | 48 | 99 | – | – |
| $Na_2S_2O_3 \cdot 5H_2O$ | Inorganic (salt hydrate) | 48 | 201 | – | 1600 s |
| 3-Heptadecanone | Organic | 48 | 218 | – | – |
| 2-Heptadecanone | Organic | 48 | 218 | – | – |
| Paraffin $C_{20}$–$C_{33}$ | Organic | 48–50 | 189 | 0.21 | 0.769 L |
| $MgSO_4 \cdot 7H_2O$ | Inorganic (salt hydrate) | 48.5 | 202 | – | – |
| Cetyl alcohol | Organic | 49.3 | 141 | – | – |
| 9-Heptadecanone | Organic | 51 | 213 | – | – |
| $Ca(NO_3)_2 \cdot 3H_2O$ | Inorganic (salt hydrate) | 51 | 104 | – | – |
| 61.5% $Mg(NO_3)_2$ + 38.5% $NH_4NO_3$ | Inorganic mixture | 52 | 125.5 | 0.494 | 1515 |
| $FeCl_3 \cdot 2H_2O$ | Inorganic (salt hydrate) | 56 | 90 | – | – |

(continued)

**Table 4.8** (continued)

| Material | Type | Melting temperature (°C) | Heat of fusion (kJ/kg) | Thermal conductivity (W/mK) | Density (kg/m$^3$) |
|---|---|---|---|---|---|
| Tristearin | Organic | 56 | 191 | – | – |
| Paraffin $C_{26}$ | Organic | 56.3 | 256 | – | – |
| $Ni(NO_3)_2 \cdot 6H_2O$ | Inorganic (salt hydrate) | 57 | 169 | – | – |
| $MnCl_2 \cdot 4H_20$ | Inorganic (salt hydrate) | 58 | 151 | – | – |
| $MgCl_2 \cdot 4H_20$ | Inorganic (salt hydrate) | 58 | 178 | – | – |
| $Na(CH_3COO) \cdot 3H_2O$ | Inorganic (salt hydrate) | 58 | 226–264 | 0.63 | 1280 L |
| 50% $Mg(NO_3)_2 \cdot 6H_2O$ + 50% $MgCl_2 \cdot 6H_2O$ | Inorganic eutectic | 58–59 | 132 | 0.510 L | 1550 L 1630 s |
| 62.5% $Mg(NO_3)_2$ + 37.5% $NH_4NO_3$ | Inorganic mixture | 58 | 267 | 0.63 s | 1.45 |
| 58.7% $Mg(NO_3)_2 \cdot 6H_2O$ + 41.3% $MgCl_2 \cdot 6H_2O$ | Inorganic eutectic | 59 | 132.2 | 0.510 (65 °C) | 1550 |
| Paraffin $C_{22}$–$C_{45}$ | Organic | 58–60 | 189 | 0.21 | 0.795 L |
| Paraffin $C_{27}$ | Organic | 58.8 | 236 | – | – |
| 80% $Mg(NO_3)_2 \cdot 6H_2O$ + 20% $MgCl_2 \cdot 6H_2O$ | Inorganic (salt hydrate) | 60 | 150 | – | – |

Data from Cabeza et al. (2011)

## 4.3 Solar Power Generation Systems with TES

There are two approaches to harness solar energy to generate power: (*i*) *direct power generation* and (*ii*) *indirect power generation.* Direct power generation stands to produce electricity from sunlight, i.e., photovoltaics. Indirect power generation produces electricity from thermal energy, i.e., thermoelectric generators (TEGs) or concentrated solar power (CSP) systems. TES units are integrated into direct systems as a passive thermal controller to prevent overheating. In indirect systems, on the other hand, TES units remove the mismatch between supply and demand. The basic concepts of each method and the implementation of heat storage units with solar power generation systems are discussed below.

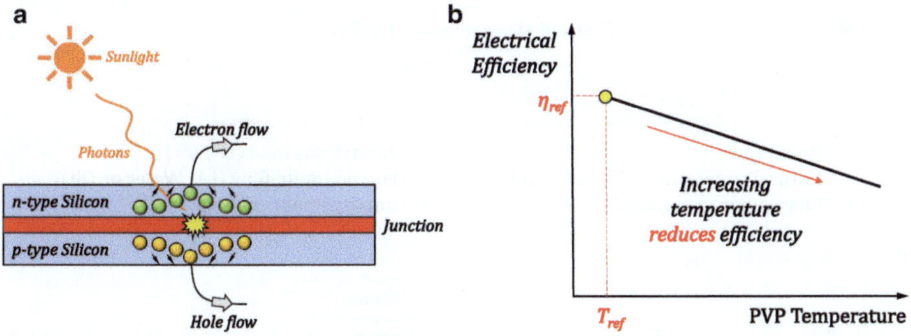


**Fig. 4.29** Schematic and efficiency variation for photovoltaic panel (PVP). (**a**) Working principle of PVP, (**b**) efficiency of PVP panel as a function of temperature

### *4.3.1 Direct Power Generation*

Sunlight is directly converted into electricity by using semiconductor cells in photovoltaic panels (PVPs). The primary barriers to widespread application of PVPs are low efficiency and high cost. Electric generation efficiencies of commercially available PVPs stand in the range of 10–20%. Only a portion of the solar spectrum is absorbed by the PVP to produce electrical power. The rest is stored in the cell as thermal energy. The cell temperature increases as the incoming solar radiation enhances. However, the electrical efficiency of a PVP is significantly affected by the temperature of the cell, and increasing the temperature reduces the panel efficiency considerably. The excess heat should be removed from the PVP to maintain the cell temperature close to the maximum efficiency temperature (or *reference temperature*). Figure 4.29 illustrates the working principle of the PVP and the relationship between panel efficiency and PVP temperature.

Among the various definitions, the following correlation (Evans and Florschuetz 1977) is widely used to evaluate PVP electricity efficiency as a function of cell temperature

$$\eta_{PVP} = \eta_{ref}\left[1 - \beta_{PVP}\left(T_{PVP} - T_{ref}\right)\right] \tag{4.1}$$

where $\eta_{ref}$ represents the reference efficiency at the reference temperature and $T_{ref}$. $\beta_{PVP}$ (in $K^{-1}$) is the temperature coefficient of the PVP. In general, 1 °C increase in the cell temperature corresponds to a reduction in electrical power generation from 0.2% to 0.5%. Krauter et al. (1999) stated that due to overheating of the building-integrated PVP (BIPVP), power generation reduces by 9.3% compared to the panel maintained outdoor. There are various thermal management tools to keep the PVP cell temperature near the reference temperature: (*i*) *hybrid PVP/thermal systems (natural or forced convection with liquid/gas medium)*, (*ii*) *heat pipes*, and (*iii*) *phase change materials* (Fig. 4.30). At higher solar radiations, PVP/thermal (PVP/T) systems can provide electrical generation and useful heat. Heat pipes improve the

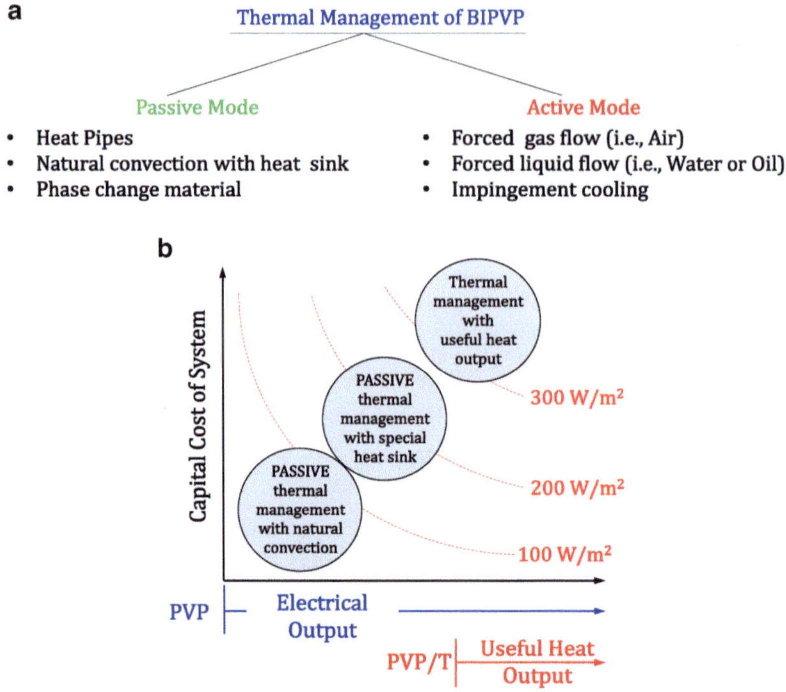


**Fig. 4.30** Thermal management of BIPVPs. (**a**) Classification of thermal management modes of BIPVPs, (**b**) thermal management options

heat transfer rate between the PVP and the heat sink. PCM, on the other hand, is a simple solution to remove excessive heat via phase change. The main advantage of the PCM-embedded thermal control is that there is no additional component, such as piping, pump, or fan, and maintenance is not required if the selected PCM possesses long-term thermal stability.

In the case of building-integrated PVPs (BIPVPs), PCM is one of the most practical solutions for thermal management of panels since PCM does not require large space on the rear side of the panels. According to the recent review of Browne et al. (2015), the first PVP/PCM system was proposed in 1983 by Ames (1983). Almost 20 years after this pioneering invention, the research on PVP/PCM systems has received increasing attention during the last two decades. A simple schematic for PCM-embedded PVP unit is given in Fig. 4.31. Here a PCM with a thickness of $t_{PCM}$ is placed on the rear side of the panel. During daytime, a portion of solar radiation is converted into electricity, and the remaining incoming energy is converted into heat. Depending on the surrounding conditions, the PVP releases heat through the ambient. In hot summer conditions, a considerable amount of heat is stored within the layers of the panel, and the PVP cell temperature increases. Incorporation of PCM increases the heat capacity of the system and provides temperature regulations. PCM stores the excess heat during daytime via melting the PCM, and the stored energy is

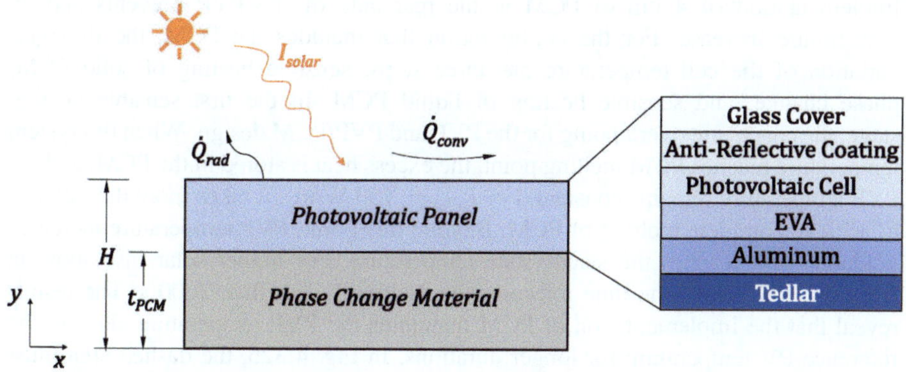


Fig. 4.31 Schematic of PVP/PCM system

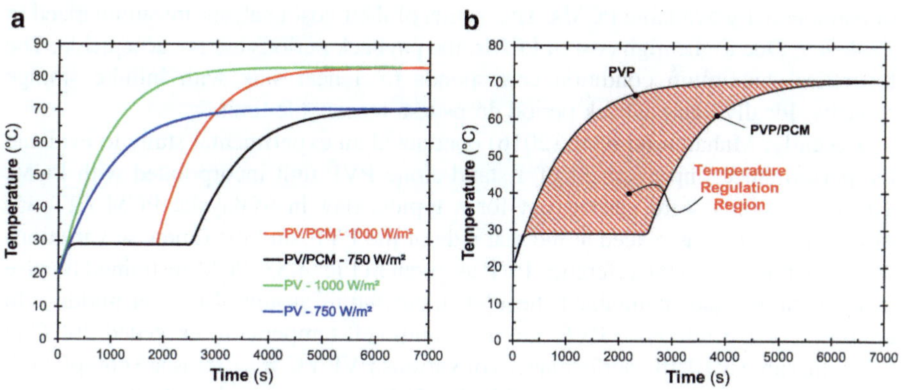


Fig. 4.32 Timewise temperature variations for PVP and PVP/PCM units. (a) Influence of PCM, (b) temperature regulation region. (Modified from Hendricks and van Sark 2011)

released through the ambient during nighttime via solidification of the PCM. The cyclic characteristic of the PVP/PCM has to be considered in detail to reveal the performance of the system accurately. Partial melting or solidification of the PCM implies inappropriate design, as excess PCM increases investment costs. In the literature, there are many numerical and experimental works on PVP/PCM units that have investigated alternative designs to improve system performance. Here we represent selected results of some recent works on PVP/PCM systems.

Hendricks and van Sark (2011) developed a one-dimensional explicit model to simulate the transient heat conduction within the PVP/PCM system. Two sets of analyses have been conducted: *(i) constant load* and *(ii) transient load*. The results for constant solar load condition are given in Fig. 4.32. Here RT-27 ($T_{melting} = 27$ °C) is used as PCM. For a PVP without PCM, the steady-state cell temperature reaches up to 82 °C and 71 °C under 1000 W/m$^2$ and 750 W/m$^2$ of solar loads, respectively.

Implementation of 4 cm of PCM at the rear side of the PVP prevents sudden temperature increase. For the configuration that includes the PCM, the timewise variation of the cell temperature has three steps: sensible heating of solid PCM, phase change, and sensible heating of liquid PCM. In the first sensible heating stage, the curves are overlapping for the PVP and PVP/PCM design. When the system temperature reaches PCM melting point, the excess heat is stored in the PCM, and the PVP temperature remains constant. For $I_{solar} = 750$ W/m$^2$, it takes more than 2500 s to achieve complete melting of PCM. Beyond this point, PVP temperature increases gradually and reaches the steady-state temperature. For higher solar insolation, at $I_{solar} = 1000$ W/m$^2$, the time for complete melting is less than 2000 s. The results reveal that the implementation of PCM maintains the PVP temperature close to the reference PV temperature for longer durations. In Fig. 4.32b, the dashed area illustrates the thermal regulation enhancement region for $I_{solar} = 750$ W/m$^2$. Referring to Eq. (4.1), reducing the PVP temperature improves the electrical efficiency of the PVP. Hendricks and van Sark (2011) also estimated the payback periods of different types of commercially available PCMs. The results of their cost analysis are summarized in Table 4.9. Due to the high cost of PCMs, the payback periods are not acceptable. The theoretical maximum condition corresponds to a heat sink with infinite storage capacity. Ideally, the payback period drops to a reasonable limit.

Recently, Mahamudul et al. (2016) conducted an experimental study to evaluate the performance improvement of a stand-alone PVP unit incorporated with PCM. The experiments were carried out for a typical day in Malaysia. PCM (RT 35) measuring 2 cm was placed at the rear side of the PVP, and the timewise variations were compared with the reference PVP as given in Fig. 4.33. PCM restrained the rise in temperature and maintained the PVP temperature around 40 °C at midday. In contrast, for the reference PVP, the maximum cell temperature exceeded 50 °C at 2 pm. In Table 4.10, the performances of various PVP/PCM systems are compiled. It is evident that PCM integration holds the PVP temperature close to the reference temperature for a more extended period.

Heat storage performance of PVP/PCM should be further enhanced by improving heat transfer within the PCM cavity to make the PVP/PCM designs more feasible. The following approaches are investigated in the literature to achieve enhanced heat transfer: (*i*) *the application of fin* and (*ii*) *the usage of macro-, micro-, or nano-sized additives inside* PCM. The latter is discussed in Section 7.2. The team of Huang conducted one of the most comprehensive studies on the PVP/PCMs with finned

**Table 4.9** Payback periods of PVP/PCM systems

| PCM type | Payback period (Year) | |
|---|---|---|
| | Malaga | Utrecht |
| Rubitherm RT-27 | 195–236 | 306–370 |
| Rubitherm RT-42 | 165–199 | 394–475 |
| Thermusol HD 35 | 182 | 324 |
| BASF micronal DS5001 | 275 | 376 |
| Theoretical maximum | 4 | 13 |

Adapted from Hendricks and van Sark (2011)

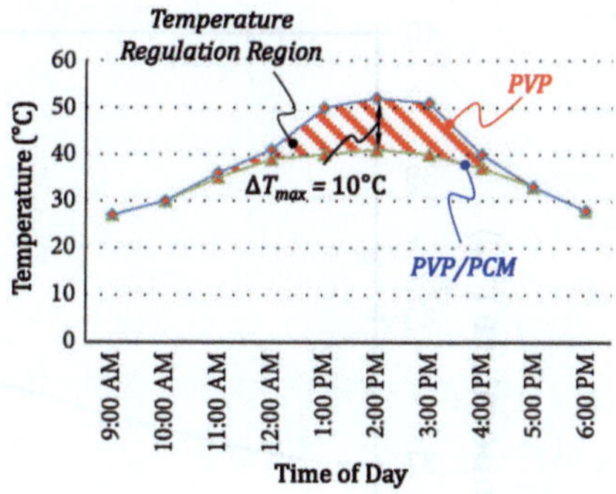


**Fig. 4.33** Timewise temperature variations for PVP and PVP/PCM units. (Modified from Mahamudul et al. 2016)

**Table 4.10** Performance comparison of different PVP/PCM systems

| Researchers | PCM type | Reference PVP temperature | PVP/PVP temperature | Temperature reduction | Constant temperature period |
|---|---|---|---|---|---|
| Huang et al. (2004) | RT-27 | 53–60 °C | 35–40 °C | 10–15 °C | 4–6 h |
| Hendricks and van Sark (2013) | RT-27 | 53–60 °C | 35–40 °C | 10–15 °C | 4–6 h |
| Hasan et al. (2014) | RT-22 | 53–60 °C | 30–35 °C | 15–18 °C | 4–6 h |
| Mahamudul et al. (2016) | RT-35 | 53–60 °C | 40–42 °C | 10–12 °C | 4–6 h |

Adapted from Mahamudul et al. (2016)

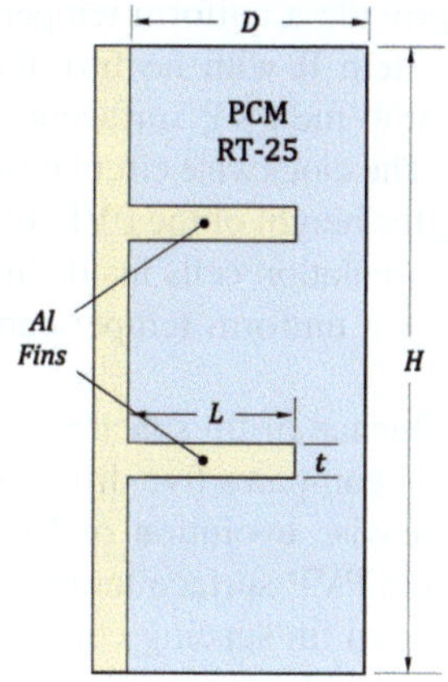


| System No | Height $H$ (mm) | Depth $D$ (mm) | Number of fins | Fin Thickness $t$ (mm) | Fin length $L$ (mm) |
|---|---|---|---|---|---|
| I | 40 | 20, 30, 50 | 0 | 0 | 0 |
| II | 132 | 20 | 0 | 0 | 0 |
| III | 132 | 20 | 2,5 | 2,3,4 | 15 |

**Fig. 4.34** Schematic and parameters of Huang et al. (2004)

cavities. Huang et al. (2004) conducted experimental and numerical studies for a PVP/PCM system to reveal the influence of various design and working conditions on the evolution of PVP temperature. Figure 4.34 shows the geometry of Huang et al. (2004) and the list of parameters that they have considered. They have obtained the spanwise and timewise variations in temperature for different fin configurations.

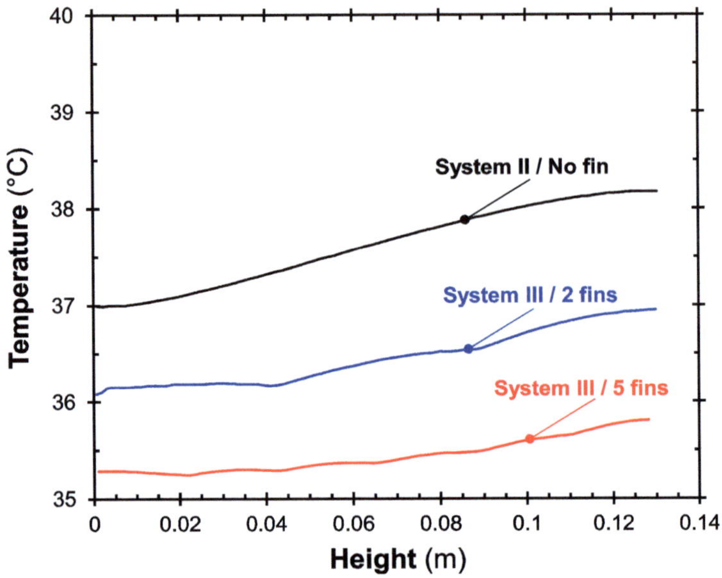


**Fig. 4.35** Variations in surface temperature along PVP height for three different PCM configurations. (Data from Huang et al. 2004)

Figure 4.35 depicts that the implementation of fins inside the PCM region increases heat transfer and reduces surface temperature. Fins also provide a uniform temperature variation along the height. In the baseline case (System II with no fin), the liquid PCM with higher temperature rises upward adjacent to the PVP surface and creates a clockwise circulation cell inside the PCM region. The clockwise circulation cell causes more than 9 °C/m temperature difference along the height of the PVP. On the other hand, in the finned system, there occur multiple circulation cells inside the liquid PCM zone, and the small fin arrangement provides a uniform temperature variation along the PVP height.

Huang et al. (2011) experimentally investigated the influence of fin spacings on temperature profiles within a PVP/PCM system. Figure 4.36 compares five different fin spacings with the baseline design considering the timewise evolution of PVP temperature. Close fin arrangement significantly reduces the PVP surface temperature. The temperature difference between the no fin and 8 mm fin spacing configuration is nearly 10 °C. Recently, Hasan et al. (2017) carried out experimental and numerical work to reveal the performance enhancement with PCM incorporation. In Fig. 4.37, the average and maximum temperature reductions are given together with the corresponding efficiency improvements for typical days for each month. Results reveal that the yearly average maximum temperature reduction is more than 10 °C. The improvement in the electrical efficiency of PVP reaches up to 7%. Hasan et al. (2017) also summarized the conditions and significant findings of the recent works on PVP/PCM systems, as listed in Table 4.11. Long-term or short-term

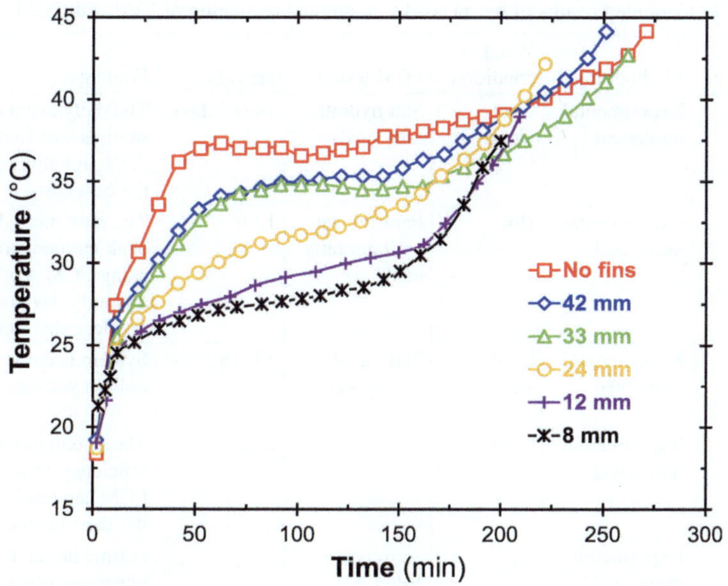


**Fig. 4.36** Influence of fin spacing on the timewise variation in PVP surface temperature. (Data from Huang et al. 2011)

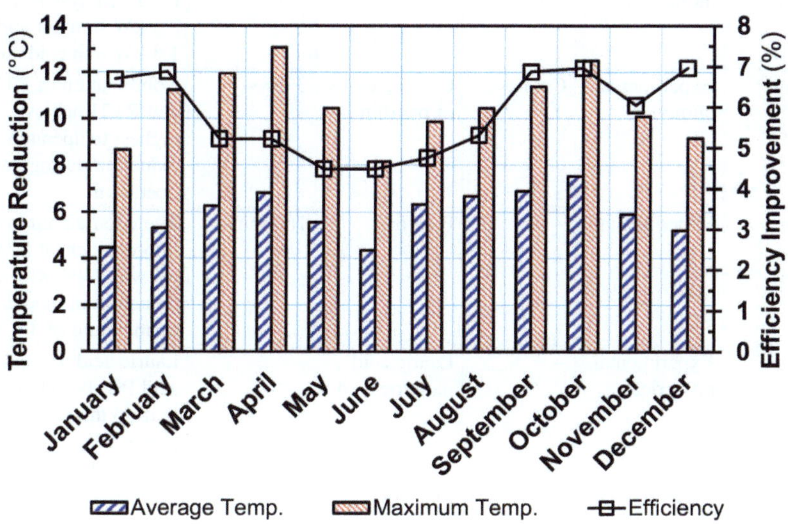


**Fig. 4.37** Annual performance of a PVP/PCM system. (Data from Hasan et al. 2017)

**Table 4.11** Compiled results of recent works on thermal regulation of PVP with PCM

| Reference | Methodology | Weather condition | PCM tested | Period | Findings |
|---|---|---|---|---|---|
| Hasan et al. (2015) | Experimental numerical | Cool warm | Salt hydrate fatty acids | 14–16 days | The daily average power savings was found to be 7.7% in hot climate in the best case |
| Hassan et al. (2014) | Experimental numerical | Hot | Paraffin, salt hydrate, fatty acids | 12 h | The maximum drop in peak temperature by using PCM was 5 °C on a cloudy day and 11 °C on clear sky conditions |
| Hasan et al. (2014) | Experimental numerical | Cool warm | Fatty acids salt hydrate | 14–16 days | System is cost-effective only in warmer climates |
| Aelenei et al. (2014) | Experimental numerical | Cool | – | 2 days | The maximum electrical efficiency of this BIPV-PCM can reach 10% and the thermal one 12% |
| Biwole et al. (2013) | Experimental numerical | – | Acronym SP/PCM paraffin | | A time lag of 1 h was observed to reach 34.9 °C, while the PCM always maintained PV temperature below 50 °C |
| Hendricks and van Sark (2013) | Experimental numerical | Moderate hot | | Year | Energy gain from PV-PCM system was 3.3 kW h in hot and 1.8 kW h in cold climate |
| Huang et al. (2011) | Experimental numerical | Cool | Combinations of paraffin | 3 days | Combination of RT27 and RT21 achieves the highest temperature reduction during daily operation |
| Sharma et al. (2016) | Experimental | – | Paraffin | 160 min | The open-circuit voltage improvement at 1000 W/m$^2$ was 4.4% with a reduction in module temperature of 3.8 °C |
| Sarwar (2012) | Experimental numerical | – | Lauric acid palmitic acid | – | Lauric acid and palmitic acid PCMs were found to drop the module temperature by 22 °C and 19.5 °C, respectively |
| Maiti et al. (2011) | Experimental numerical | Indoor outdoor | Paraffin wax metal-wax composite | 7 h | The PV-PCM system with embedded metal shreds maintains the temperature around 65–68 °C for 3 h compared with paraffin only, which reached 84 °C in less than 50 min |

(continued)

**Table 4.11** (continued)

| Reference | Methodology | Weather condition | PCM tested | Period | Findings |
|---|---|---|---|---|---|
| Lillo et al. (2011) | Experimental | 12 sites | PCMs with different melting points | – | CPV-PCM system yields 37.2% more power than CPV-only system in hot climate and 18% in cold climate |
| Wu (2009) | Experimental | – | Paraffin | – | A temperature drop of 18 °C with 10% increase in electrical efficiency is achieved by the CPV-PCM system at high intensity of 672 $W/m^2$ |

Data from Hasan et al. (2017)

**Table 4.12** Positive and negative aspects of thermal management of PVP with PCM

| Advantages | Disadvantages |
|---|---|
| Higher heat transfer rates compared to both forced air circulation and forced water circulation | High PCM cost |
| Higher heat absorption due to latent heating | Some PCMs are toxic |
| Isothermal nature of heat removal | Some PCMs have fire safety issues |
| No electricity consumption | Some PCMs are strongly corrosive |
| Passive heat exchange | PCMs may have disposal problem after their cycle is complete |
| No maintenance cost | |

Data from Makki et al. (2015)

experimental/numerical studies reveal that PCM usage improves the electrical efficiency of a PVP panel as much as 10% and such a thermal regulation design is a practical solution for hot climate conditions or concentrated PVPs (CPVPs). In Table 4.12, the pros and cons of PVP thermal regulation with PCM are listed. It is certain that the selection of an appropriate PCM type is crucial. A long-term stable, safe, and nontoxic material with a low cost should be found to be used as a PCM.

### 4.3.2 *Indirect Power Generation*

Alternatively, solar energy is collected to produce *electricity* from *heat*. There are two practical approaches to produce electricity from the solar heat: (*i*) *thermoelectric conversion* and (*ii*) *mechanical conversion.*

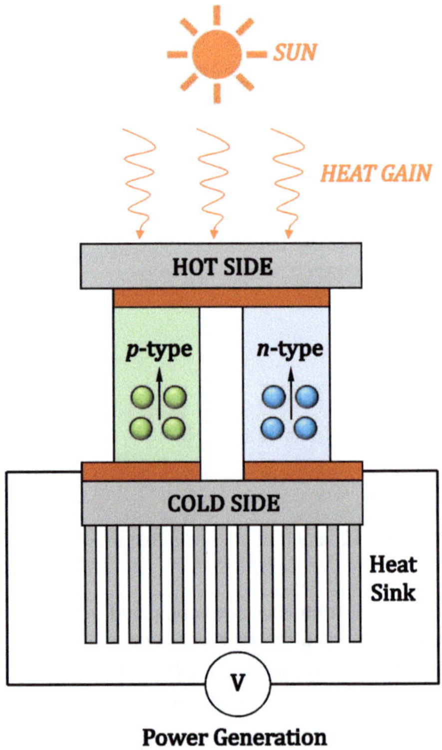


**Fig. 4.38** Schematic of a thermoelectric generator

**Thermoelectric Generators**

Thermoelectric generators (TEGs) are used to convert thermal energy into electricity. TEGs are composed of hundreds of thermoelectric couples that are connected electrically in series and thermally in parallel. The temperature difference between the junctions of the TEG generates an open-circuit voltage. TES units are integrated with the TEGs to limit the fluctuations in the hot/cold end temperatures/thermal loads. A schematic is given in Fig. 4.38 to illustrate the main components of a TEG. A TEG is composed merely of n-type and p-type semiconductors. The principle behind thermal energy to electricity conversion is the Seebeck effect. Several factors affect the electricity generation performance of a TEG, such as materials that are used as semiconductors, the amount of supplied thermal energy, and the design of the heat sink. TEGs are recently used in various applications to harvest waste heat and produce electricity. In the case of solar energy, TEGs are commonly used as stand-alone electricity generators or used coupled with PVPs. In a PVP-TEG unit, the TEG is placed on the rear side of the PVP. The TEG extracts thermal energy from the PVP to produce electricity. Such a configuration provides two advantages. The heat rejection from the PVP improves the electricity conversion efficiency of the cell. Further achievement is obtained by producing electricity from the TEG. On the other hand, stand-alone TEG units directly generate electricity from solar energy.

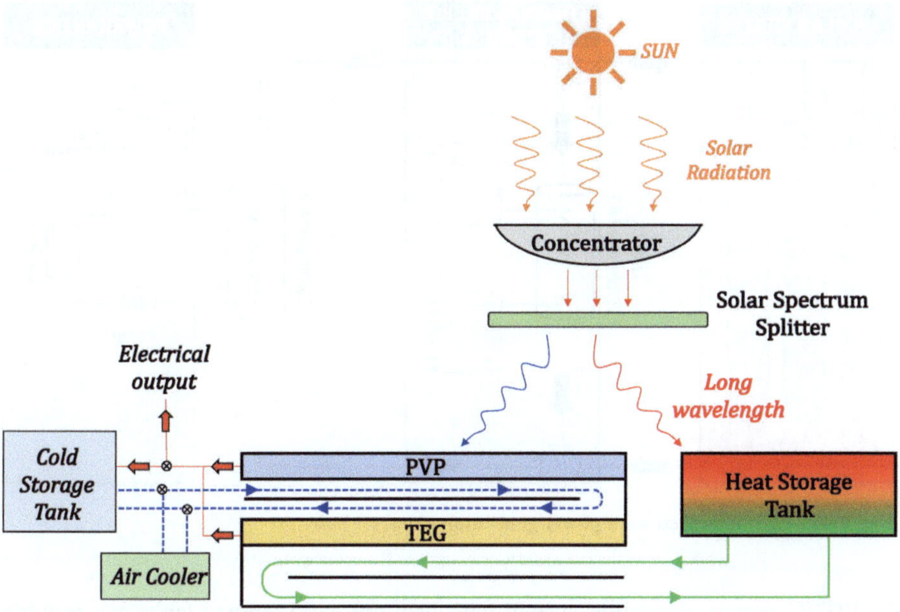


**Fig. 4.39** PVP/TEG combined electric generation system with heat and cold storage tanks. (Modified from Li et al. 2014)

However, fluctuations in solar energy make it difficult to achieve a continuous output from the TEG. PCM integration to a TEG unit makes the system more flexible. The solar energy that is stored within the PCM container during daytime provides electricity generation at nighttime.

Li et al. (2014) proposed a unique system to produce electricity from a PVP/TEG unit with heat and cold storage tanks. The schematic of the proposed system is given in Fig. 4.39. Solar spectrum splitter directs the short wavelength to PVP and long wavelength to the heat storage tank. Such a design provides lower PVP temperatures and higher temperatures with enhanced thermal energy within the storage tank. A closed-loop circulation transfers the thermal energy from heat storage tank to the hot side of the TEG (*solid green lines*). There is a secondary flow circulation on the heat sink of the TEG (*dashed blue lines*). The secondary loop rejects the heat from both heat sink of the TEG and the PVP, so this region is called as a cooling chamber. During off-peak times, the ambient air is circulated to remove the excess heat from the system, and a portion of the electrical output from the PVP/TEG system is used to obtain low-temperature cold storage inside the tank. The stored cold energy is then utilized during peak hours to regulate the PVP and TEG temperatures. Reducing the PVP temperature improves the electrical output of the cell. The temperature difference between the hot and cold sides of the TEG also tends to improve the electrical output of the TEG. Consequently, it is reported that such a novel combined

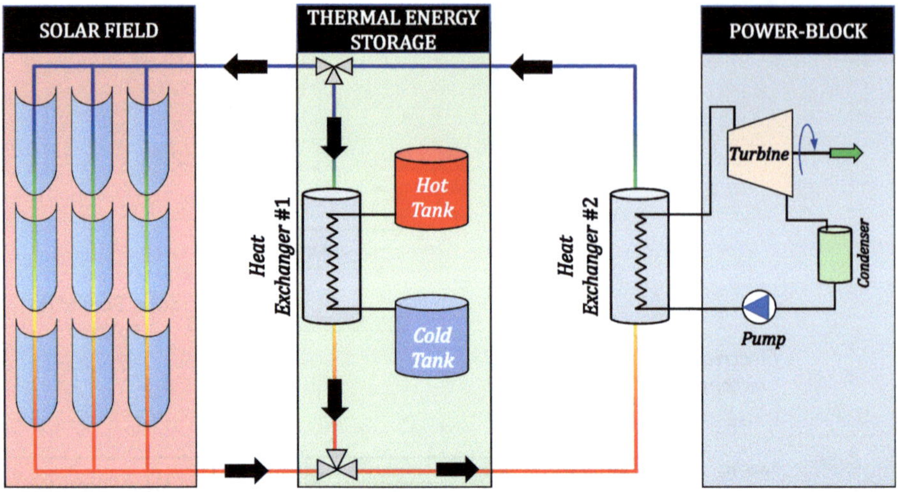


**Fig. 4.40** TES-integrated solar power generation plant

PVP/TEG power generation system with heat and cold storage tanks can provide 30% improvement in electrical output.

***Concentrated Solar Power***

Concentrated solar power (CSP) plants receive solar energy in a concentrated small collector surface. The working fluid extracts heat from the collector surface and transfers it to a conventional power generation block, i.e., Rankine cycle, to generate electricity. The integration of TES units to CSP plants provides continuous power production by storing solar energy during daytime. Figure 4.40 schematically represents the blocks of a TES-integrated CSP unit. The plant consists of three main blocks: (*i*) *solar field*, (*ii*) *thermal storage*, and (*iii*) *power block*. In this particular case, the thermal storage block includes hot and cold storage tanks. A portion of thermal energy that is collected in the solar field is transferred to Heat Exchanger #1 to store excess heat within the storage tank. The secondary circulation from cold tank to hot tank allows storing thermal energy during daytime. The circulation is reversed when the solar energy is not accessible. Flow through hot tank to cold tank releases the thermal energy that is supplied to the power block. Heat Exchanger #2 transfers the thermal energy that is extracted either from the solar field or thermal storage unit to the power block.

Figure 4.41 shows the variation in the capacity of the hybrid solar power plant throughout the day. The power plant is deployed as a fossil fuel backup. During nighttime, the plant is driven by the fuel backup system. From sunrise to sunset, solar energy is directly supplied to the power block, and the excess heat is directed to the TES unit for future use. After sunset, the stored thermal energy is released to run the plant for a limited duration. Such a hybrid and TES-integrated power plant provides

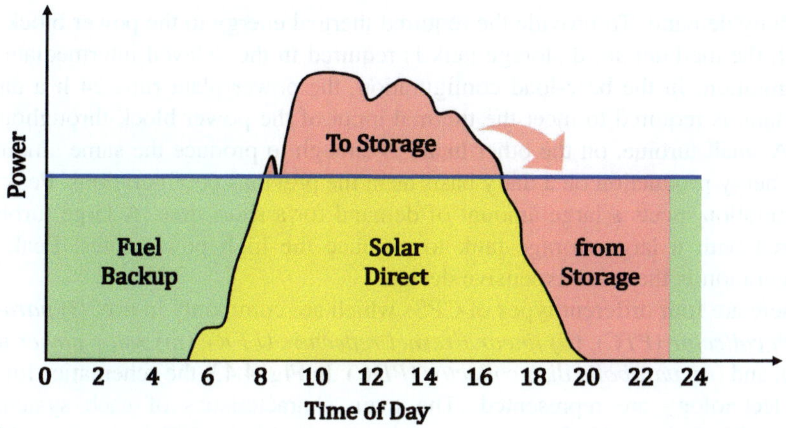


**Fig. 4.41** Load shaving for a solar power generation plant

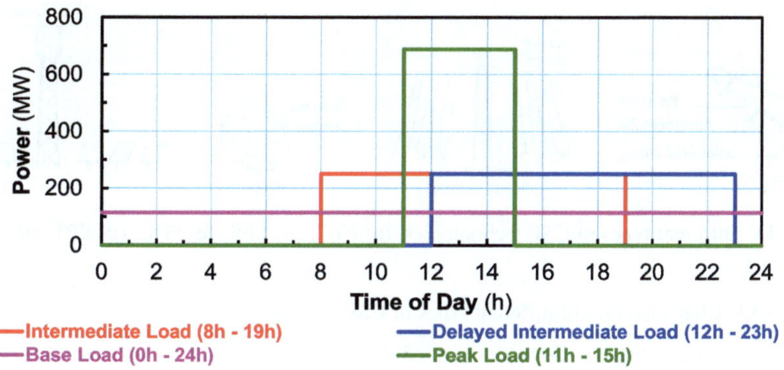


**Fig. 4.42** Power rates and production durations for solar power plants with different TES configurations. (Adapted from IEA International Energy Agency 2018)

flexible and environmentally friendly power generation. There are various alternative approaches for TES implementation to a solar power plant.

Figure 4.42 schematically illustrates four different configurations for TES-integrated CSPs that proposed in the recent report by IEA. The configurations are labeled as (*i*) *intermediate load*, (*ii*) *delayed intermediate*, (*iii*) *based load*, and (*iv*) *peak load*. In each configuration, the solar field size is identical, and the turbine produces the same amount of electricity. However, depending on the charging and discharging scenarios, the power rates and production times vary. In the intermediate load mode, electricity is produced when sunlight is available (i.e., from 8 am to 19 pm). Such a configuration requires a small-sized storage tank to regulate the fluctuations in the solar load. For intermediate load configuration, the investment cost is low as the storage tank is small. The delayed intermediate load design produces the electricity from noon to midnight. Such a design can meet the peak electricity demand. To provide the required thermal energy to the power block after sunset, the medium-sized storage tank is required in the delayed intermediate load configuration. In the base-load configuration, the power plant runs 24 h a day. A large tank is required to meet the thermal input of the power block throughout the day. A small turbine, on the other hand, is enough to produce the same amount of total energy production on a daily basis as in the previous configurations. Peak load configuration meets a large amount of demand for a short time. A large turbine is required with a large storage tank to produce the high power rates. Peak load configuration is the most expensive design.

There are four different types of CPSs which are commonly in use: (*i*) *parabolic trough collector* (*PTC*), (*ii*) *linear Fresnel reflectors* (*LFR*), (*iii*) *solar power tower* (*SPT*), and (*iv*) *parabolic dish collector* (*PDC*). In Fig. 4.43, the schematics for each CSP technology are represented. The main characteristics of each system are compared in Table 4.13. Storage integration is possible in CSPs, except the PDCs.

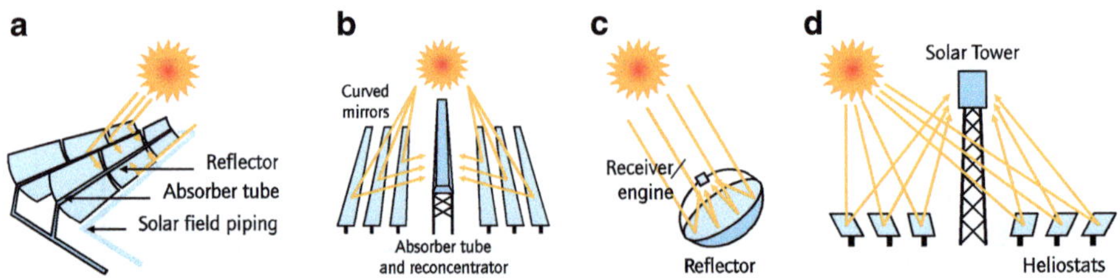


**Fig. 4.43** Different types of CSP technologies. (**a**) PTC, (**b**) LFR, (**c**) PDC, (**d**) SPT. (IEA 2018)

**Table 4.13** Characteristics of different types of CSP technologies

| Technology | Operating temperature (°C) | Solar concentration ratio | Storage integration possibility | Advantages | Disadvantages |
|---|---|---|---|---|---|
| PTC | 20–400 | 15–45 | Possible | Low installation cost | Large area occupied |
| | | | | Large experimental feedback | Low thermal efficiency |
| LFR | 50–300 | 10–40 | Possible | Low installation cost | Low thermal efficiency |
| SPT | 300–1000 | 150–1500 | Highly possible | High thermal efficiency | Large area occupied |
| | | | | | High installation cost |
| PDC | 120–1500 | 100–1000 | Difficult | Small area occupied | High installation cost |
| | | | | High thermal efficiency | Little experimental feedback |

Data from Pelay et al. (2017)

*PTC* parabolic trough collector, *LFR* linear Fresnel reflectors, *SPT* solar power tower, *PDC* parabolic dish collector

**Table 4.14** Evolution of power, hybridization, and TES properties of CSP power plants

| Build status | Average power of plants (MW) | Hybridization (%) | TES integration (%) | Storage capacity (hours) |
|---|---|---|---|---|
| *Before* 2000 | 20.5 | 79.8 | 52.5 | 3.0 |
| 2000–2010 | 23.2 | 64.3 | 49.8 | 3.2 |
| *After* 2010 | 41.9 | 80.7 | 44.0 | 7.0 |
| Under construction | 64.3 | 38.3 | 72.6 | 6.0 |
| Planned | 125.5 | 56.0 | 77.4 | 7.6 |

Data from Pelay et al. (2017)

According to the recent statistics provided by Pelay et al. (2017), PTC and SPT are the most widely used CSP technologies. Power plants which involve PTC and SPT correspond to 70% of the CSPs which are currently in use or under construction. The total capacity of CSPs dramatically increased from 21 $MW_{electric}$ to 120 $MW_{electric}$ (by six times) beyond 2000. TES integration is also planned to be improved. Currently, nearly half of the CSP plants are integrated with TES systems, and the plan is to increase the ratio by 77%. The storage durations for the TES-integrated CSPs are also increased dramatically from 3 h to 7 h. The increment of storage duration is achieved by implementation of new storage technologies with highly efficient methods. Table 4.14 summarizes some statistical data for current and planned CSP plants.

## 4.4 Closing Remarks

In the current chapter, an overview of the TES applications is represented. The content is mainly focused on the implementations of sensible and latent heat storage units for heating/cooling, hot water supply, and solar power generation systems. It is revealed that the TES is a promising technique to provide thermal comfort in buildings either in active or passive heating/cooling systems. The implementation of TES in conventional HVAC units not only reduces temperature fluctuations but also reduces operational and investment costs. PCMs are used in a wide range of thermal applications to maintain system temperature in the desired range. There is increasing attention on the development of PCM-incorporated photovoltaic and thermoelectric power generation systems. TES is also integrated with CSP power generation systems to maintain the thermal source in various load scenarios to improve effectiveness and power generation durations.

## Nomenclature

| | |
|---|---|
| $c$ | Specific heat, J/kgK |
| $C$ | Volumetric heat capacity, J/m$^3$K |
| $D$ | Depth, m |
| $\Delta H$ | Reaction enthalpy, J/mol |
| $h$ | Specific enthalpy, J/kg |
| $I_{solar}$ | Incident solar radiation, W/m$^2$ |
| $L$ | Length, m |
| $\dot{Q}$ | The rate of heat transfer, W |
| $t$ | Thickness |
| $T$ | Temperature, K or °C |
| $x, y$ | Cartesian coordinate, m |

## Greek Letters

| | |
|---|---|
| $\beta$ | The temperature coefficient of a PVP, K$^{-1}$ |
| $\eta$ | Efficiency |

## Subscripts

| | |
|---|---|
| *conv* | Convection |
| *m* | Melting |
| *pvp* | Photovoltaic panel |
| *rad* | Radiation |
| *ref* | Reference |
| *sf* | Solid to liquid |

## Abbreviations

| | |
|---|---|
| ATES | Aquifer thermal energy storage |
| BIPVP | Building-integrated photovoltaic panel |
| BTES | Borehole thermal energy storage |
| COP | Coefficient of performance |
| CSP | Concentrated solar power |
| CTES | Cold thermal energy storage |
| HTES | Heat thermal energy storage |
| HTF | Heat transfer fluid |
| HVAC | Heating ventilation air conditioning |
| IEA | International Energy Agency |
| LHTES | Latent heat thermal energy storage system |
| LFR | Linear Fresnel reflectors |
| PCM | Phase change material |
| PDC | Parabolic dish collector |
| PTC | Parabolic trough collector |
| PVP | Photovoltaic panel |
| PVP/T | Photovoltaic thermal system |
| SDHW | Solar domestic hot water |
| SPT | Solar power tower |
| TEG | Thermoelectric generator |
| TES | Thermal energy storage |

# References

Aelenei, L., Pereira, R., Gonçalves, H., & Athienitis, A. (2014). Thermal performance of a hybrid BIPV-PCM: Modeling, design and experimental investigation. *Energy Procedia, 48*, 474–483.

Al-Sanea, S. A., Zedan, M. F., & Al-Hussain, S. N. (2012). Effect of thermal mass on performance of insulated building walls and the concept of energy savings potential. *Applied Energy, 89*(1), 430–442.

Ames, W. D. (1983). *Photovoltaic devices for producing electrical and heat energy*. United States of America. Patent number US4389533.

Biwole, P. H., Eclache, P., & Kuznik, F. (2013). Phase-change materials to improve solar panel's performance. *Energy and Buildings, 62*, 59–67.

Browne, M. C., Norton, B., & McCormack, S. J. (2015). Phase change materials for photovoltaic thermal management. *Renewable and Sustainable Energy Reviews, 47*, 762–782.

Cabeza, L. F., Mehling, H., Hiebler, S., & Ziegler, F. (2002). Heat transfer enhancement in water when used as PCM in thermal energy storage. *Applied Thermal Engineering, 22*(10), 1141–1151.

Cabeza, L. F., Ibanez, M., Sole, C., Roca, J., & Nogués, M. (2006). Experimentation with a water tank including a PCM module. *Solar Energy Materials and Solar Cells, 90*(9), 1273–1282.

Cabeza, L. F., Castell, A., Barreneche, C. D., De Gracia, A., & Fernández, A. I. (2011). Materials used as PCM in thermal energy storage in buildings: A review. *Renewable and Sustainable Energy Reviews, 15*(3), 1675–1695.

Evans, D. L., & Florschuetz, L. W. (1977). Cost studies on terrestrial photovoltaic power systems with sunlight concentration. *Solar Energy, 19*(3), 255–262.

Ezan, M. A. (2006). *Design and optimization of Ice-on-Coil latent thermal storage system*. MSc thesis, Graduate School of Natural and Applied Sciences of Dokuz Eylul University, Izmir.

Ezan, M. A. (2011). *Experimental and numerical investigation of cold thermal energy storage systems*. PhD thesis, Graduate School of Natural and Applied Sciences of Dokuz Eylul University, Izmir.

Ezan, M. A., Cetin, L., & Erek, A. (2011). Ice thickness measurement method for thermal energy storage unit. *Journal of Thermal Science and Technology, 31*, 1–10.

Hasan, A., McCormack, S. J., Huang, M. J., & Norton, B. (2014). Energy and cost saving of a photovoltaic-phase change materials (PV-PCM) system through temperature regulation and performance enhancement of photovoltaics. *Energies, 7*(3), 1318–1331.

Hasan, A., McCormack, S. J., Huang, M. J., Sarwar, J., & Norton, B. (2015). Increased photovoltaic performance through temperature regulation by phase change materials: Materials comparison in different climates. *Solar Energy, 115*, 264–276.

Hasan, A., Sarwar, J., Alnoman, H., & Abdelbaqi, S. (2017). Yearly energy performance of a photovoltaic-phase change material (PV-PCM) system in hot climate. *Solar Energy, 146*, 417–429.

Hasnain, S. M. (1998). Review on sustainable thermal energy storage technologies, part I: Heat storage materials and techniques. *Energy Conversion and Management, 39*(11), 1127–1138.

Hassan, A., Nouman, H., Assi, A., & Norton, B. (2014). *Temperature regulation and thermal energy storage potential of phase change materials layer contained at the back of a building integrated photovoltaic panel*. Proceedings of the 30th International PLEA Conference (pp. 16–18), CEPT University, Ahmedabad.

Heier, J., Bales, C., & Martin, V. (2015). Combining thermal energy storage with buildings–a review. *Renewable and Sustainable Energy Reviews, 42*, 1305–1325.

Hendricks, J. H. C., & Sark, W. G. J. H. M. (2011). Annual performance enhancement of building integrated photovoltaic modules by applying phase change materials. *Progress in Photovoltaics: Research and Applications, 21*(4), 620–630.

Huang, M. J., Eames, P. C., & Norton, B. (2004). Thermal regulation of building-integrated photovoltaics using phase change materials. *International Journal of Heat and Mass Transfer, 47*(12), 2715–2733.

Huang, M. J., Eames, P. C., Norton, B., & Hewitt, N. J. (2011). Natural convection in an internally finned phase change material heat sink for the thermal management of photovoltaics. *Solar Energy Materials and Solar Cells, 95*(7), 1598–1603.

Icesynergy. www.icesynergy.com. Accessed 12 Jan 2018.

IEA International Energy Agency. https://www.iea.org/newsroom/news/2017/january/making-freshwater-from-the-sun.html. Accessed 12 Jan 2018.

Iten, M., Liu, S., & Shukla, A. (2016). A review on the air-PCM-TES application for free cooling and heating in the buildings. *Renewable and Sustainable Energy Reviews, 61*, 175–186.

Kenisarin, M., & Mahkamov, K. (2007). Solar energy storage using phase change materials. *Renewable and Sustainable Energy Reviews, 11*(9), 1913–1965.

Khalifa, A. J. N., Suffer, K. H., & Mahmoud, M. S. (2013). A storage domestic solar hot water system with a back layer of phase change material. *Experimental Thermal and Fluid Science, 44*, 174–181.

Krauter, S., Araújo, R. G., Schroer, S., Hanitsch, R., Salhi, M. J., Triebel, C., & Lemoine, R. (1999). Combined photovoltaic and solar thermal systems for facade integration and building insulation. *Solar Energy, 67*(4), 239–248.

Kürklü, A., Özmerzi, A., & Bilgin, S. (2002). Thermal performance of a water-phase change material solar collector. *Renewable Energy, 26*(3), 391–399.

Li, Y., Witharana, S., Cao, H., Lasfargues, M., Huang, Y., & Ding, Y. (2014). Wide spectrum solar energy harvesting through an integrated photovoltaic and thermoelectric system. *Particuology, 15*, 39–44.

Lillo, B. I., Silva, P. M., & Larraneta, G. C. M. (2011). *Use of phase change materials in photovoltaic modules with solar concentration of up to 2*. 26th European photovoltaic solar energy conference and exhibition, Hamburg.

Mahamudul, H., Rahman, M. M., Metselaar, H. S. C., Mekhilef, S., Shezan, S. A., Sohel, R., et al. (2016). Temperature regulation of photovoltaic module using phase change material: A numerical analysis and experimental investigation. *International Journal of Photoenergy, 5917028*, 1–8.

Maiti, S., Banerjee, S., Vyas, K., Patel, P., & Ghosh, P. K. (2011). Self regulation of photovoltaic module temperature in V-trough using a metal–wax composite phase change matrix. *Solar Energy, 85*(9), 1805–1816.

Makki, A., Omer, S., & Sabir, H. (2015). Advancements in hybrid photovoltaic systems for enhanced solar cells performance. *Renewable and Sustainable Energy Reviews, 41*, 658–684.

Mehling, H., Cabeza, L. F., Hippeli, S., & Hiebler, S. (2003). PCM-module to improve hot water heat stores with stratification. *Renewable Energy, 28*(5), 699–711.

Oró, E., De Gracia, A., Castell, A., Farid, M. M., & Cabeza, L. F. (2012). Review on phase change materials (PCMs) for cold thermal energy storage applications. *Applied Energy, 99*, 513–533.

Pelay, U., Luo, L., Fan, Y., Stitou, D., & Rood, M. (2017). Thermal energy storage systems for concentrated solar power plants. *Renewable and Sustainable Energy Reviews, 79*, 82–100.

Rabin, Y., Bar-Niv, I., Korin, E., & Mikic, B. (1995). Integrated solar collector storage system based on a salt-hydrate phase-change material. *Solar Energy, 55*(6), 435–444.

Rosen, M. A. (2001). The exergy of stratified thermal energy storages. *Solar Energy, 71*(3), 173–185.

Sarwar, J. (2012). *Experimental and numerical investigation of thermal regulation of photovoltaic and concentrated photovoltaic using phase change materials*. PhD thesis. Dublin Institute of Technology.

Sayyar, M., Weerasiri, R. R., Soroushian, P., & Lu, J. (2014). Experimental and numerical study of shape-stable phase-change nanocomposite toward energy-efficient building constructions. *Energy and Buildings, 75*, 249–255.

Seddegh, S., Wang, X., Henderson, A. D., & Xing, Z. (2015). Solar domestic hot water systems using latent heat energy storage medium: A review. *Renewable and Sustainable Energy Reviews, 49*, 517–533.

Sharif, M. A., Al-Abidi, A. A., Mat, S., Sopian, K., Ruslan, M. H., Sulaiman, M. Y., & Rosli, M. A. M. (2015). Review of the application of phase change material for heating and domestic hot water systems. *Renewable and Sustainable Energy Reviews, 42*, 557–568.

Sharma, S., Tahir, A., Reddy, K. S., & Mallick, T. K. (2016). Performance enhancement of a building-integrated concentrating photovoltaic system using phase change material. *Solar Energy Materials and Solar Cells, 149*, 29–39.

Shinde, G. D., Suresh, P. R., & Sancheti, S. D. (2015). Influence of fins on solidification of phase change material in rectangular capsule. *International Journal of Modern Trends in Engineering and Research (IJMTER), 2*(6), 351–360.

Tatsidjodoung, P., Le Pierrès, N., & Luo, L. (2013). A review of potential materials for thermal energy storage in building applications. *Renewable and Sustainable Energy Reviews, 18*, 327–349.

U.S. Energy Information Administration. https://www.eia.gov/energyexplained/index.cfm?page=us_energy_homes#tab1. Accessed 12 Jan 2018.

Wu, Y. (2009). *Thermal management of concentrator photovoltaics*. PhD thesis, University of Warwick.

Yamagishi, Y., Sugeno, T., Ishige, T., Takeuchi, H., & Pyatenko, A. T. (1996, August). An evaluation of microencapsulated PCM for use in cold energy transportation medium. *Energy Conversion Engineering Conference, 1996. IECEC 96. Proceedings of the 31st Intersociety* (Vol. 3, pp. 2077–2083). IEEE.

# Chapter 5
# System Modeling and Analysis

## 5.1 Introduction

To build a thermal system for either heating or cooling purposes, the design engineer must consider various alternatives such as capacities of equipment, working scenarios, the input of controllers, the working conditions of heat transfer fluids, and so on. Developing a numerical model, either in component or system scale, makes it possible to determine the performance of the alternatives before building up the system. Such an approach to design appropriate thermal systems which fit the demands of the consumer is useful. Systems mostly interact with the surrounding, hence considering the local climatic conditions in the numerical models also makes it possible to optimize the system correctly. Thermodynamic analyses provide a better understanding to compare the systems that have different working or design configurations. The first and second laws of thermodynamics should be used to conduct the quantitative and qualitative considerations of a system. Heat transfer analysis allows evaluating the steady or unsteady behavior of thermal systems by reducing the heat transfer mechanisms into empirical or numerical correlations. Computational fluid dynamics (CFD), on the other hand, is resolving the set of governing equations to obtain the transport variables at a finite number of volumes. CFD is a powerful tool to resolve the flow field and temperature distributions inside complex geometries (or complicated problems) to obtain an in-depth understanding. In this chapter, the foundations of the modeling of thermal systems are represented, and the solution steps of the integral and differential forms of the governing equations are discussed with simple illustrative examples for heat storage applications.



I. Dincer, M. A. Ezan, *Heat Storage: A Unique Solution For Energy Systems*, Green Energy and Technology, https://doi.org/10.1007/978-3-319-91893-8_5

## 5.2 Thermodynamic Analysis

Thermodynamic analyses, either the first law or the second law, deal with the integral form of the balance equation to determine the performance of the system, i.e., efficiency. The thermodynamic balance equations for the sensible heat and latent heat storage tanks differ from each other. These are discussed in detail in the following subsections.

### 5.2.1 *Energy and Exergy Analyses of Sensible Heat TES Unit*

Consider a closed system that is filled with a fluid storage medium (Fig. 5.1). The heat transfer fluid is circulated inside a pipeline to transfer heat to/from the storage unit. Fluids A and B correspond to the fluids within the tank and heat transfer fluid, respectively.

As the storage tank is a thermodynamically closed system (Dincer et al. 1997), the energy balance equation (*conservation of energy*) for a low-temperature cold thermal energy storage unit (CTS) can be written as follows:

***Energy Balance:***

$$\text{Cold Charged} - [\text{Cold Discharged} + \text{Cold Loss}] = \text{Cold Accumulation}$$

or

$$Q_C - Q_D - Q_L = \Delta E \quad (\text{in J or kJ}) \tag{5.1}$$

$Q_C$ and $Q_D$ indicate, respectively, the total heat transfer between the storage medium (fluid A) and the heat transfer fluid (fluid B) during the charging and discharging periods. $Q_L$ is the total heat gain from the surroundings during the overall process. $\Delta E$ is the internal energy variation of the storage medium. If the storage system undergoes a complete cycle, the final temperature will be identical to the initial

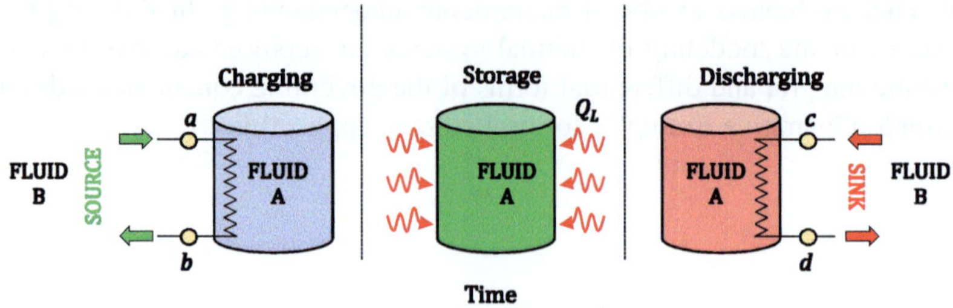


**Fig. 5.1** Charging, storage, and discharging processes of a sensible heat storage (SHS) unit

temperature of the storage medium, so the variation of the internal energy is zero, $\Delta E = 0$. The energy balance equation can be written regarding the enthalpies of the flowing fluid at the points $a$, $b$, $c$, and $d$ in Fig. 5.1 as

$$\underbrace{(H_b - H_a)}_{Q_C} - \left[\underbrace{(H_c - H_d)}_{Q_D} + Q_L\right] = \Delta E \tag{5.2}$$

The total energy transfer during the charging and discharging periods are defined as follows:

$$Q_C = \int_{t=0}^{t_{ch}} \dot{m}_a c_a (T_b(t) - T_a) dt \tag{5.3}$$

$$Q_D = \int_{t=0}^{t_{dis}} \dot{m}_c c_c (T_c(t) - T_d) dt \tag{5.4}$$

where $\dot{m}$ (in kg/s) and $c$ (in J/kg°C) are the mass flow rate and the specific heat of the heat transfer fluid, respectively. $t_{ch}$ and $t_{dis}$ refer to the charging and discharging durations. In the case of the constant inlet and outlet temperatures during the charging/discharging periods, the equations can be reduced to the following forms:

$$Q_C = H_b - H_a = m_a c_a (T_b - T_a) \tag{5.5}$$

$$Q_D = H_c - H_d = m_c c_c (T_c - T_d) \tag{5.6}$$

where $m$ (in kg) is the total mass of the streaming fluid during the charging or discharging processes. Energy variation of the system is defined as the difference between the final state ($f$) and the initial state ($i$) as

$$\Delta E = E_f - E_i = m(u_f - u_i) \tag{5.7}$$

where $m$ is the mass of the storage medium. $u_f$ and $u_i$ denote the internal energies of the storage medium at the final and initial states, respectively. For the sensible heat storage unit without phase change, the variation of the internal energy can be defined in terms of the temperature difference as

$$\Delta E = m\bar{c}(\bar{T}_f - \bar{T}_i) \tag{5.8}$$

Here the overbars indicate the arithmetic average quantities. For a linearly stratified storage tank, the average temperature of the storage medium is defined as follows by Rosen and Hooper (1991):

$$\bar{T} = \frac{T_{top} + T_{bottom}}{2} \tag{5.9}$$

$T_{top}$ and $T_{bottom}$ are the fluid temperatures at the top and bottom of the linearly stratified sensible heat storage unit, respectively. The first law efficiency (or the energy efficiency) of the system is defined as the ratio of the product energy output to the total energy input. In general, for a TES system that undergoes a cyclic process, the overall energetic efficiency (*or* the energy efficiency) is defined as follows:

***Energy efficiency:***

$$\eta = \frac{\text{Product Energy Recovered}}{\text{Energy Input}} \tag{5.10}$$

For simplicity, in the case of a CTS or HTS, the word "energy" can be replaced with "cold" or "heat," respectively. The overall energy efficiency of the process in Fig. 5.1 becomes

***Overall efficiency:***

$$\eta_{overall} = \frac{Q_D}{Q_C} = \frac{H_c - H_d}{H_b - H_a} \tag{5.11}$$

Moreover, the energy efficiencies of the sub-processes, i.e., charging, storing, and discharging, can be written as

***Charging efficiency:***

$$\eta_{CH} = \frac{\Delta E_{CH}}{H_b - H_a} \tag{5.12}$$

***Storage efficiency:***

$$\eta_{ST} = \frac{\Delta E_{CH} + Q_L}{\Delta E_{CH}} \tag{5.13}$$

***Discharging efficiency:***

$$\eta_{DIS} = \frac{H_c - H_d}{\Delta E_{DIS}} \tag{5.14}$$

Unlike the first law of thermodynamics, the second law of thermodynamics quantifies the "quality of energy." The quality of energy is measured from the state of the system in relation to its surrounding conditions (Jegadheeswaran et al. 2010). The quality or usefulness of energy is termed as *availability* or *exergy*. Exergy represents the measure of a system's "capacity to perform work" (Kestin 1980). Since the second law of thermodynamics introduces exergy, the analysis based on the second law is widely called *exergy analysis* (Jegadheeswaran et al. 2010). Unlike the energy, exergy is not a conserved property. For a real system exergy is destroyed. However, exergy can be conserved if the process reversibly takes place. It is well known that all real processes are irreversible, and irreversibilities related to the process are exposed exergy destruction. Hence, exergy balance of a system with heat gain can be expressed as follows:

***Exergy balance:***

$$\text{Exergy Input} - [\text{Exergy Output} + \text{Exergy Loss/Gain}] - \text{Exergy Destruction} = \text{Exergy Accumulation}$$

or

$$Ex_C - Ex_D - Ex_L - Ex_{Dest} = \Delta Ex \quad (\text{in J}) \tag{5.15}$$

In the exergy balance equation, the exergy transfers associated with the $Q_C$, $Q_D$, and $Q_L$ are denoted with $Ex_C$, $Ex_D$, and $Ex_L$, respectively. $Ex_{Dest}$ is the exergy destruction due to the irreversibilities, and $\Delta Ex$ is the variation of exergy within the system. The exergy of the flow at the inlet and outlet sections of the storage tank can be expressed in a general form as follows:

$$ex_k = (H_k - H_o) - T_o(S_k - S_o) \tag{5.16}$$

where *k* represents the inlet/outlet ports as $k = a, b, c$, or, $d$ in Fig. 5.1. $ex_k$, $H_k$, and $S_k$ denote the exergy, enthalpy, and entropy of the flowing fluid at state *k*, respectively. $H_o$ and $S_o$, on the other hand, correspond to the enthalpy and entropy values of flowing fluid at the environmental (*or* dead state) conditions. The exergy transfer associated with the charging process can be written as the difference between inlet and outlet as

$$Ex_C = ex_b - ex_a \tag{5.17}$$

or

$$Ex_C = ex_b - ex_a = (H_b - H_a) - T_o(S_b - S_a) \tag{5.18}$$

Assuming that the specific heat of the HTF remains constant from the inlet to the outlet of the system, entropy and enthalpy variations can be obtained by using the following fundamental thermodynamic relations:

$$dh = cdT \rightarrow \int_{in}^{out} dh = c\int_{in}^{out} dT \rightarrow (h_{in} - h_{out})_{HTF} = c_{HTF}(T_{HTF,in} - T_{HTF,out}) \tag{5.19}$$

$$ds = \frac{dh}{T} \rightarrow \int_{in}^{out} ds = c\int_{in}^{out} \frac{dT}{T} \rightarrow (s_{in} - s_{out})_{HTF} = c_{HTF}\ln\left(\frac{T_{HTF,in}}{T_{HTF,out}}\right) \tag{5.20}$$

Consequently, the exergy transfer during the charging can be expressed as follows:

$$Ex_C = ex_b - ex_a = m_a c_a \left\{ (T_b - T_a) - T_o \ln\left(\frac{T_b}{T_a}\right) \right\} \tag{5.21}$$

Similarly, the exergy transfer during the discharging process can be written as

$$Ex_D = ex_c - ex_d = m_c c_c \left\{ (T_c - T_d) - T_o \ln\left(\frac{T_c}{T_d}\right) \right\} \tag{5.22}$$

Exergy loss due to heat transfer with the surroundings, $Q_L$, during the storage period is defined as

$$Ex_L = \left[1 - \frac{T_o}{T_s}\right] Q_L \tag{5.23}$$

where $T_s$ is the temperature of the boundary at which the heat transfer takes place between the system and the surroundings. The $Ex_L$ term has a negative sign when the boundary temperature of the system is less than the surroundings ($T_s < T_o$). Such a condition takes place for low-temperature thermal energy storage systems (or CTES). One should note that the direction of the heat transfer is opposite to the exergy transfer associated with the heat transfer in the case of CTES.

Exergy variation of the system is defined as the difference between exergy content at the final state ($f$) and the initial state ($i$) as

$$\Delta Ex = Ex_f - Ex_i = m\left[(u_f - u_i) - T_o(s_f - s_i)\right] \tag{5.24}$$

where $s$ is the specific entropy of the storage medium. For the sensible heat storage unit without phase change, the variation of the exergy can be defined with regard to the temperature difference as

$$\Delta Ex = Ex_f - Ex_i = mc\left[(T_f - T_i) - T_o \ln\left(\frac{T_f}{T_i}\right)\right] \tag{5.25}$$

Similar to the first law efficiency (*or* energy efficiency) of the system, the second law efficiency (*or* exergy efficiency) is defined as the ratio of the product exergy output to the total exergy input. For a TES system that undergoes a cyclic process, the overall exergetic efficiency (*or* the exergy efficiency) is defined as follows:

***Exergy efficiency:***

$$\psi = \frac{\text{Product Exergy Recovered}}{\text{Exergy Input}} \tag{5.26}$$

The overall exergy efficiency of the process in Fig. 5.1 becomes

***Overall efficiency:***

$$\psi_{overall} = \frac{Ex_D}{Ex_C} = \frac{ex_c - ex_d}{ex_a - ex_b} \tag{5.27}$$

Moreover, the exergy efficiencies of the sub-processes, i.e., charging, storing, and discharging, can be written as

***Charging efficiency:***

$$\psi_{CH} = \frac{\Delta Ex_{CH}}{ex_a - ex_b} \tag{5.28}$$

***Storage efficiency:***

$$\psi_{ST} = \frac{\Delta Ex_{CH} + \Delta Ex_{ST}}{\Delta Ex_{CH}} \tag{5.29}$$

***Discharging efficiency:***

$$\psi_{DIS} = \frac{Ex_D}{\Delta Ex_{CH} + \Delta Ex_{ST}} \tag{5.30}$$

In the following worked example, the results of thermodynamic analyses for two sensible heat TES systems, Systems 1 and 2, are compared by carrying out analyses that are based on first law and second law aspects.

**Illustrative Example 1: Sensible Heat Storage Unit**

Consider two TES systems that are used for sensible heat storage at high temperatures. Water is used as heat transfer fluid. During the charging process, the TES system receives 104,650 kJ from the water stream of 500 kg. The inlet and outlet temperatures of the water stream are given as 80 °C and 30 °C, respectively. The heat input ($Q_C$) through the storage medium can be obtained by using the temperature drop of the HTF during the charging period as

*Energy stored in Systems 1 and 2:*

$Q_C = (mc)_{HTF}(T_{in} - T_{out})_{HTF} = 500 \times 4.186 \times (80 - 30) = 104,650$ kJ.

***(a) First Law Analysis of System 1***

For System 1, 1 day after the charging process, 94,185 kJ of heat is recovered from the storage tank. During the discharging period, a stream of 4500 kg of water is heated from 30 °C to 35 °C. Hence, the rejected thermal energy from System 1 can be evaluated as

*Energy recovered from System 1:*

$Q_{D,1} = (mc)_{HTF}(T_{out} - T_{in})_{HTF} = 4500 \times 4.186 \times (35 - 30) = 94,185$ kJ.

(continued)

As a result, the overall efficiency of System 1 can be evaluated from Eq. (5.11) as

*The efficiency of System 1:* $\eta_{sys,1} = \frac{Q_{D,1}}{Q_C} = \frac{94,185}{104,650} = 0.9$ or 90%.

The heat loss through the surrounding during the storage period is merely the difference between the charged and discharged energy amount as follows:

*The total amount of heat loss in System 1:*
$Q_{L,1} = Q_C - Q_{D,1} = 104,650 - 94,185 = 10,465$ kJ.

***(b) First Law Analysis of System 2***

In the second system, System 2, the stored energy is rejected from the tank 90 days after the charging process. During the discharging period, a stream of 500 kg of water is heated from 30 °C to 75 °C. Hence, the rejected thermal energy from System 2 can be evaluated as

*Energy recovered from System 2:*
$Q_{D,2} = (mc)_{HTF}(T_{out} - T_{in})_{HTF} = 500 \times 4.186 \times (75 - 30) = 94,185$ kJ.

Since the rejected heat from the system is the same as in System 1, the overall efficiency of System 2 will be identical as in System 1 as follows:

*The efficiency of the System 2:* $\eta_{sys,2} = \frac{Q_{D,2}}{Q_C} = \frac{94,185}{104,650} = 0.9$ or 90%.

The heat loss through the surrounding during the storage period is also identical to System 1 as follows:

*The total amount of heat loss in System 2:*
$Q_{L,2} = Q_C - Q_{D,2} = 104,650 - 94,185 = 10,465$ kJ.

***(c) Second Law Analysis of System 1 and System 2***

Considering the first law of thermodynamics, these two TES systems have the same energetic efficiencies even though the HTF outlet temperatures are not the same during the discharging period. System 1 provides thermal energy after 1 day at a temperature of 35 °C. In System 2, on the other hand, the same amount of thermal energy is obtained after 90 days with a higher temperature, 75 °C. As System 2 provides thermal energy at a higher temperature, it has a higher performance. However, the first law of thermodynamics does not take into account the *quality* or economic value of the thermal energy. Instead, it deals only with the quantity.

The true usefulness of a TES system could be achieved by introducing exergy efficiency. In the following, second law analysis is conducted for

(continued)

Systems 1 and 2. First, we should evaluate the exergy transfer during charging as

$$Ex_C = (mc)_{HTF}\left\{(T_{in} - T_{out}) - T_o \ln\left(\frac{T_{in}}{T_{out}}\right)\right\}_{HTF}$$

which becomes

$$Ex_C = 500 \times 4.186 \times \left\{(353 - 303) - 298 \times \ln\left(\frac{353}{303}\right)\right\} = 9386.88 \text{ kJ}$$

The exergy transfers during the discharging process for the System 1 and System 2 are evaluated by using Eq. (5.22) as follows:

$$\begin{aligned} \textit{System 1}: \; Ex_{D,1} &= 4500 \times 4.186 \times \left\{(308 - 303) - 298 \times \ln\left(\frac{308}{303}\right)\right\} \\ &= 2310.18 \text{ kJ} \\ \textit{System 2}: \; Ex_{D,2} &= 500 \times 4.186 \times \left\{(348 - 303) - 298 \times \ln\left(\frac{348}{303}\right)\right\} \\ &= 7819.52 \text{ kJ} \end{aligned}$$

Then, from Eq. (5.27), the exergy efficiencies of Systems 1 and 2 become

$$\textit{System 1}: \; \psi_{system,1} = \frac{Ex_{D,1}}{Ex_C} = \frac{2310.18}{9386.88} = 0.25 \quad \text{or} \quad 25\%$$

$$\textit{System 2}: \; \psi_{system,2} = \frac{Ex_{D,2}}{Ex_C} = \frac{7819.52}{9386.88} = 0.83 \quad \text{or} \quad 83\%$$

In this illustrative example, the exergy efficiencies of the Systems 1 and 2 are obtained as 25% and 83%. System 2 has greater exergy as the temperature of the HTF is higher than System 1. Consequently, even though the two systems provide different outlet temperatures, the analysis based on the first law aspect gives the same energy efficiencies as 90%. It is clear that considering only the first law aspect is not an appropriate tool to evaluate the performance of a TES unit properly. Some different measures, such as exergy-based, should be considered to determine the performance of the systems more realistically by considering not only the amount of thermal energy (*quantity*) but also the usefulness (*quality*). In the case of thermal science, the measure of the usefulness is related to the temperature. Higher temperature means greater availability of thermal energy.

**Discussions on the Example**

The worked example highlights the importance of the exergy-based thermodynamic analysis to evaluate the performance of a TES unit more realistically. Dincer (2002) lists the following advantages of the exergy-based thermodynamic analysis:

- It provides a more proper accounting of the loss of availability of heat in a TES system using the conservation of mass and energy principles together with the second law of thermodynamics for the goals of design and analysis.
- It gives more meaningful and useful information than energy analysis regarding the efficiency, loss, and performance of TES systems.
- It is more correct, reflecting the thermodynamic and economic values of the operation of TES systems.
- It is an efficient technique and reveals whether or not and by how much it is possible to design more efficient TES systems by reducing inefficiencies in the existing units.

## 5.2.2 *Energy and Exergy Analyses of Latent Heat TES Unit*

In the case of a latent heat TES unit, the variation of internal energy in the system is related to sensible and latent cooling or heating. For the charging process, the storage domain (or PCM) with an initial temperature of $T_i$ cools down via HTF. First superheated liquid PCM (phase change material) cools down to the melting temperature, $T_m$, and then turns into the solid phase. Enthalpy variation of this process can be illustrated as in Fig. 5.2. Here, $T_{final}$ designates the final temperature of subcooled solid PCM.

***Energy Analysis:***

As the system consists of a mixture of the solid and liquid PCM, the total energy variation of the system is defined as the difference between the final $(f)$ and the initial $(i)$ states of each phase as

$$\Delta E = E_f - E_i = m\left\{(1 - F)\left(h_{solid,f} - h_i\right) + F\left(h_{liquid,f} - h_i\right)\right\} \tag{5.31}$$

where $m$ is the total mass of the storage medium. $h_{solid}$ and $h_{liquid}$ denote the internal energies of the solid and liquid portions of the PCM, respectively, and $F$ is the liquid fraction.

***Exergy Analysis:***

Exergy variation of the system is defined as the difference between exergy content at the final state $(f)$ and the initial state $(i)$ as

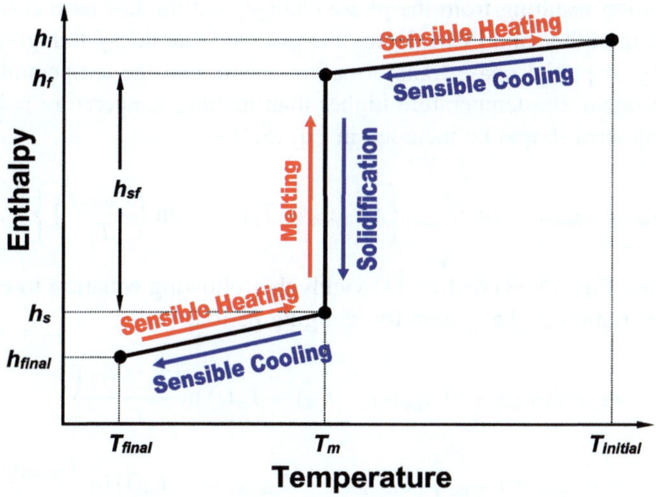


**Fig. 5.2** Variation of enthalpy with temperature for a latent heat TES

$$\left(\frac{dEx}{dt}\right)_{system} = \left(\frac{\Delta Ex}{\Delta t}\right)_{system} = \left(\frac{H^n - H^{n-1}}{\Delta t}\right)_{system} - T_o\left(\frac{S^n - S^{n-1}}{\Delta t}\right)_{system} \tag{5.32a}$$

where

$$\left(\frac{H^n - H^{n-1}}{\Delta t}\right)_{system} = \frac{1}{\Delta t}\left\{(mc)_{liquid}(T_m - T_i) - h_{sf}\Delta m_{solid} + (mc)_{solid}(T_f - T_m)\right\} \tag{5.32b}$$

$$\left(\frac{S^n - S^{n-1}}{\Delta t}\right)_{system} = \frac{1}{\Delta t}\left\{(mc)_{liquid}\ln\left(\frac{T_m}{T_i}\right) - h_{sf}\frac{\Delta m_{solid}}{T_m} + (mc)_{solid}\ln\left(\frac{T_f}{T_m}\right)\right\} \tag{5.32c}$$

Hence, exergy variation in the system can be obtained as

$$\begin{aligned}\Delta Ex_{system} = (mc)_{liquid}\left\{(T_m - T_i) - T_o\ln\left(\frac{T_m}{T_i}\right)\right\} + \Delta m_{solid}h_{sf}\left\{-1 + \frac{T_o}{T_m}\right\} \\ + (mc)_{solid}\left\{(T_f - T_m) - T_o\ln\left(\frac{T_f}{T_m}\right)\right\}\end{aligned} \tag{5.33}$$

where the first term on the right-hand side of the equation represents the exergy variation due to sensible cooling of liquid PCM, the second term designates the exergy variation resulting from the phase change, and the last term corresponds to the subcooling of the solid PCM. Here, it is assumed that the system goes into solid state entirely. If partial phase transformation occurs and the temperature of liquid PCM drops down to a temperature higher than melting temperature point ($T_{liquid}$), the following term should be included in Eq. (5.33):

$$\Delta Ex_{sensible,liquid} = (mc)_{liquid}\left\{ \left(T_{liquid} - T_i\right) - T_o \ln\left(\frac{T_{liquid}}{T_i}\right)\right\} \tag{5.34}$$

Combining Eqs. (5.33) and (5.34) yields the following equation to evaluate the net exergy variation of the system for charging period:

$$\begin{aligned}\Delta Ex_{system} = {} & m_{solid}(t)c_{solid}\left\{ \left(\bar{T}_{solid}(t) - T_m\right) - T_o(t)\ln\frac{\bar{T}_{solid}(t)}{T_m}\right\} \\ & + m_{liquid}(t)c_{liquid}\left\{ \left(\bar{T}_{liquid}(t) - \bar{T}_{liquid,i}\right) - T_o(t)\ln\frac{\bar{T}_{liquid}(t)}{\bar{T}_{liquid,i}}\right\} \\ & + m_{solid}(t)c_{liquid}\left\{ \left(T_m - \bar{T}_{liquid,i}\right) - T_o(t)\ln\frac{T_m}{\bar{T}_{liquid,i}}\right\} \\ & + m_{solid}(t)\left\{ -h_{sf} + T_o(t)\frac{h_{sf}}{T_m}\right\}\end{aligned} \tag{5.35}$$

where the first two terms signify the exergy variations inside the solid and liquid phases of water due to sensible cooling. The third term represents the exergy change that arises from the cooling of superheated water that transformed into ice. The last term indicates the exergy variation that arises from phase change. On the other hand, the net exergy variation of the system for discharging period can be written as follows:

$$\begin{aligned}\Delta Ex_{system} = {} & m_{solid}(t)c_{solid}\left\{ \left[\bar{T}_{solid}(t) - \bar{T}_{solid,i}\right] - T_o(t)\ln\frac{\bar{T}_{solid}(t)}{\bar{T}_{solid,i}}\right\} \\ & + m_{liquid}(t)c_{liquid}\left\{ \left[\bar{T}_{liquid}(t) - T_m\right] - T_o(t)\ln\frac{\bar{T}_{liquid}(t)}{T_m}\right\} \\ & + m_{liquid}(t)c_{solid}\left\{ \left[T_m - \bar{T}_{solid,i}\right] - T_o(t)\ln\frac{T_m}{\bar{T}_{solid,i}}\right\} \\ & + m_{liquid}(t)\left\{ h_{sf} + T_o(t)\frac{h_{sf}}{T_m}\right\}\end{aligned} \tag{5.36}$$

where the first two terms signify the exergy variation inside the solid and liquid phases of PCM due to sensible heating. The third term represents the exergy change that arises from the heating of subcooled solid PCM that transformed into a liquid state. The last term indicates the exergy variation that arises from phase change. The energy and exergy efficiencies that are defined for the sensible heat TES in the previous section are identical in latent heat TES. Further details on energy and exergy analyses of latent heat TES systems can be found in MacPhee and Dincer (2009) and Jegadheeswaran et al. (2010).

In the following worked example, the results of thermodynamic analyses for sensible and latent heat TES systems are compared by carrying out analyses that are based on the first law and second law aspects.

**Illustrative Example 2: Sensible Heat and Latent Heat Storage Units**

Consider two TES systems that are used for low-temperature cold storage. Water is used as a storage medium in two systems. The systems work in a cyclic manner, which means that the initial and final states of the systems are identical at the end of the charging, storage, and discharging sequences. For simplicity, in the sensible and latent heat storage tanks, it is assumed that the storage medium, water, is thoroughly mixed. Two cases of Dincer and Rosen (2011) are considered here. The first case is thoroughly mixed sensible heat storage; the second case is latent heat storage at melting temperature. In the analyses, the following simplifications are considered:

- During charging and discharging, the tanks are assumed to be adiabatic, which corresponds to no heat gain from the surroundings.
- The outer surface of the storage tank is 2 °C higher than the mean temperature of the storage medium.
- The thermophysical properties of the heat transfer fluid (HTF) are constant. The mass flow rate of the HTF is varied in such a way that the inlet and outlet temperatures of the HTF remain constant during the charging and discharging periods.
- The ambient temperature remains constant throughout the cycle at 20 °C.
- There is no work interaction and negligible variation in the kinetic and potential energy terms.

The specific heat of the storage medium and heat transfer fluid is identical to 4.18 kJ/kgK. The melting temperature of the water is 0 °C with a latent heat of 333.43 kJ/kg. The mean temperature of the storage medium at the beginning of the charging (*Ch*), storage (*St*), and discharging (*Dis*) is given in Table 5.1. In the latent heat thermal energy storage tank, the letters *l* and *s* indicate the liquid and solid states. Due to heat gain during the storage period, a part of PCM turns into a liquid state. Here the liquid fraction (*F*) is assumed to be 10%. The schematic representation of the current system is similar to the one in Fig. 5.1. The inlet temperature of the HTF is kept constant during the charging and discharging periods. For the charging $T_a = -10$ °C and for the discharging $T_c = 20$ °C. The outlet temperatures of the HTF during the charging and discharging periods are given in Table 5.1. The storage tank has an inner diameter of 2 m with an internal height of 5 m. The total mass of

(continued)

**Table 5.1** Working conditions for the cold storage systems

| Temperature (°C) | Sensible heat storage unit | Latent heat storage unit |
|---|---|---|
| $T_b$ *(HTF outlet of charging)* | 4 | −1 |
| $T_b$ *(HTF outlet of during discharging)* | 11 | 10 |
| $T_{Ch}$ | 10.5 | 0 *(liquid)* |
| $T_{St}$ | 5 | 0 *(solid)* |
| $T_{dis}$ | 6 | 0 *(liquid and solid)* |

the storage medium is 15,708 kg. Note that in the calculations, for simplicity, the absolute energy and exergy quantities are given in MJ.

***Unit 1 – Sensible Heat Storage***

During the charging process, the variation of the energy content of the storage medium can be evaluated as

$$\Delta E_{Ch} = (mc)_{water}(T_{St} - T_{Ch})_{water} = 15708 \times 4.18 \times (5 - 10.5)$$
$$= 361.1 \text{ MJ}$$

As the tank is assumed to be adiabatic during the charging period, there is no heat gain from the surroundings; the energy efficiency is 100%. From the energy balance, *mc* of the HTF is obtained as

$$\Delta E_{Ch} = Q_{HTF,Ch} = (mc)_{HTF,Ch}(T_b - T_a)_{HTF} \rightarrow (mc)_{HTF,Ch} = 25.8 \text{ MJ/K}$$

From Eq. (5.25), the exergy variation of the system is evaluated as

$$\Delta Ex_{Ch} = 15708 \times 4.18\left[(278 - 283.5) - 293 \ln\left(\frac{278}{283.5}\right)\right]$$
$$\rightarrow \Delta Ex_{Ch} = 15.8 \text{ MJ}$$

The exergy input of the HTF is, on the other hand, obtained from Eq. (5.21) as

$$Ex_{Ch} = 25.8\left\{(277 - 263) - 293 \ln\left(\frac{277}{263}\right)\right\} \rightarrow Ex_{Ch} = 30.9 \text{ MJ}$$

As the system is adiabatic, the difference between the exergy input and the exergy variation of the system gives the exergy destruction that arises from the internal irreversibilities:

$$I_{Ch} = 15.1 \text{ MJ}$$

(continued)

The exergy efficiency of the charging process is simply defined as the ratio of the exergy variation of the storage medium to the exergy input by the HTF:

$$\psi_{Ch} = \frac{15.8}{30.9} = 0.51 \quad \rightarrow \quad \psi_{Ch} = 51\%$$

During the storage period, due to heat gain from surroundings, the temperature of the storage medium rises. As a first step, the energy balance equation can be written as

$$\Delta E_{St} = Q_{gain,St} = 15708 \times 4.18 \times (6 - 5) \quad \rightarrow \quad Q_{gain,St} = 65.7 \text{ MJ}$$

The energy efficiency of the system during the storage period is evaluated as

$$\eta_{St} = \frac{361.1 - 65.7}{361.1} = 0.82 \quad \rightarrow \quad \eta_{St} = 82\%$$

From Eq. (5.25), exergy variation of the system during the storage period is evaluated as

$$\Delta Ex_{St} = 15708 \times 4.18 \left[ (279 - 278) - 293 \ln \left( \frac{279}{278} \right) \right]$$
$$\rightarrow \quad \Delta Ex_{St} = 3.4 \text{ MJ}$$

The exergy loss due to heat gain is evaluated from Eq. (5.23) as

$$Ex_{St,L} = \left[ 1 - \frac{293}{280.5} \right] \times 65.7 \quad \rightarrow \quad Ex_{St,L} = 2.9 \text{ MJ}$$

The internal irreversibilities during the storage period are obtained from the exergy balance equation as follows:

$$I_{St} = \Delta Ex_{St} - Ex_{St,L} \quad \rightarrow \quad I_{St} = 0.5 \text{ MJ}$$

The exergy efficiency of the storage period is evaluated from Eq. (5.29)

$$\psi_{St} = \frac{15.8 - 3.4}{15.8} = 0.78 \quad \rightarrow \quad \psi_{St} = 78\%$$

Thermodynamic analysis of the discharging process is similar to the charging one. The variation of the energy content of the storage medium can be evaluated as

$$\Delta E_{Dis} = 15708 \times 4.18 \times (10.5 - 6) = 295.5 \text{ MJ}$$

(continued)

From the energy balance, *mc* of the HTF is obtained as

$$(mc)_{HTF,Dis} = 32.8\ \text{MJ/K}$$

From Eq. (5.25) exergy variation of the system is evaluated as

$$\Delta Ex_{Dis} = 12.4\ \text{MJ}$$

The exergy recovered by the HTF is, on the other hand, obtained from Eq. (5.22) as

$$Ex_{Dis} = 4.6\ \text{MJ}$$

As the system is adiabatic, the difference between the exergy recovered and the exergy variation of the system gives the exergy destruction that arises from the internal irreversibilities

$$I_{Dis} = 7.7\ \text{MJ}$$

The exergy efficiency of the discharging process is evaluated as

$$\psi_{Dis} = \frac{4.6}{12.4} = 0.38 \rightarrow \psi_{Dis} = 38\%$$

The overall energy and exergy efficiencies of the sensible heat thermal energy storage system are defined as follows:

$$\eta_{overall} = \frac{\text{Recovered Energy by HTF during Discharging}}{\text{Input Energy by HTF during Charging}} = \frac{295.5}{361.1} = 0.82 \rightarrow \eta_{overall} = 82\%$$

$$\psi_{overall} = \frac{\text{Recovered Exergy by HTF during Discharging}}{\text{Input Exergy by HTF during Charging}} = \frac{4.6}{30.9} = 0.15 \rightarrow \psi_{overall} = 15\%$$

***Unit 2 – Latent Heat Storage***

For the latent heat storage unit, during the charging process, the liquid water completely turns into ice. Hence the variation of the energy content of the storage medium can be evaluated as

$$\Delta E_{Ch} = m_{water} L_{water} = 15708 \times 333.431 = 5237.5\ \text{MJ}$$

The tank is assumed to be adiabatic during the charging period, so there is no heat gain from the surroundings, and the energy efficiency of the charging process is 100%. From the energy balance, *mc* of the HTF is obtained as

(continued)

$$\Delta E_{Ch} = Q_{HTF,Ch} = (mc)_{HTF,Ch}(T_b - T_a)_{HTF} \quad \rightarrow \quad (mc)_{HTF,Ch} = 581.9\ \text{MJ/K}$$

During the charging, there is no temperature variation within the storage medium, so Eq. (5.33) is simplified into

$$\Delta Ex_{Ch} = 15708 \times 333.431\left\{-1 + \frac{293}{273}\right\} \quad \rightarrow \quad \Delta Ex_{Ch} = 383.7\ \text{MJ}$$

The exergy input of the HTF is, on the other hand, obtained from Eq. (5.21) as

$$Ex_{Ch} = 581.9\left\{(272 - 263) - 293\ln\left(\frac{272}{263}\right)\right\} \quad \rightarrow \quad Ex_{Ch} = 499.8\ \text{MJ}$$

The difference between the exergy input and the exergy variation of the system gives the exergy destruction that arises from the internal irreversibilities (see Eq. 5.15):

$$I_{Ch} = 116.1\ \text{MJ}$$

The exergy efficiency of the charging process is simply defined as the ratio of the exergy variation of the storage medium to the exergy input by the HTF:

$$\psi_{Ch} = \frac{383.7}{499.8} = 0.77 \quad \rightarrow \quad \psi_{Ch} = 77\%$$

The heat gain from the surroundings during the storage period can be evaluated from the energy balance as

$$\Delta E_{St} = Q_{gain,St} = Fm_{water}L_{water} = 0.1 \times 15708 \times 333.431$$
$$\rightarrow \quad Q_{gain,St} = 523.8\ \text{MJ}$$

The energy efficiency of the system during the storage period is evaluated as

$$\eta_{St} = \frac{5327.5 - 523.8}{532.8} = 0.9 \quad \rightarrow \quad \eta_{St} = 90\%$$

From Eq. (5.33), the exergy variation of the system during the storage period is evaluated as

$$\Delta Ex_{St} = 15708 \times 333.431\left\{-1 + \frac{293}{273}\right\} \times 0.1 \quad \rightarrow \quad \Delta Ex_{Ch} = 38.4\ \text{MJ}$$

Exergy loss due to heat gain is evaluated from Eq. (5.23) as

(continued)

$$Ex_{St,L} = \left[1 - \frac{293}{275}\right] \times 523.8 \rightarrow Ex_{St,L} = 34.3 \text{ MJ}$$

The internal irreversibilities during the storage period are obtained from the exergy balance equation as follows:

$$I_{St} = \Delta Ex_{St} - Ex_{St,L} \rightarrow I_{St} = 4.1 \text{ MJ}$$

The exergy efficiency of the storage period is evaluated from Eq. (5.29):

$$\psi_{St} = \frac{383.7 - 38.4}{383.7} = 0.9 \rightarrow \psi_{St} = 90\%$$

The variation of the energy content of the PCM during the discharging period can be evaluated as

$$\Delta E_{Dis} = 15708 \times 333.431 \times 0.9 = 4713.8 \text{ MJ}$$

From the energy balance, *mc* of the HTF is obtained as

$$(mc)_{HTF,Dis} = 277.3 \text{ MJ/K}$$

From Eq. (5.33), the exergy variation of the system is evaluated as

$$\Delta Ex_{Dis} = 345.3 \text{ MJ}$$

Exergy recovered by the HTF is, on the other hand, obtained from Eq. (5.21) as

$$Ex_{Dis} = 142.3 \text{ MJ}$$

The exergy destruction that arises from the internal irreversibilities for the discharging process is

$$I_{Dis} = 203.1 \text{ MJ}$$

The exergy efficiency of the discharging process is evaluated as

$$\psi_{Dis} = \frac{142.3}{345.3} = 0.41 \rightarrow \psi_{Dis} = 41\%$$

The overall energy and exergy efficiencies of the latent heat thermal energy storage system are defined as follows:

(continued)

$$\eta_{overall} = \frac{\text{Recovered Energy by HTF during Discharging}}{\text{Input Energy by HTF during Charging}}$$

$$= \frac{4713.8}{5237.5} = 0.9 \rightarrow \eta_{overall} = 90\%$$

$$\psi_{overall} = \frac{\text{Recovered Exergy by HTF during Discharging}}{\text{Input Exergy by HTF during Charging}}$$

$$= \frac{142.3}{499.8} = 0.28 \rightarrow \psi_{overall} = 28\%$$

***Discussion***

In this illustrative example, thermodynamic analyses of low-temperature CTES systems are conducted. Results reveal the difference between the energetic and exergetic point of views. In the energy-based analyses, only the energy quantities are considered by disregarding the usefulness of the "cold energy." Exergy-based analyses make it possible to discuss the internal or external irreversibilities so that an in-depth understanding of the system could be achieved.

## 5.3 Heat Transfer Analysis

The thermodynamic analysis makes it possible to evaluate the system performance when the data set for the system is obtained from experimental work. However, during a design process, it is required to conduct some numerical studies to reveal the performance of the system before building up the system. In such a case, the heat transfer analysis is a useful tool to predict the timewise variations within the system. In this part, the thermal resistance network technique (Jekel et al. 1993; Drees and Braun 1995; Neto and Krarti 1997) is discussed to carry out energy and entropy-based analyses for the charging period of the ice-on-coil latent heat thermal energy storage (LHTES) system. In Fig. 5.3, a schematic representation of the experimental setup is given for the charging process. The solidification mode is illustrated in Fig. 5.3b. During solidification, the HTF is cooled down at the evaporator (1) side of the chiller and passes through the flow meter (3) and Pt-100 (4) type temperature sensor before reaching the storage tank (5). Inside the tank, HTF is distributed into the coil (6) to perform outward solidification around tubes. Timewise variations of the interface positions and the temperature values of the solid and liquid phases of water are observed with the aid of measurement cards (7). At the outlet section of the tank, the temperature value of the HTF is measured by Pt-100 (4) type temperature sensor. A suction pump (8) is placed at the return pipeline to circulate HTF inside the system.

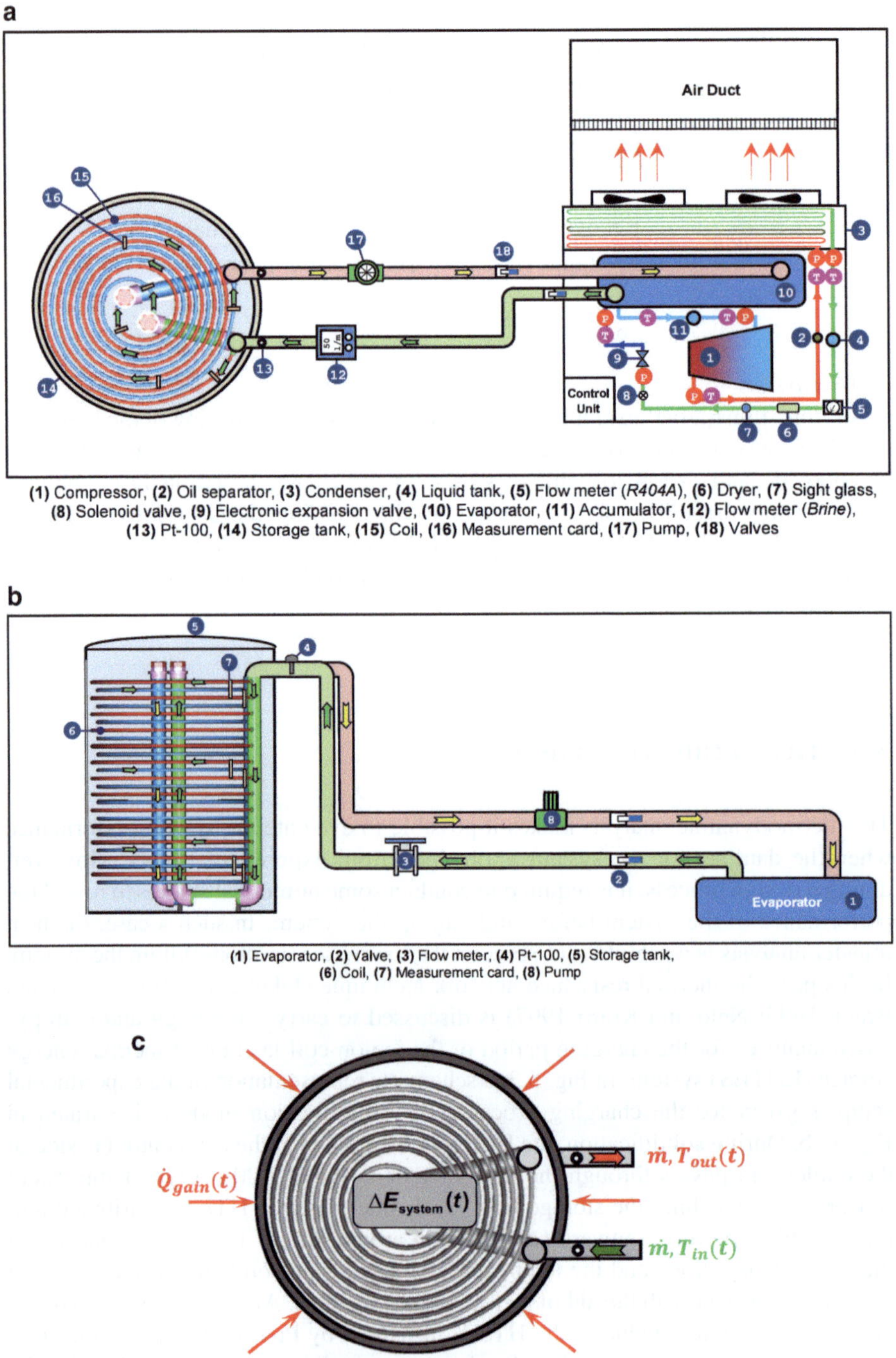


**Fig. 5.3** The schematic representation of the charging process: (**a**) top view, (**b**) front view, (**c**) energy balance for the storage tank. (Ezan 2011)

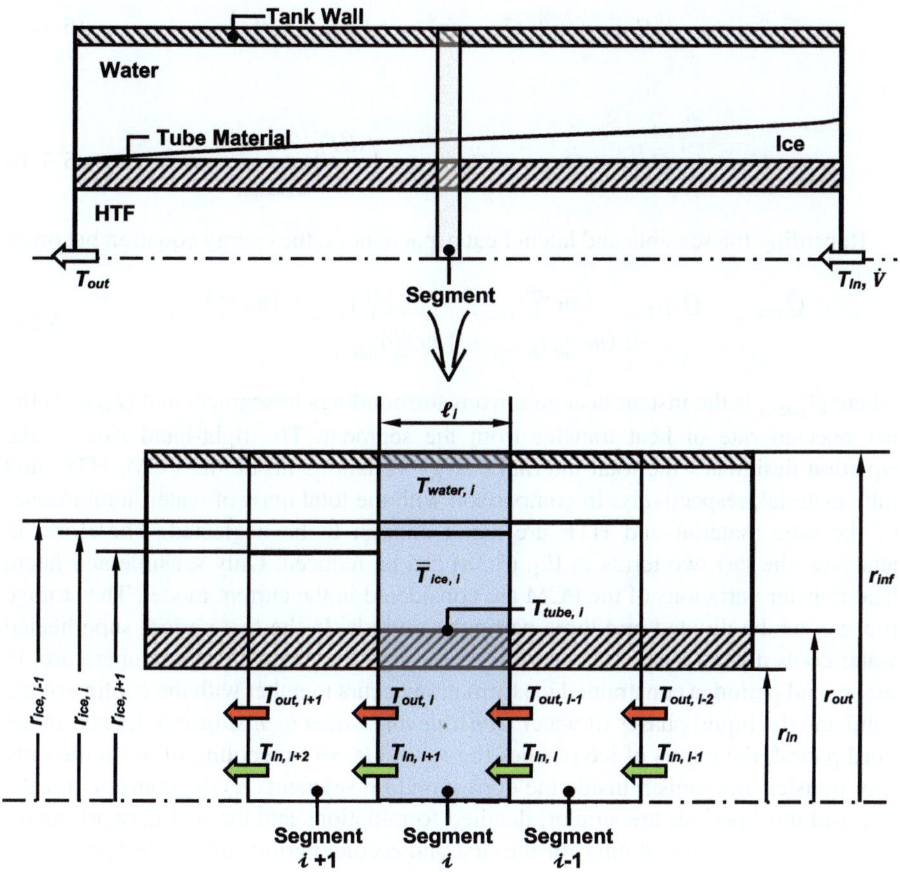


**Fig. 5.4** The mathematical model of the ice-on-coil LHTES system. (Ezan 2011)

To develop a simplified thermal resistance model for the storage tank, the following basic assumptions are considered:

1. Quasi-steady-state condition is valid for each segment.
2. Heat gain from surroundings to the tank is uniformly distributed into the system.
3. Sensible heat storage inside the tube material and the heat transfer fluid (HTF) are small enough to be neglected.
4. Thermophysical properties of ethylene glycol/water mixture are defined as functions of the inlet temperature of the HTF. On the other hand, for water, only density and thermal expansion coefficient are selected as temperature dependent.

In the model, the coil is divided into small segments as seen in Fig. 5.4. Quasi-steady-state energy balance is applied to each segment and time step to simulate the solidification process around the coil. First, the radius of ice is determined iteratively for each segment. The temperature of the HTF at the outlet section of each segment is then calculated and set as the inlet temperature of the next segment. The energy balance for an individual segment ($i$) can be written in general form as

$$E_{HTF,i}(t) \pm Q_{gain,i}(t) = \Delta E_{segment,i}(t) \tag{5.37a}$$

or

$$\dot{Q}_{in,i} - \dot{Q}_{out,i} + \dot{Q}_{gain,i} = \left(\frac{dH}{dt}\right)_{segment,i} \tag{5.37b}$$

Regarding the sensible and latent heat capacitances, the energy equation becomes

$$\begin{aligned} \dot{Q}_{gain,i} - \dot{Q}_{HTF,i} = \left(mc\tfrac{dT}{dt}\right)_{water,i} + h_{sf}\left(\tfrac{dm}{dt}\right)_{ice,i} + \left(mc\tfrac{dT}{dt}\right)_{ice,i} \\ + \left(mc\tfrac{dT}{dt}\right)_{HTF,i} + \left(mc\tfrac{dT}{dt}\right)_{tube,i} \end{aligned} \tag{5.38}$$

where $\dot{Q}_{gain,i}$ is the instant heat gain from surroundings to segment and $\dot{Q}_{HTF,i}$is the net rejected rate of heat transfer from the segment. The right-hand side of the equation designates the total internal energy variations inside the PCM, HTF, and tube material, respectively. In comparison with the total mass of water, total masses of the tube material and HTF are small enough to be neglected. Therefore, in analyses, the last two terms in Eq. (5.38) can be reduced. Only sensible and latent heat transfer variations of the PCM are considered in the current model. The storage process can be divided into three particular periods. In the first period, superheated water cools down (*sensible heat storage*) through the phase change temperature. In the second period, a constrained ice formation occurs together with the cooling of the solid and the liquid phases of water (*sensible and latent heat storage*). Lastly, in the third period, the radius of ice reaches the $r_{inf}$ value, so subcooling of ice is the only heat transfer mechanism inside the corresponding segment. As the concepts for the first and third periods are similar, detailed formulations and thermal network representations are illustrated only for the first and second periods of the storage.

### 5.3.1 Period 1: Sensible Heat Storage

In Fig. 5.5, the components of Eq. (5.38) are represented as thermal resistance network in general form. For simplicity, internal energy variations in the tube material and HTF can be neglected. Hence, energy equation becomes

$$\dot{Q}_{gain,i} - \dot{Q}_{HTF,i} = \left(mc\frac{dT}{dt}\right)_{water,i} \tag{5.39a}$$

or

$$\dot{Q}_{gain,i} - \dot{Q}_{HTF,i} = \dot{Q}_{water,sen,i} \tag{5.39b}$$

The heat transfer rate between the HTF and water can be defined using the enthalpy variation of the HTF as

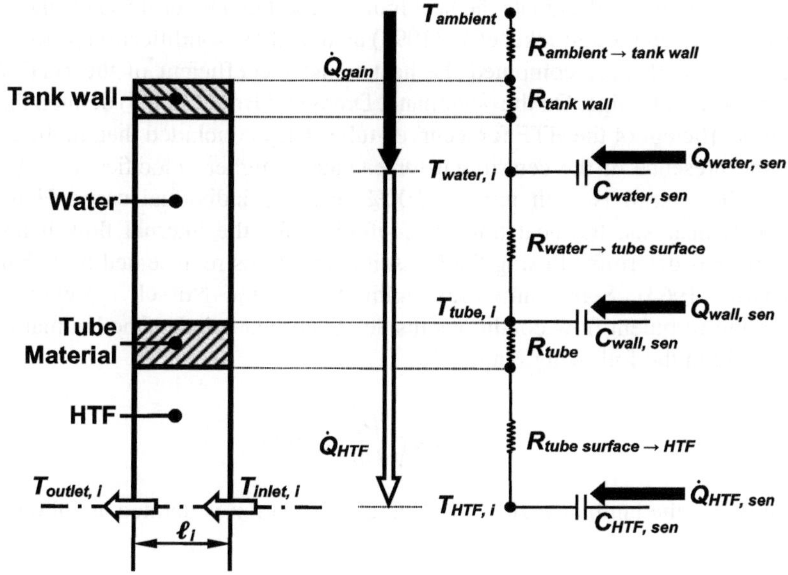


**Fig. 5.5** Thermal resistance network for Period 1: sensible heat storage. (Ezan 2011)

$$\dot{Q}_{HTF,i} = \left(\dot{m}c\right)_{HTF}\left[T_{out,i} - T_{in,i}\right] \tag{5.40}$$

and, alternatively, using the overall heat transfer coefficient definition as follows:

$$\dot{Q}_{HTF,i} = (UA)_{water\rightarrow HTF}\left[T_{water,i} - \bar{T}_{HTF,i}\right] \tag{5.41}$$

where $\bar{T}_{HTF,i}$ represents the mean temperature value of the HTF, $(T_{in,i} + T_{out,i})/2$. Overall heat transfer coefficient, $(UA)_{water\rightarrow HTF}$, can be obtained by using the thermal resistance network given in Fig. 5.5, as

$$(UA)_{water\rightarrow HTF} = \left[R_{tube\ surface\rightarrow HTF} + R_{tube} + R_{water\rightarrow tube\ surface}\right]^{-1} \tag{5.42}$$

The thermal resistances of $R_{tube\ surface\rightarrow HTF}$, $R_{tube}$, and $R_{water\rightarrow tube\ surface}$ are calculated using the following equations:

$$R_{tube\ surface\rightarrow HTF} = \left[2\pi r_i \ell_i h_{HTF}\right]^{-1} \tag{5.42a}$$

$$R_{tube} = \frac{\ln\left(r_o/r_i\right)}{2\pi k_{tube}\ell_i} \tag{5.42b}$$

$$R_{water\rightarrow tube\ surface} = \left[2\pi r_o \ell_i h_{water}\right]^{-1} \tag{5.42c}$$

where $h_{HTF}$ and $h_{water}$ designate the heat transfer coefficients for the *HTF* and *water*, respectively. Even though Jekel et al. (1993) analyzed the solidification process for a spiral-shaped coil, they computed the heat transfer coefficient of the HTF, $h_{HTF}$, assuming a straight pipe. On the other hand, Drees and Braun (1995) defined the heat transfer coefficient of the HTF for a curved tube. They concluded that, in the curved tubes, the presence of the centrifugal forces causes higher velocities, and this can increase the overall Nusselt number 500% in some individual cases. Hence, in the current analysis, the heat transfer coefficient for the internal flow inside the curved pipe is determined using the Nusselt correlations represented by Seban and McLauglin (1963). Seban and McLauglin defined the Nusselt correlations for laminar and turbulent flow conditions inside the curved spirals. The laminar condition is valid in the following range:

$$11.6 < \mathrm{Re}\sqrt{\frac{D_i}{D_H}} < 2000 \tag{5.43}$$

Moreover, for the laminar region, the Nusselt correlation is defined as follows:

$$\overline{\mathrm{Nu}} = 0.13\left[\frac{f}{8}\mathrm{Re}^2\right]^{1/3}\mathrm{Pr}^{1/3} \tag{5.44}$$

where $f$ is the Darcy-Weisbach friction factor and defined as

$$\frac{f}{f_s} = \left\{1 - \left[1 - \left(\frac{11.6}{\mathrm{Re}\sqrt{D_i/D_H}}\right)^{0.45}\right]^{2.22}\right\}^{-1} \tag{5.45}$$

with $f_s = 64/\mathrm{Re}$. Besides, the initial point of the turbulent flow determined by the critical Reynolds number as

$$\mathrm{Re}_{cr} = 2 \times 10^4\left(\frac{D_i}{D_H}\right)^{0.32} \tag{5.46}$$

Moreover, the following Nusselt correlation is valid for $\mathrm{Re}(D_i/D_H)^2 > 6$:

$$\overline{\mathrm{Nu}} = \frac{f}{8}\mathrm{Re}\mathrm{Pr}^{0.4} \tag{5.47}$$

On the other hand, the heat transfer coefficient of water at the outer surface of the tube is determined using the Churchill and Chu correlation, which can be found in Incropera and DeWitt (2002):

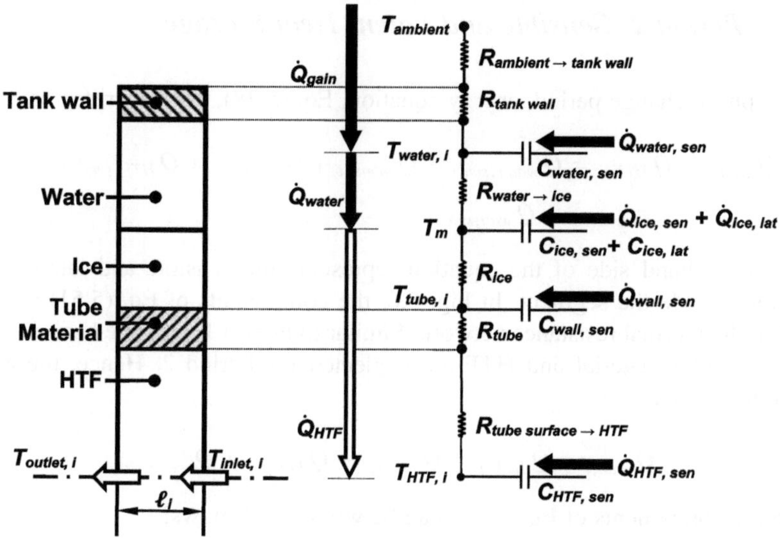


**Fig. 5.6** Thermal resistance network for Period 2: sensible and latent heat storage (Ezan 2011)

$$\overline{\mathrm{Nu}}_D = \left\{ 0.60 + \frac{0.387\mathrm{Ra}_D^{1/6}}{\left[1 + (0.559/\mathrm{Pr})^{9/16}\right]^{8/27}} \right\}^2 \tag{5.48}$$

An iterative procedure can be utilized for Eqs. (5.40) and (5.41) to obtain the outlet temperature value of the HTF from each segment. Afterward, the outer surface temperature values of the tube can be updated by defining the following energy balance equation between the tube surface and the HTF as

$$\dot{Q}_{HTF,i} = (UA)_{tube\ surface \to HTF}\left[T_{tube,i} - \bar{T}_{HTF,i}\right] \tag{5.49}$$

where the overall heat transfer coefficient between the HTF and the outer surface of tube, $(UA)_{tube\ surface \to HTF}$, can be obtained using the thermal resistance network given in Fig. 5.5. On the other hand, the variation of the mean temperature value of water can be obtained by rearranging Eq. (5.39a) to be

$$T_{water,i} = T^0_{water,i} + \frac{\left(\dot{Q}_{gain} - \dot{Q}_{HTF}\right)\Delta t}{(\rho \forall c)_{water}} \tag{5.50}$$

where $T_{water,\ i}$ and $T^0_{water,i}$ indicate mean temperature values of water for the current and previous time steps, respectively.

### 5.3.2 Period 2: Sensible and Latent Heat Storage

For the phase change period, energy equation, Eq. (5.38), can be written as

$$\dot{Q}_{gain,i} - \dot{Q}_{HTF,i} = \dot{Q}_{water,sen,i} + \dot{Q}_{latent,i} + \dot{Q}_{ice,sen,i} + \dot{Q}_{HTF,sen,i} + \dot{Q}_{wall,sen,i} \tag{5.51}$$

The right-hand side of the equation represents the sensible and latent energy variations inside the segment. In Fig. 5.6, the components of Eq. (5.51) are illustrated in the thermal resistance network. Similar to Period 1, sensible heat variations inside the tube material and HTF are neglected for Period 2. Hence, the energy equation reduced as

$$\dot{Q}_{gain,i} - \dot{Q}_{HTF,i} = \dot{Q}_{water,i} + \dot{Q}_{latent,i} + \dot{Q}_{ice,sen,i} \tag{5.52}$$

where the components of Eq. (5.52) can be written as follows:

$$\dot{Q}_{HTF,i} = (UA)_{ice \to HTF}\left[T_m - \bar{T}_{HTF,i}\right] \tag{5.53a}$$

$$\dot{Q}_{water,i} = (2\pi r_{ice,i}\ell_i)h_{water}\left[T_{water,i} - T_m\right] \tag{5.53b}$$

$$\dot{Q}_{latent,i} = h_{sf}\frac{m_{ice,i} - m^0_{ice,i}}{\Delta t} \tag{5.53c}$$

$$\dot{Q}_{ice,sen,i} = (mc)_{ice,i}\frac{\bar{T}_{ice,i} - \bar{T}^0_{ice,i}}{\Delta t} \tag{5.53d}$$

The heat transfer coefficient between the heat transfer fluid and the surface of ice, $(UA)_{ice \to HTF}$, can be obtained using the thermal resistance network given in Fig. 5.6. On the other hand, for quasi-steady-state conditions, the mean temperature value of ice can be calculated as

$$\bar{T}_{ice} = T_{tube,i} + (T_{tube,i} - T_m)\left[\frac{1}{2\ln\left(r_{ice,i}/r_o\right)} - \frac{r^2_{ice,i}}{r^2_{ice,i} - r^2_o}\right] \tag{5.54}$$

The iterative solution procedure can be applied to Eqs. (5.52) and (5.53) to obtain the radius values of ice for each segment. Afterward, the surface temperature of the tube can be updated by defining the energy balance between the tube surface and the HTF, as in Eq. (5.49). The mean temperature of the water can be obtained by defining the following energy balance:

$$T_{water,i} = T^0_{water,i} + \frac{\left(\dot{Q}_{gain} - \dot{Q}_{water,i}\right)\Delta t}{(\rho \forall c)_{water}} \tag{5.55}$$

Internal energy variation within the tank can be obtained by Eq. (5.38) as

$$\Delta E_{system} = \sum_{t=0}^{t}\sum_{i=1}^{n}\left\{\left[mc\left(T-T^0\right)\right]_{water,i} + h_{sf}\left(m-m^0\right)_{ice,i} + \left[mc\left(T-T^0\right)\right]_{water,i}\right\} \tag{5.56}$$

where superscript "0" designate the previous time step and $n$ indicates the number of segments. On the other hand, the total energy delivered by the heat transfer fluid is determined with the integration of Eq. (5.40) to be

$$E_{HTF}(t) = \left(\dot{m}\,c\right)_{HTF}\int_{t=0}^{t}\left[T_{out} - T_{in}\right]dt \tag{5.57}$$

The heat transfer analysis makes it possible to predict the temperature variation within the system. After evaluating the temperatures within the system and the outlet temperature of the HTF, the thermodynamic assessment could be conducted by using Eqs. (5.26), (5.27), (5.28), (5.29), and (5.30).

## 5.4 Computational Fluid Dynamics (CFD) Analysis

In CFD analysis, transport equations are resolved by discretizing the differential equations into the algebraic sets of equations to evaluate the spatial velocity and temperature distributions. In this section, first the governing equations are given in 2D orthogonal coordinate systems. After that, the fundamental aspects of the CFD solution methods are discussed, and some CFD applications on numerical modeling of sensible and latent heat storage systems are represented.

### 5.4.1 *Governing Equations*

For Cartesian, cylindrical, and spherical coordinate systems, two-dimensional *mass*, *momentum*, and *energy* equations are written as follows.

#### 5.4.1.1 Cartesian Coordinate System (*x*–*y*)

*Mass:*

$$\frac{\partial}{\partial t}(\rho) + \frac{\partial}{\partial x}(\rho u) + \frac{\partial}{\partial y}(\rho v) = 0 \tag{5.58}$$

*Momentum:*

$$\text{x-direction:} \quad \frac{\partial}{\partial t}(\rho u) + \frac{\partial}{\partial x}(\rho uu) + \frac{\partial}{\partial y}(\rho vu) = -\frac{\partial p}{\partial x} + \frac{\partial}{\partial x}\left[\frac{\partial}{\partial x}(\mu u)\right] + \frac{\partial}{\partial y}\left[\frac{\partial}{\partial y}(\mu u)\right] + F_x \tag{5.59}$$

$$\text{y-direction:} \quad \frac{\partial}{\partial t}(\rho v) + \frac{\partial}{\partial x}(\rho uv) + \frac{\partial}{\partial y}(\rho vv) = -\frac{\partial p}{\partial y} + \frac{\partial}{\partial x}\left[\frac{\partial}{\partial x}(\mu v)\right] + \frac{\partial}{\partial y}\left[\frac{\partial}{\partial y}(\mu v)\right] + F_y \tag{5.60}$$

*Energy:*

$$\frac{\partial}{\partial t}(H) + \frac{\partial}{\partial x}(uH) + \frac{\partial}{\partial y}(vH) = \frac{\partial}{\partial x}\left[\frac{\partial}{\partial x}(kT)\right] + \frac{\partial}{\partial y}\left[\frac{\partial}{\partial y}(kT)\right] \tag{5.61}$$

#### 5.4.1.2 Cylindrical Coordinate System (*r*–*θ*)

*Mass:*

$$\frac{\partial}{\partial t}(\rho) + \frac{1}{r}\frac{\partial}{\partial r}(r\rho u_r) + \frac{1}{r}\frac{\partial}{\partial \theta}(\rho u_\theta) = 0 \tag{5.62}$$

*Momentum:*

$$\text{r-direction:} \quad \frac{\partial}{\partial t}(\rho u_r) + \frac{\partial}{\partial r}(\rho u_r u_r) + \frac{1}{r}\frac{\partial}{\partial \theta}(\rho u_\theta u_r) + \frac{u_\theta^2}{r} = -\frac{\partial p}{\partial r} + \frac{1}{r}\frac{\partial}{\partial r}\left[r\frac{\partial}{\partial r}(\mu u_r)\right] - \mu\frac{u_r}{r^2} + \frac{1}{r^2}\frac{\partial}{\partial \theta}\left[\frac{\partial}{\partial \theta}(\mu u_r)\right] - \frac{2}{r^2}\frac{\partial}{\partial \theta}(\mu u_\theta) + F_r \tag{5.63}$$

$$\theta\text{-direction:} \quad \frac{\partial}{\partial t}(\rho u_\theta) + \frac{\partial}{\partial r}(\rho u_r u_\theta) + \frac{1}{r}\frac{\partial}{\partial \theta}(\rho u_\theta u_\theta) + \frac{u_\theta u_r}{r} = -\frac{1}{r}\frac{\partial p}{\partial \theta} + \frac{1}{r}\frac{\partial}{\partial r}\left[r\frac{\partial}{\partial r}(\mu u_\theta)\right] - \mu\frac{u_\theta}{r^2} + \frac{1}{r^2}\frac{\partial}{\partial \theta}\left[\frac{\partial}{\partial \theta}(\mu u_\theta)\right] - \frac{2}{r^2}\frac{\partial}{\partial \theta}(\mu u_r) + F_\theta \tag{5.64}$$

*Energy:*

$$\frac{\partial}{\partial t}(H) + \frac{\partial}{\partial r}(u_r H) + \frac{1}{r}\frac{\partial}{\partial \theta}(u_\theta H) = \frac{1}{r}\frac{\partial}{\partial r}\left(kr\frac{\partial T}{\partial r}\right) + \frac{1}{r^2}\frac{\partial}{\partial \theta}\left(k\frac{\partial T}{\partial \theta}\right) \tag{5.65}$$

#### 5.4.1.3 Spherical Coordinate System (*r*–*θ*)

*Mass:*

$$\frac{\partial}{\partial t}(\rho) + \frac{1}{r^2}\frac{\partial}{\partial r}(r^2 u_r) + \frac{1}{r\sin\theta}\frac{\partial}{\partial \theta}(u_\theta \sin\theta) = 0 \tag{5.66}$$

*Momentum:*

$$\begin{aligned} r\text{-direction:}\quad & \frac{\partial}{\partial t}(\rho u_r) + \frac{\partial}{\partial r}(\rho u_r u_r) + \frac{1}{r}\frac{\partial}{\partial \theta}(\rho u_\theta u_r) - \frac{\rho u_\theta{}^2}{r} = \\ & -\frac{\partial p}{\partial r} + \frac{1}{r^2}\frac{\partial}{\partial r}\left(\mu r^2\frac{\partial u_r}{\partial r}\right) + \frac{1}{r^2\sin\theta}\frac{\partial}{\partial \theta}\left(\mu\sin\theta\frac{\partial u_r}{\partial \theta}\right) \\ & -\mu\left[\frac{2u_r}{r^2} + \frac{2}{r^2}\frac{\partial u_\theta}{\partial \theta} + \frac{2}{r^2}u_\theta\frac{\cos\theta}{\sin\theta}\right] + F_r \end{aligned} \tag{5.67}$$

$$\begin{aligned} \theta\text{-direction:}\quad & \frac{\partial}{\partial t}(\rho u_\theta) + \frac{\partial}{\partial r}(\rho u_r u_\theta) + \frac{1}{r}\frac{\partial}{\partial \theta}(\rho u_\theta u_\theta) + \rho\frac{u_r u_\theta}{r} = \\ & -\frac{1}{r}\frac{\partial p}{\partial \theta} + \frac{1}{r^2}\frac{\partial}{\partial r}\left(\mu r^2\frac{\partial u_\theta}{\partial r}\right) + \frac{1}{r^2\sin\theta}\frac{\partial}{\partial \theta}\left(\mu\sin\theta\frac{\partial u_\theta}{\partial \theta}\right) \\ & -\mu\left[\frac{2}{r^2}\frac{\partial u_r}{\partial \theta} - \frac{u_\theta}{r^2\sin^2\theta}\right] + F_\theta \end{aligned} \tag{5.68}$$

*Energy:*

$$\frac{\partial}{\partial t}(H) + \frac{\partial}{\partial r}(u_r H) + \frac{1}{r}\frac{\partial}{\partial \theta}(u_\theta H) = \frac{1}{r^2}\frac{\partial}{\partial r}\left(kr^2\frac{\partial T}{\partial r}\right) + \frac{1}{r^2\sin\theta}\frac{\partial}{\partial \theta}\left(k\sin\frac{\partial T}{\partial \theta}\right) \tag{5.69}$$

On the right-hand side of the momentum equations, *F*s indicate the external forces, such as buoyancy, magnetic, or electrical. In TES systems, the influence of natural convection becomes important when the temperature difference within the cavity is higher. In such a case, the external force term is defined as $F = \rho g$. Here $g$ is the gravitational acceleration and $\rho$ is the density of the fluid as a function of temperature. As a common approach, the variation of density as a function of temperature is defined by using the Boussinesq approach. That is, the buoyancy force acting on a control volume is defined as

$$F = \rho g \beta \left(T - T_{ref}\right) \tag{5.70}$$

Energy equations are defined in terms of the volumetric enthalpy ($H$ in J/m$^3$). The temperature transformation method could be applied into the energy equation to convert enthalpy terms into the temperature-based form to evaluate a general formulation which is applicable to simulate mathematical models of sensible and latent heat storage unit. The enthalpy value of a material can be computed as the sum of the sensible and the latent heat components by

$$H = h + \rho h_{sf} f \qquad \left(\text{in J/m}^3\right) \tag{5.71}$$

where $h$ is the sensible enthalpy, $f$ is the liquid fraction, and $h_{sf}$ is the latent heat. Sensible enthalpy can be defined in terms of the specific heat as follows:

$$\frac{dh}{dT} = \rho c \tag{5.72}$$

Hence, the sensible enthalpy can be derived as

$$h = h_{ref} + \int_{T_{ref}}^{T} (\rho c) dT \tag{5.73}$$

The temperature transformation method is based on the equivalent heat capacity method (Morgan 1981). To account for the latent heat effect on the liquid-solid interface, equivalent heat capacity is introduced, assuming that the phase change process occurs over a temperature range. The equivalent heat capacity method has the advantage of being simple for programming but also has many difficulties in the selection of the time step size, mesh size, and the phase change temperature range (Cao and Faghri 1990). Cao and Faghri (1990) proposed a new temperature-based fixed grid formulization to overcome the drawbacks of the former method. Similar to the heat capacity method, the proposed method of Cao and Faghri (1990) also assumes that the phase change takes place over a range of phase change temperature from $T_m - \delta T_m$ to $T_m + \delta T_m$, rather than a fixed temperature, and the enthalpy variation of the material is assumed to be linear in the mushy region. Here, $T_m - \delta T_m$ and $T_m + \delta T_m$ designate the phase transformation temperatures for the solid and the liquid states of the material, respectively. Hence, in addition to the *solid* and *liquid* phases, there is a transition phase that takes place called *mushy*, as shown in Fig. 5.7.

The relationship between the enthalpy and temperature can be obtained by assuming linear variations. For three different phase regions, the relationship between total enthalpy and temperature can be obtained as follows:

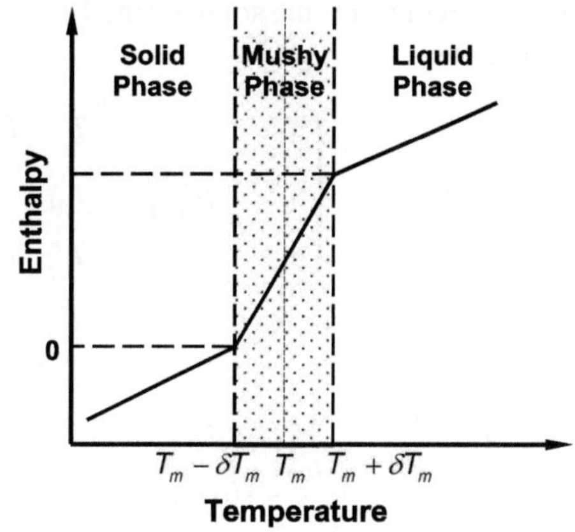


**Fig. 5.7** Illustration of enthalpy-temperature relationship

*Solid phase:* $T < T_m - \delta T_m$

$$H(T) = (\rho c)_s(T - T_m + \delta T_m) \tag{5.74}$$

*Mushy phase:* $T_m - \delta T_m \le T \le T_m + \delta T_m$

$$H(T) = (\rho c)_m(T - T_m) + \rho\frac{h_{sf}}{2\delta T_m}(T - T_m) + (\rho c)_m\delta T_m + \rho\frac{h_{sf}}{2} \tag{5.75}$$

*Liquid phase:* $T > T_m + \delta T_m$

$$H(T) = (\rho c)_l(T - T_m) + (\rho c)_s\delta T_m + \rho h_{sf} \tag{5.76}$$

Hence, the relationship between the enthalpy and temperature can be expressed as

$$H(T) = \begin{cases} C_s(T - T_m) + C_s\delta T_m & T < T_m - \delta T_m \\ \left(C_m + \rho\frac{h_{sf}}{2\delta T_m}\right)(T - T_m + \delta T_m) & T_m - \delta T_m \le T \le T_m + \delta T_m \\ C_l(T - T_m) + C_s\delta T_m + \rho h_{sf} & T > T_m + \delta T_m \end{cases} \tag{5.77}$$

where $C$ represents the volumetric heat capacity ($C = \rho c$) and $C_m$ is the volumetric heat capacity of the mushy region. $C_m$ is defined as the average of the solid and liquid phase values, $C_m = 0.5(C_s + C_l)$. Cao and Faghri (1990) introduced a linear temperature-dependent function to define the enthalpy as

$$H = CT + S \tag{5.78}$$

where $S$ represents the source term. Enthalpy can be written in terms of $C$ and $S$ terms as follows:

$$C = C(T) = \begin{cases} C_s & T < T_m - \delta T_m & \text{Solid Phase} \\ C_m + \rho \dfrac{h_{sf}}{2\delta T_m} & T_m - \delta T_m \le T \le T_m + \delta T_m & \text{Mushy Phase} \\ C_l & T > T_m + \delta T_m & \text{Liquid Phase} \end{cases} \tag{5.79}$$

$$S = S(T) = \begin{cases} C_s(\delta T_m - T_m) & T < T_m - \delta T_m & \text{Solid Phase} \\ \left(C_m + \rho \dfrac{h_{sf}}{2\delta T_m}\right)(\delta T_m - T_m) & T_m - \delta T_m \le T \le T_m + \delta T_m & \text{Mushy Phase} \\ C_s \delta T_m - C_l T_m + \rho h_{sf} & T > T_m + \delta T_m & \text{Liquid Phase} \end{cases} \tag{5.80}$$

For simplicity, the temperature transforming method is applied to energy equation for the Cartesian coordinate system (Eq. 5.61). A similar approach can quickly be followed for the cylindrical and spherical coordinate systems. Eq. (5.61) can be rearranged to obtain the energy equation in the temperature-based form as follows:

$$\begin{aligned} \frac{\partial}{\partial t}(CT) + \frac{\partial}{\partial x}(uCT) + \frac{\partial}{\partial y}(vCT) &= \frac{\partial}{\partial x}\left(k\frac{\partial T}{\partial x}\right) + \frac{\partial}{\partial y}\left(k\frac{\partial T}{\partial y}\right) \\ &\quad - \frac{\partial}{\partial t}(S) - \frac{\partial}{\partial x}(uS) - \frac{\partial}{\partial y}(vS) \end{aligned} \tag{5.81}$$

The last three terms on the right-hand side are named source terms. As mentioned by Wang et al. (2010), the $S$ term is constant inside the liquid phase ($-\partial S/\partial x = 0, -\partial S/\partial y = 0$), and moreover, velocity components are both zero in the solid and mushy regions, so the last two terms drop, and only the time-dependent term remains as the source term of the energy equation:

$$\frac{\partial}{\partial t}(CT) + \frac{\partial}{\partial x}(uCT) + \frac{\partial}{\partial y}(vCT) = \frac{\partial}{\partial x}\left(k\frac{\partial T}{\partial x}\right) + \frac{\partial}{\partial y}\left(k\frac{\partial T}{\partial y}\right) - \frac{\partial}{\partial t}(S) \tag{5.82}$$

The $C$ and $S$ terms are evaluated by using Eqs. (5.79) and (5.80). An iterative solution method should be followed to predict the spatial and temporal temperatures within the computational domain. Fundamental aspects of the CFD are briefly introduced in the following subsection.

### 5.4.2 *Fundamental Aspects of CFD and Finite Volume Method*

For a two-dimensional transient convection-diffusion process, transport equations can be written as in Eqs. (5.58), (5.59), (5.60), (5.61), (5.62), (5.63), (5.64), (5.65), (5.66), (5.67), (5.68), and (5.69). Analytical solutions for these partial differential equations can be obtained for only some simplified problems. Analytical solutions make it possible to express the field variables, e.g., $u$, $v$, $T$, etc., as functions of spatial locations, e.g., $x$, $y$. Nevertheless, in real fluid flows, because of the two- or three-dimensional nature of problems, analytical relationships are not readily achievable. Even when the problem is reduced to a two-dimensional problem, it is difficult to obtain analytical results for convection-diffusion problems, except for some simplified cases. Rather than attaining closed-form analytical expressions, using computational fluid dynamics (CFD) methods, transport equations can be solved for the discrete locations. In the CFD methodology, first partial differential equations are replaced into algebraic equations, and then, the discrete values of the flow field variables are computed by solving the sets of algebraic equations or matrices. There are many computational methods for discretization of the transport equations for a computational domain. Commonly, the following methods are preferred for CFD applications (Tu et al. 2008): finite difference method, finite element method, spectral methods, and finite volume (*control volume*) method. Detailed discussions about the pros and cons of each approach can be found elsewhere (Versteeg and Malalasekera 2007; Tu et al. 2008). As an overview, the solution procedure for these CFD methods is illustrated in Fig. 5.8.

Tu et al. (2008) designated that nowadays the majority of commercial CFD codes are based on the finite volume method. The finite volume approach (or *control volume method*) is a useful tool for discretizing the differential equations (Patankar 1980). The most attractive feature of the control volume formulation is that the resulting solution would imply that the integral conversation of the quantities such as mass, momentum, and energy is precisely satisfied over any group of control volumes and, naturally, over the whole calculation domain (Patankar 1980). In this method, the calculation domain is divided into some non-overlapping control volumes (Fig. 5.9) such that there is one control volume surrounding each grid point.

All transport equations can be written in terms of the generic variable of $\phi$ as follows:

$$\frac{\partial}{\partial t}(\rho\phi) + \mathrm{div}(\rho\phi\mathbf{u}) = \mathrm{div}(\Gamma \mathrm{grad}\phi) + S_{\phi} \tag{5.83}$$

Alternatively, in words,

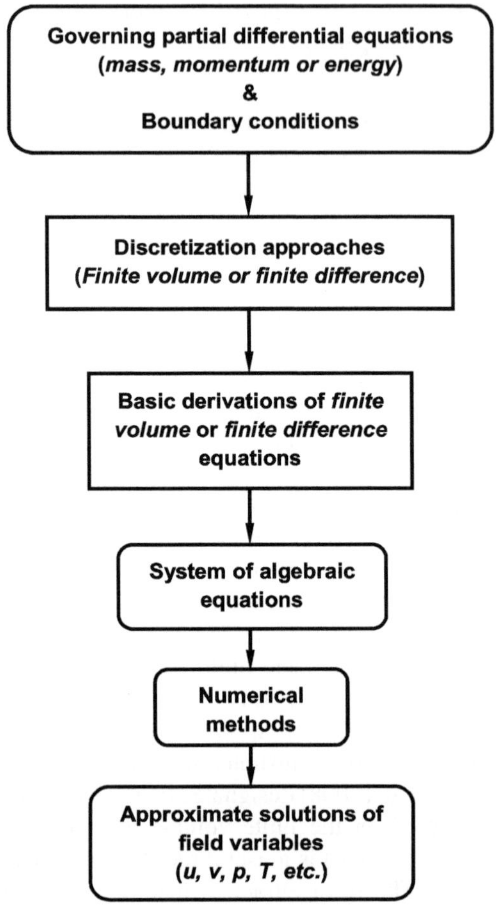


**Fig. 5.8** Overview of the computational solution procedure for CFD problems. (Adapted from Tu et al. 2008)

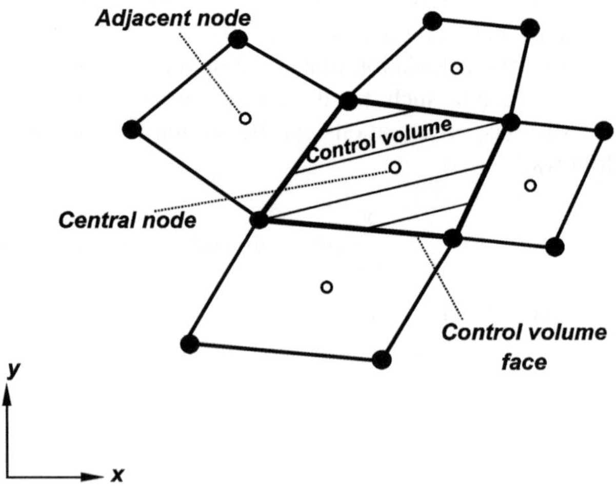


**Fig. 5.9** Representation of the structured grid arrangement (open symbols at the center of the control volumes denote computational node). (Adapted from Tu et al. 2008)

$$\begin{bmatrix}\text{Rate of change of} \\ \phi \text{ in the control volume} \\ \text{with respect time}\end{bmatrix} + \begin{bmatrix}\text{Net flux of } \phi \text{ due} \\ \text{to convection out of} \\ \text{the control volume}\end{bmatrix} = \begin{bmatrix}\text{Net flux of } \phi \text{ due} \\ \text{to diffusion into} \\ \text{the control volume}\end{bmatrix} + \begin{bmatrix}\text{Net rate of creation of} \\ \phi \text{ inside the control} \\ \text{volume}\end{bmatrix}$$

Equation (5.83) includes various transport processes, such as the rate of change terms and the convective term on the left-hand side; on the other hand, the diffusive term ($\Gamma$, *diffusion coefficient*) and the source term are on the right-hand side (Versteeg and Malalasekera 2007). The transport equation can be integrated over each control volume to achieve the discretization equation as follows:

$$\int_{CV} \frac{\partial}{\partial t}(\rho\phi)dV + \int_{CV} \text{div}(\rho\phi\mathbf{u})dV = \int_{CV} \text{div}(\Gamma \text{grad}\phi)dV + \int_{CV} S_\phi dV \tag{5.84}$$

Piecewise profiles expressing the variation of $\phi$ between the grid points are used to evaluate the required integrals. The result is the discretization equation containing the values of $\phi$ for a group of grid points. The discretization equation obtained in this manner expresses the conservation principle for $\phi$ for the finite control volume, just as the differential equation expresses it for an infinitesimal control volume (Patankar 1980). The general form of the discretized equation can be written as

$$a_P\phi_P = a_W\phi_W + a_E\phi_E + a_N\phi_N + a_S\phi_S + b \tag{5.85}$$

where the coefficients of $\phi$'s are the unknowns and $a$'s are the coefficients. $b$ includes the source terms and the boundary conditions. For a laminar 2D CFD problem which involves heat transfer, there are four unknowns as $u$, $v$, $P$, and $T$. Even though there are three equations for flow field, there is no dedicated equation to evaluate the unknown pressure field. That is, resolving the velocity field is one of the biggest challenges in CFD problems. There are some alternative solution methods, such as stream-function/vorticity approach, in which the pressure terms are omitted. There are several solution algorithms to predict the velocity field by resolving the *mass* and *momentum* equations without omitting the pressure terms. One of the most popular solution algorithm is the SIMPLE algorithm. SIMPLE (*Semi-Implicit Method for Pressure-Linked Equations*) algorithm was developed initially by Patankar and Spalding (1972). The real difficulty in the calculation of the velocity field lies in the unknown pressure field (Patankar 1980). Governing equations can be discretized for the domain by utilizing the finite volume approach. The first step of the SIMPLE algorithm is defining the proper control volumes for scalar variables, such as pressure, temperature, density, and so on. Among many grid arrangements, the staggered arrangement is the most popular and comprehensive accepted method to obtain realistic pressure and velocity field inside the computational domain. In staggered grid arrangement, while the velocity components are defined at the control volume faces, the rest of the variables, or *scalars*, are stored at the central node of the control volume.

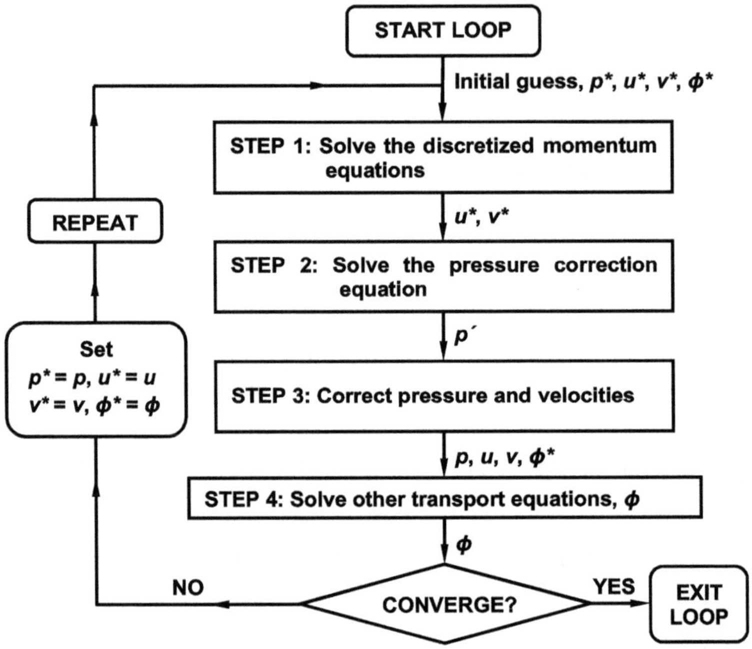


**Fig. 5.10** The flowchart of the SIMPLE algorithm

Solution procedure for the SIMPLE algorithm is illustrated in Fig. 5.9. The method is based on an iterative solution procedure. Here each governing equation is solved, *decoupled*, or *segregated* from other equations; hence, this solution algorithm is known as segregated. The segregated algorithm is memory efficient since the discretized equations need only be stored in the memory one at a time. However, the solution convergence is relatively slow, in as much as the equations are solved in a decoupled manner (ANSYS Inc. 2009).

### 5.4.3 CFD Applications on Thermal Energy Storage

CFD tools are widely used to predict velocity fields and temperature distributions within the TES units. In this part of the book, results of some selected recent works are presented. Transient modeling of complex SHTES or LHTES tanks requires discretizing the domain into a significant number of control volumes. That is, in most cases, it is better to use commercial CFD tools to develop the numerical model. Commercial CFD packages (i.e., ANSYS-FLUENT or COMSOL) includes built-in tools to simulate conjugate heat transfer within a storage tank. Using software with parallel processing significantly reduces the required simulation time and provides to resolve more complex problems.

#### 5.4.3.1 Sensible Heat TES Systems

In the sensible heat thermal energy storage (SHTES) systems with the liquid storage medium, the thermal stratification phenomenon has great importance to design a storage tank with high thermal efficiency. The thermal gradient along the height of the tank could be obtained by using CFD analyses, and numerous numerical works deal with optimization of the thermal stratification within the SHTES tanks. Table 5.2 shows a list of selected works on modeling the storage tanks with CFD. As a commonly followed approach, the mode of analyses is transient. The effects of the type and position of the baffles and entrance are determined.

In the following, two selected numerical works from the literature are reviewed. Altuntop et al. (2005) developed a three-dimensional tank model in FLUENT software to evaluate the influence of obstacles within the tank on the thermal stratification. The schematic of the storage tank and the obstacle configurations that are studied are represented in Fig. 5.11. In the model, the hot water from the solar collector enters the sensible heat storage tank with a temperature of $T_2$. The cold water enters from the bottom of the tank with a temperature of $T_4$. The hot-water and cold-water outlet temperatures, on the other hand, are indicated with $T_3$ and $T_1$, respectively. The primary purpose of the study is to obtain higher thermal stratification between the hot and cold sides of the tank.

Altuntop et al. (2005) carried out transient simulations for each design to evaluate temperature distribution within the sensible heat storage (SHS) tank and the temperature difference between the hot and cold sides of the tank. In Fig. 5.12, the temperature distributions within the storage tank are given on a selected plane for three different obstacle designs.

Figure 5.13 compares the temperature differences between the cold and hot side of the tank. Altuntop et al. (2005) stated that to achieve a better thermal stratification, $(T_3 - T_4)$ should be higher but $(T_2 - T_3)$ and $(T_1 - T_4)$ should be lower. According to these criteria, the configuration in which obstacle 11 is used gives a better thermal stratification.

Abdelhak et al. (2015) investigated the transient flow behavior inside a water tank with an electrical heater. They simulated two different configurations as given in

**Table 5.2** Some selected CFD studies on modeling of SHTES tanks

| Reference | Software | Primary parameter | Mode of analysis |
|---|---|---|---|
| Yee and Lai (2001) | In-house code | Location of the baffle and thickness of the porous tube | Transient |
| Shah and Furbo (2003) | FLUENT 5.5 | Entrance effect | Transient |
| Altuntop et al. (2005) | FLUENT 6.1.22 | Type of obstacle | Transient |
| Abdelhak et al. (2015) | FLUENT 6.3 | Horizontal and vertical configurations | Transient |
| Bouhal et al. (2017) | FLUENT 15 | Position and type of the baffles | Transient |
| Cascetta et al. (2016) | ANSYS FLUENT | Development of a model for packed bed | Transient |

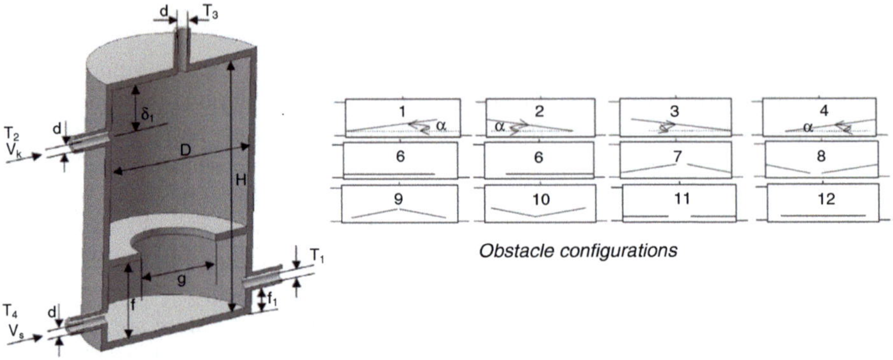


**Fig. 5.11** Sensible heat storage model of Altuntop et al. (2005)

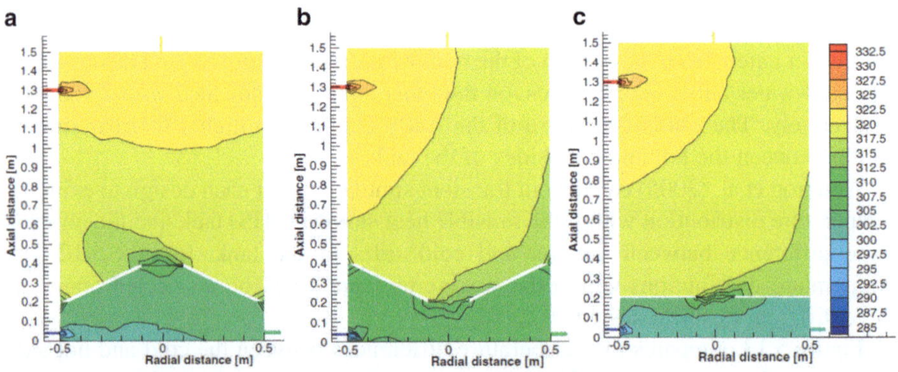


**Fig. 5.12** Temperature distribution within the SHS tanks: (**a**) obstacle 7, (**b**) obstacle 9, (**c**) obstacle 11. (Altuntop et al. 2005)

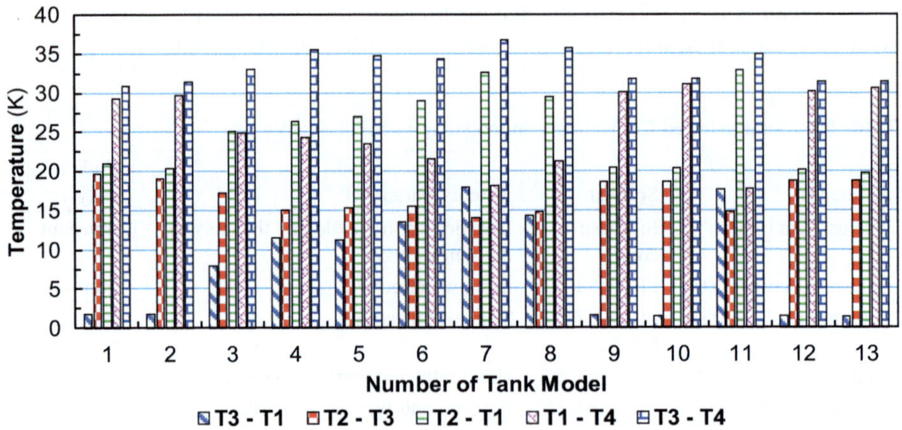


**Fig. 5.13** Temperature differences for each configuration with obstacles. (Data from Altuntop et al. 2005)

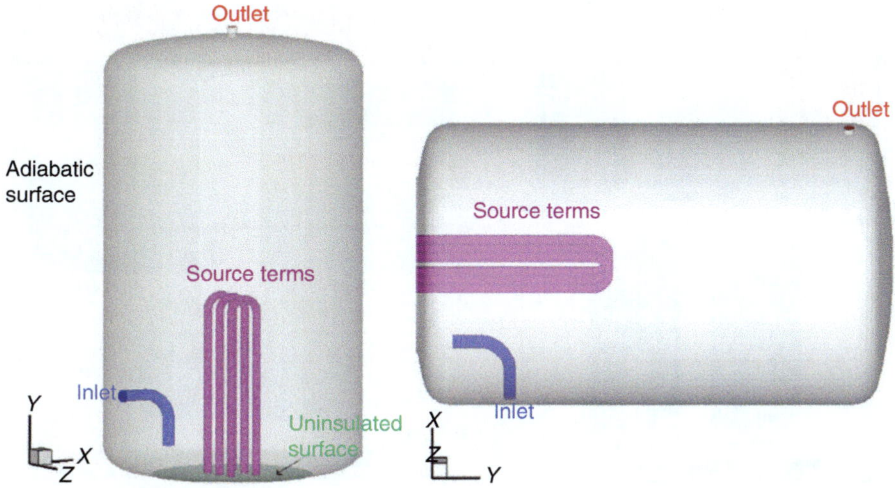


**Fig. 5.14** Vertical and horizontal tank configurations with an electrical heater. (Abdelhak et al. 2015)

Fig. 5.14. In the first model, the electrical heaters are placed in a vertical position and cold fluid flows parallel to the electrical heater. In the second model, the heater is placed in a horizontal position and the fluid flows in a vertical direction. The temperature distributions within the tanks are evaluated by resolving the governing equations in FLUENT. Figure 5.15 shows the temperature distributions within the tanks for vertical and horizontal heater configurations. The timewise variation of the nondimensional stratification number is also evaluated to compare the performances of vertical and horizontal configurations. It is concluded that the stratification efficiency of the horizontal tank is lower than the vertical one.

### 5.4.3.2 Latent Heat TES Systems

In LHTES systems, various types of encapsulation techniques are used to provide heat transfer between the heat transfer fluid and the PCM. The transient response of the storage unit depends on the working parameters, design parameters, and thermophysical properties of the materials. Modeling of the PCM domain requires high computational cost due to its complexity when the natural convection within the liquids PCM is considered. In some works, to simplify the long-term simulations, the heat transfer mechanism within the PCM domain is reduced into conduction. However, in some cases, neglecting the natural convection inside the liquid PCM may cause unrealistic prediction due to the enhanced heat transfer for the convection dominated phase change. The effective thermal conductivity definition is used to incorporate the enhanced heat transfer inside the liquid PCM. Such an approach allows achieving reasonable predictions with lower computational costs. In the literature, planar, cylindrical, or spherical capsules are used in various configurations

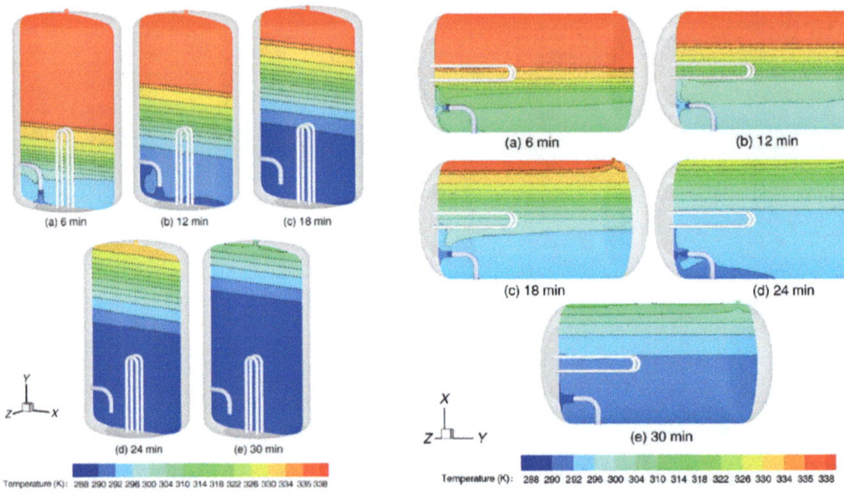


**Fig. 5.15** Evolution of temperature distributions. (Abdelhak et al. 2015)

**Table 5.3** Some selected CFD studies on modeling of LHTES tanks

| Reference | Software | Geometry | Primary parameters | Mode of heat transfer inside PCM |
|---|---|---|---|---|
| Guo and Zhang (2008) | FLUENT 6.2 | Shell-and-tube-type heat exchanger (HEX) with PCM | Geometric and working parameters of the HEX | Conduction |
| Xia et al. (2010) | FLUENT 6.2 | Packed bed latent heat storage tank with PCM | The capsule material and wall thickness | Effective thermal conductivity is defined to consider the natural convection |
| Tay et al. (2013) | ANSYS CFX | Cylindrical tank with PCM | Influence of pinned, finned, and plane tubes | Conduction |
| Fornarelli et al. (2016) | N/A | Shell-and-tube-type heat exchanger | Influence of mushy zone constant and natural convection | Convection and conduction |
| Allouche et al. (2016) | ANSYS FLUENT | Tube-in-tank PCM | Flow rate of the HTF | Convection and conduction |
| Promoppatum et al. (2017) | COMSOL | Cylindrical tubes with PCM in cross flow | Tube arrangement and the influence of aluminum loading into PCM | Conduction |

to achieve higher charging and discharging performance for an LHTES with PCM. Some of the recent works that deal with CFD modeling of LHTES tanks are listed in Table 5.3. In the following, two selected numerical works from the literature are reviewed.

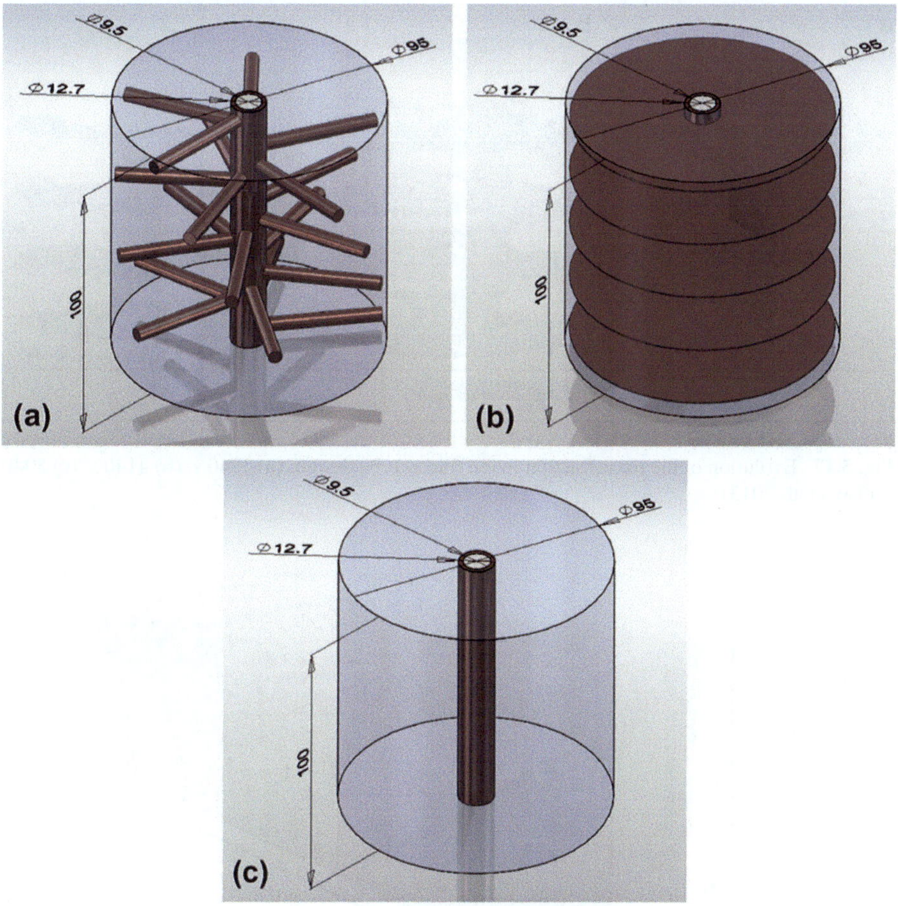


**Fig. 5.16** LHTES tank design configurations: (**a**) pinned tube, (**b**) finned tube, and (**c**) plain tube. (Tay et al. 2013)

Tay et al. (2013) simulated the transient heat transfer and fluid flow problem within a PCM-filled storage tank with three different tube arrangements. The schematic of the pinned tube, finned tube, and plain tube designs are shown in Fig. 5.16. In the pinned and finned tube designs, authors varied the geometric parameters to introduce the influence of different configurations in LHTES tanks. Three-dimensional mathematical models are developed in ANSYS-CFX software, and transient analyses are conducted. Nondimensional compactness factor and the effectiveness of the tanks with various tube configurations are obtained from the timewise variations of the liquid fraction of the PCM. In Fig. 5.17, mass fraction distributions are given at three different flow times inside the LHTES tank with finned tube arrangement. It is concluded that the finned tube design provides 40% better effectiveness.

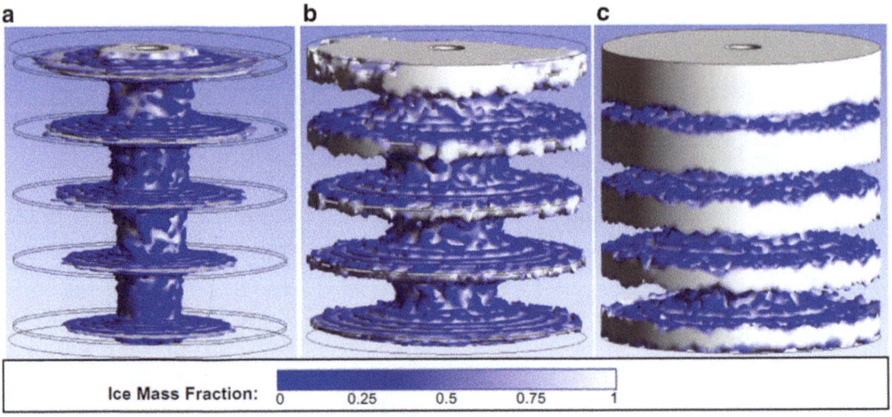


**Fig. 5.17** Evolution of the mass fraction in the finned tube design: (**a**) 1740 s, (**b**) 4140 s, (**c**) 5940 s. (Tay et al. 2013)

**Fig. 5.18** Cross-flow heat exchanger-type LHTES. (Promoppatum et al. 2017)

Promoppatum et al. (2017) considered a cross-flow heat exchanger with vertical tubes. The PCM-filled tubes are placed in a staggered arrangement to improve the convective heat transfer between the PCM and the working fluid. In Fig. 5.18, the isometric view of the storage unit is given. The discharging period of the storage tank is simulated in which the cold air from the building passes through the tube array to reject the stored thermal energy within the liquid PCM. Notice that the colors of the tubes vary in the flow direction of the air in Fig. 5.18. Varying colors indicate the PCM-filled tubes with different melting temperatures. A two-dimensional numerical model is considered to simulate the phase change problem within the PCM. The temperature distributions within the PCM domain are evaluated under various

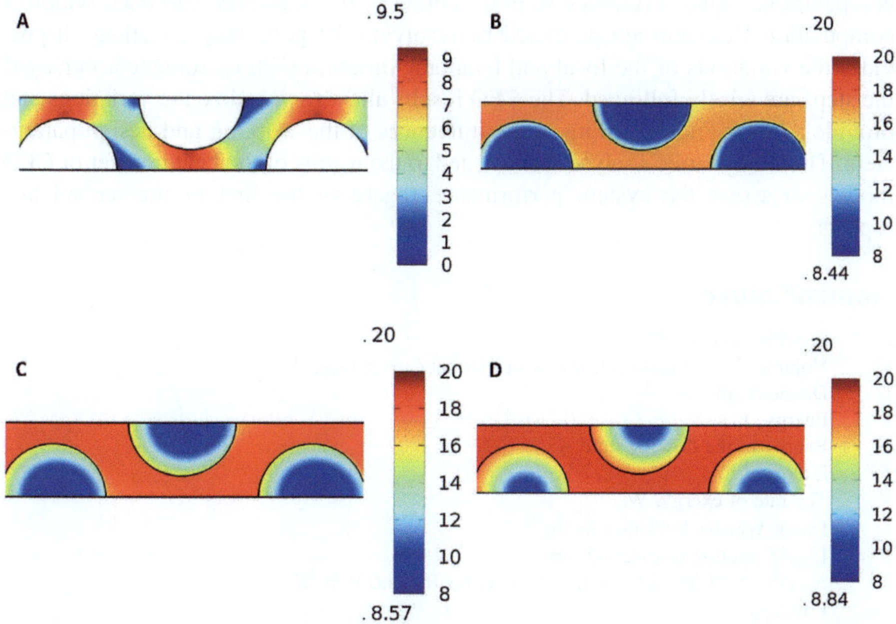


**Fig. 5.19** Velocity and temperature distributions inside cross-flow heat exchanger: (**a**) velocity (m/s), (**b**) temperature (°C) at 900 s, (**c**) temperature (°C) at 1800 s, (**d**) temperature (°C) at 2700 s. (Promoppatum et al. 2017)

working and design conditions. Figure 5.19 represents the velocity and temperature contours at selected instants. The authors also investigated the influence of aluminum insertion into the pure PCM to improve the thermal conductivity of the material. It is concluded that the incorporation of aluminum, even at minimal volumetric ratios, significantly improves the thermal performance of the storage tank.

## 5.5 Closing Remarks

In this chapter, various modeling and analysis studies of the TES units are presented and discussed. The thermodynamic analyses allow assessing the performance both quantitatively and qualitatively. Illustrative examples show that the first and second law analyses provide an understanding of the performance of the system. Energy and exergy analyses should be considered together to achieve the usefulness of each process of the TES unit. Heat transfer analyses are used to determine the variations of the temperature or interface front within the storage unit under varying design and working conditions. In the analyses, the heat transfer mechanisms are reduced to RC (resistance/capacitance) thermal networks. Empirical or numerical correlations are used to evaluate the heat transfer coefficients at various flow conditions. In the computational fluid dynamics (CFD) approach, the transport variables within a computational domain are predicted by resolving the governing equations. It provides the variations of the local and temporal variables with reasonable accuracy if the steps are wisely followed. The CFD results allow to visualize the variations and provide a better understanding of the influences of the working and design parameters. Thermodynamic analyses are applied to the results of the heat transfer or CFD models to assess the system performance regarding the first or the second law aspects.

## Nomenclature

| | |
|---|---|
| $c$ | Specific heat, J/kgK |
| $C$ | Volumetric heat capacity, J/m$^3$K or thermal capacitance, W |
| $D$ | Diameter, m |
| $E$ | Energy, J |
| $ex$ | Exergy of the flowing fluid, J |
| $Ex$ | Exergy, J |
| $\dot{E}x$ | The rate of exergy, W |
| $f$ | Darcy-Weisbach friction factor |
| $F$ | Liquid fraction or external force |
| $h$ | Specific enthalpy, J/kg or heat transfer coefficient, W/m$^2$K |
| $H$ | Enthalpy, J |
| $H$ | Volumetric enthalpy, J/m$^3$ |
| $I$ | Irreversibility, J |
| $k$ | Thermal conductivity, W/mK |
| $\ell$ | Length of the tube, m |
| $m$ | Mass, kg |
| $\dot{m}$ | Mass flow rate, kg/s |
| Nu | Nusselt |
| Pr | Prandtl number |
| $Q$ | Total heat transfer, J |
| $\dot{Q}$ | Heat transfer rate, W |
| $r, x$ | Radial and axial coordinates, m |
| $r$ | Radius, m |
| $R$ | Thermal resistance, K/W |
| Ra | Rayleigh number |
| Re | Reynolds number |
| $s$ | Specific entropy, J/kgK |
| $S$ | Total entropy, J/K, or source term in energy equation, J/m$^3$ |
| $t$ | Time, s |
| $T$ | Temperature, K or °C |
| $\bar{T}$ | Mean temperature, °C *or* K |
| $u, v$ | Velocity components, m/s |
| $U$ | Overall heat transfer coefficient, W/m$^2$K |
| $x, y$ | Cartesian coordinates, m |

## Greek Letters

| | |
|---|---|
| $\beta$ | Thermal expansion coefficient, 1/K |
| $\Delta$ | Difference |

| | |
|---|---|
| $2\delta T_m$ | Phase change temperature range, °C *or* K |
| $\eta$ | Energy efficiency |
| $\phi$ | Generic variable |
| $\Gamma$ | Diffusion coefficient |
| $\rho$ | Density ($kgm^{-3}$) |
| $\mu$ | Dynamic viscosity ($kgm^{-1}\ s^{-1}$) |
| $\theta$ | Polar coordinate |
| $\psi$ | Exergy efficiency |

## Subscripts

| | |
|---|---|
| *CH, C* | Charging |
| *DIS, D* | Discharging |
| *dest* | Destruction |
| *f* | Final |
| *H* | Hydraulic |
| *i* | Initial or inner |
| *in* | Inlet |
| *L* | Lost |
| *m* | Melting |
| *o* | Dead state |
| *out* | Outlet |
| *s* | Surface |
| *S* | Solid |
| *sf* | Solid to liquid |
| *ST* | Storage |

## References


Abdelhak, O., Mhiri, H., & Bournot, P. (2015). CFD analysis of thermal stratification in domestic hot water storage tank during dynamic mode. *Building Simulation, 8*(4), 421–429 Tsinghua University Press.

Allouche, Y., Varga, S., Bouden, C., & Oliveira, A. C. (2016). Validation of a CFD model for the simulation of heat transfer in a tubes-in-tank PCM storage unit. *Renewable Energy, 89*, 371–379.

Altuntop, N., Arslan, M., Ozceyhan, V., & Kanoglu, M. (2005). Effect of obstacles on thermal stratification in hot water storage tanks. *Applied Thermal Engineering, 25*(14–15), 2285–2298.

ANSYS Inc. (2009). *ANSYS FLUENT user's guide, version 12*. ANSYS Inc.

Bouhal, T., Fertahi, S., Agrouaz, Y., El Rhafiki, T., Kousksou, T., & Jamil, A. (2017). Numerical modeling and optimization of thermal stratification in solar hot water storage tanks for domestic applications: CFD study. *Solar Energy, 157*, 441–455.

Cao, Y., & Faghri, A. (1990). A numerical−analysis of phase−change problems including natural convection. *ASME Journal of Heat Transfer, 112*, 812–816.

Cascetta, M., Cau, G., Puddu, P., & Serra, F. (2016). A comparison between CFD simulation and experimental investigation of a packed-bed thermal energy storage system. *Applied Thermal Engineering, 98*, 1263–1272.

Dincer, I. (2002). On thermal energy storage systems and applications in buildings. *Energy and Buildings, 34*(4), 377–388.

Dincer, I., Dost, S., & Li, X. (1997). Performance analyses of sensible heat storage systems for thermal applications. *International Journal of Energy Research, 21*(12), 1157–1171.

Dincer, I., & Rosen, M. (2011). *Thermal energy storage: Systems and applications* (2nd ed.). Hoboken: Wiley.

Drees, K. H., & Braun, J. E. (1995). Modeling of area–constrained ice storage tanks. *HVAC&R Research, 1*, 143–158.

Ezan, M. A. (2011). *Experimental and numerical investigation of cold thermal energy storage systems*. PhD thesis, Graduate School of Natural and Applied Sciences of Dokuz Eylul University, Izmir.

Fornarelli, F., Camporeale, S. M., Fortunato, B., Torresi, M., Oresta, P., Magliocchetti, L., et al. (2016). CFD analysis of melting process in a shell-and-tube latent heat storage for concentrated solar power plants. *Applied Energy, 164*, 711–722.

Guo, C., & Zhang, W. (2008). Numerical simulation and parametric study on new type of high temperature latent heat thermal energy storage system. *Energy Conversion and Management, 49* (5), 919–927.

Incropera, F. P., & DeWitt, P. D. (2002). *Fundamentals of heat and mass transfer*. New York: Wiley.

Jegadheeswaran, S., Pohekar, S. D., & Kousksou, T. (2010). Exergy based performance evaluation of latent heat thermal storage system: A review. *Renewable and Sustainable Energy Reviews, 14* (9), 2580–2595.

Jekel, T. B., Mitchell, J. W., & Klein, S. A. (1993). Modeling of ice–storage tanks. *ASHRAE Transactions, 99*, 1016–1024.

Kestin, J. (1980). Availability: The concept and associated terminology. *Energy, 5*(8-9), 679–692.

MacPhee, D., & Dincer, I. (2009). Thermodynamic analysis of freezing and melting processes in a bed of spherical PCM capsules. *Journal of Solar Energy Engineering, 131*(3), 031017.

Morgan, K. (1981). A numerical analysis of freezing and melting with convection. *Computer Methods in Applied Mechanics and Engineering, 28*, 275–284.

Neto, J. H. M., & Krarti, M. (1997). Deterministic model for an internal melt ice-on-coil thermal energy storage tank. *ASHRAE Transactions, 103*, 113–124.

Patankar, S. V. (1980). *Numerical heat transfer and fluid flow*. New York: Hemisphere.

Patankar, S. V., & Spalding, D. B. (1972). A calculation procedure for heat, mass and momentum transfer in three–dimensional parabolic flows. *International Journal of Heat Mass Transfer, 15*, 1787–1806.

Promoppatum, P., Yao, S. C., Hultz, T., & Agee, D. (2017). Experimental and numerical investigation of the cross-flow PCM heat exchanger for the energy saving of building HVAC. *Energy and Buildings, 138*, 468–478.

Rosen, M. A., & Hooper, F. C. (1991). A general method for evaluating the energy and exergy contents of stratified thermal energy storages for linear-based storage fluid temperature distributions. *Proceedings of the 17th Annual Conference of Solar Energy Society of Canada*, Toronto, pp. 182–187.

Seban, R. A., & McLaughlin, E. F. (1963). Heat transfer in tube coils with laminar and turbulent flow. *International Journal of Heat and Mass Transfer, 6*, 387–395.

Shah, L. J., & Furbo, S. (2003). Entrance effects in solar storage tanks. *Solar Energy, 75*(4), 337–348.

Tay, N. H. S., Bruno, F., & Belusko, M. (2013). Comparison of pinned and finned tubes in a phase change thermal energy storage system using CFD. *Applied Energy, 104*, 79–86.

Tu, J., Yeoh, G. H., & Liu, C. (2008). *Computational fluid dynamics: A practical approach*. Butterworth: Heinemann.

Versteeg, H., & Malalasekera, W. (2007). *An introduction to computational fluid dynamics: The finite volume method* (2nd ed.). Harlow: Prentice Hall.

Wang, S. M., Faghri, A., & Bergman, T. L. (2010). A comprehensive numerical model for melting with natural convection. *International Journal of Heat and Mass Transfer, 53*, 1986–2000.

Xia, L., Zhang, P., & Wang, R. Z. (2010). Numerical heat transfer analysis of the packed bed latent heat storage system based on an effective packed bed model. *Energy, 35*(5), 2022–2032.

Yee, C. K., & Lai, F. C. (2001). Effects of a porous manifold on thermal stratification in a liquid storage tank. *Solar Energy, 71*(4), 241–254.

# Chapter 6
# System Optimization

## 6.1 Introduction

The design of a thermal energy storage system includes many aspects. The primary goal of a design engineer is to build a device or system that meets the minimum requirements of a facility, such as a building or a plant. However, designing a system that *works* does not mean that the design process of the system is completed. Several factors should be considered, such as safety, environmental issues, and cost, to find a better design. A better design may be the one that has the highest efficiency, lowest cost, and minimum harmful effects. The optimization procedure starts with selecting the main output variables that will be made minimum or maximum. Such variables are known as objective functions. In the case of a thermal system, the following quantities are defined as the objective functions: profit, cost, and efficiency. An engineer may design an individual component of a TES system or the entire system that includes several components. Consider a sensible heat storage unit that is used in a building heating unit. The system may include a storage tank, pump, heat exchanger, and controllers. Various alternative designs may provide the required heat output through the building. Each alternative design may satisfy the requirements or constraints. In such a case, an optimization procedure could be followed to find a proper design that maximizes the overall efficiency of the system and minimizes investment costs. Besides, optimization may be conducted for each component of the system for the same goals.

Thermodynamic-based optimization aims to design thermal systems considering the energetic, exergetic, environmental, and economic aspects. Optimization of thermal systems has great potential in using the current energy resources of the world more efficiently. The aim of the current chapter is to present the basic definitions and methods of the optimization, with some illustrative examples of the optimization processes of thermal energy storage systems for sensible and latent heat storage and their applications.

 
I. Dincer, M. A. Ezan, *Heat Storage: A Unique Solution For Energy Systems*, Green Energy and Technology, https://doi.org/10.1007/978-3-319-91893-8_6

## 6.2 Optimization

Optimization is the process of maximizing or minimizing a function subject to several constraints (Dincer et al. 2017). In engineering design, optimization seeks the best possible configuration for a given problem. An optimization problem may involve one (*single*) or more than one (*multiple*) objective functions. Single optimization deals with finding a better solution considering a single criterion. However, we already know that in any real-world problem, there is always more than one constraint that should be considered while designing a system. In the case of a thermal system design process, a design engineer should consider many aspects such as economic, environmental, and operational. In the following, some essential optimization terms or concepts are defined (Dincer et al. 2017).

***Objective Functions and System Criteria*** An objective function is based on the purpose of the decision-maker. The objective function can be either maximized or minimized. Optimization criteria, on the other hand, can vary widely. For instance, optimization criteria can be based on economic purposes (e.g., total capital investment, total annual levelized costs, cost of exergy destruction, cost of environmental impact), efficiency aims (e.g., energy, exergy, and others), other technological goals (production rate, production time, total weight), environmental impact objectives (reduced pollutant emissions), and other objectives (Dincer et al. 2017). Note that a multi-objective optimization technique lets us consider more than one objective function for an optimization problem.

***Decision Variables*** In an optimization problem, selecting the appropriate decision variables is of vital importance to achieve the desired goal more wisely. Dincer et al. (2017) pointed out some critical points that should be kept in mind while selecting the decision variables: (*i*) include all critical variables that can affect the performance and cost-effectiveness of the system, (*ii*) do not include variables of minor importance, and (*iii*) distinguish among independent variables whose values are amenable to change.

***Constraints*** Constraints build merely the borders of a design problem. Constraints can be physical variables that are defined by the design engineer or some physical equations, such as mass conservation or energy conservation. Dincer et al. (2017) listed some possible restrictions on variables that may arise due to the limitations of space, equipment, or material, such as (*i*) restriction of the physical dimensions, (*ii*) temperature limits (high and/or low), (*iii*) maximum/minimum allowed pressure, (*iv*) maximum/minimum flow rate, and (*v*) maximum/minimum force. In a thermal system, there are many additional constraints that arise from conservation laws, i.e., mass, momentum or energy, and balance equations, i.e., entropy and exergy.

According to Dincer et al. (2017), optimization techniques that are widely used in thermal system designs are categorized as follows:

***Classical Optimization*** Such techniques are used for continuous and differentiable functions. However, classical methods may not be useful in practical applications as they mostly involve objective functions that are not continuous and/or differentiable.

***Numerical Optimization Methods*** In this category, the following techniques are used: (*i*) linear programming, (*ii*) integer programming, (*iii*) quadratic programming, (*iv*) nonlinear programming, (*v*) stochastic programming, (*v*) dynamic programming, (*vi*) combinatorial optimization, and (*vii*) evolutionary algorithm.

***Evolutionary Algorithms*** Such techniques are based on biological evolution, production, mutation, recombination, and selection (Dincer et al. 2017). Well-known evolutionary algorithms are (*i*) genetic algorithm (GA), (*ii*) artificial neural networks (ANN), and (*iii*) fuzzy logic. Details of these methods could be found elsewhere (Dincer et al. 2017).

## 6.3 Second Law-Based Optimization of Sensible and Latent Heat TES Systems

As pointed out in the rest of the book, a complete assessment of a thermal system could only be achieved by conducting first law and second law analyses. Following this route for a storage system provides useful information to a design engineer about the quantity and the quality of thermal energy during the charging and discharging periods. The second law-based optimization technique aims to minimize irreversibilities within a system by reducing entropy generation. Its primary purpose is to store useful work given by thermodynamic availability or exergy (Badar et al. 1993).

**Illustrative Example 1: Entropy Generation Minimization of a Sensible Heat TES System**

Bejan (1978) considered a sensible heat TES unit that involves a liquid storage tank and an immersed heat exchanger within the tank. Hot gas flows through the heat exchanger during the charging period. The schematic of the system is illustrated in Fig. 6.1a. The tank is covered with an insulation material to minimize heat loss through the ambient. Initially, the temperature of the liquid tank is equal to the ambient temperature $T_o$. The temperature of the storage material within the storage tank gradually increases as the hot gas flows through the heat exchanger. The evolution of the tank temperature ($T_{tank}$) is illustrated in Fig. 6.1b together with the timewise variations in the inlet ($T_{g,in}$) and outlet ($T_{g,out}$) temperatures of the hot gas flow. Bejan (1978) developed a heat transfer model for a sensible heat storage tank before going through thermodynamic optimization. A lumped model is assumed to be valid by considering a well-mixed liquid, disregarding spatial temperature variations

(continued)

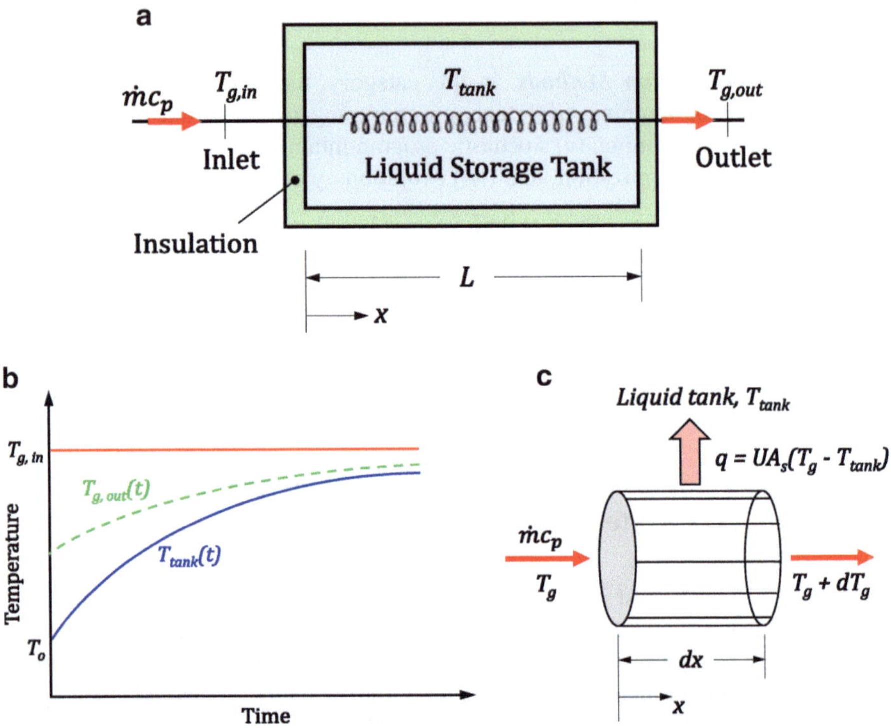


**Fig. 6.1** Geometry and temperature variation of a sensible heat TES unit. (**a**) Sensible heat storage tank with immersed gas heat exchanger, (**b**) evolution of the temperature, (**c**) heat transfer between gas and tank. (Adapted from Bejan 1978)

and natural convection, with a uniform temperature of $T_{tank}$ which depends only time.

The heat transfer between the hot gas and the storage material, liquid, could be evaluated by considering the energy balance for a differential control volume as illustrated in Fig. 6.1c. The energy balance yields

$$UA_s\left(T_g(x) - T_{\text{tank}}\right) = \dot{m}\, c_p dT_g \quad (\text{in W}) \tag{6.1}$$

where $U$ is the overall heat transfer coefficient between the liquid and the hot fluid flow within the tube. Notice that the heat loss through the surrounding is assumed to be negligible. $A_s$ is the heat transfer surface area and is defined as $A_s = Pdx$. $P$ stands for the perimeter of the tube, and $dx$ is the length of the differential control volume. Integrating Eq. (6.1) from inlet ($x = 0$) to outlet ($x = L$) of the heat exchanger yields

(continued)

$$\frac{T_{g,out}(t) - T_{\text{tank}}(t)}{T_{g,in} - T_{g,out}(t)} = \exp\left(-\frac{UP}{\dot{m}c_p}L\right) \tag{6.2}$$

where $T_{g,in}$ and $T_{g,out}$ indicate the inlet and outlet temperatures of the hot gas flow. Notice that the model assumes constant inlet temperature of hot gas throughout the charging (Bejan 1978). For simplicity, the dimensionless *number of heat transfer unit* (NTU) is defined as

$$\text{NTU} = \frac{UP}{\dot{m}c_p}L \tag{6.3}$$

Considering the liquid tank as a transient closed system, the conservation of energy is written in the following form:

$$\left(mc\frac{dT}{dt}\right)_{\text{tank}} = \dot{m}c_p\left[T_{g,out}(t) - T_{g,in}\right] \tag{6.4}$$

The left-hand side of the equation is the sensible internal energy variation in the storage material, i.e., liquid, and the right-hand side is the rate of heat transfer provided from the hot flow. Bejan (1978) combined Eqs. (6.2), (6.3), and (6.4), and the integration yields

$$\frac{T_{\text{tank}}(t) - T_o}{T_{g,in} - T_o} = 1 - \exp(-y\theta) \tag{6.5}$$

$$\frac{T_{\text{out}}(t) - T_{\text{tank}}(t)}{T_{g,in} - T_{\text{tank}}(t)} = 1 - y\exp(-y\theta) \tag{6.6}$$

where $y$ and $\theta$ are dimensionless groups. Bejan (1978) defined $y$ and $\theta$ as

$$y = 1 - \exp(-\text{NTU}) \tag{6.7}$$

$$\theta = \frac{(\dot{m}c_p)_{gas}}{(mc)_{\text{tank}}}t \tag{6.8}$$

Figure 6.2a illustrates the variation of $y$ as a function of NTU. $y$ approaches unity beyond NTU = 10. Further increasing NTU does not change the $y$-parameter. Figure 6.2b, c, on the other hand, show the variations of tank temperature and outlet temperature, respectively, as a function of NTU. Here the initial tank temperature and the inlet temperature of the hot gas are selected as $T_o = 25$ °C and $T_{g,in} = 60$ °C, respectively. Both tank temperature and the outlet temperature of the gas asymptotically approach the inlet temperature. Note that increasing the NTU improves the speed of heat transfer. The amount

(continued)

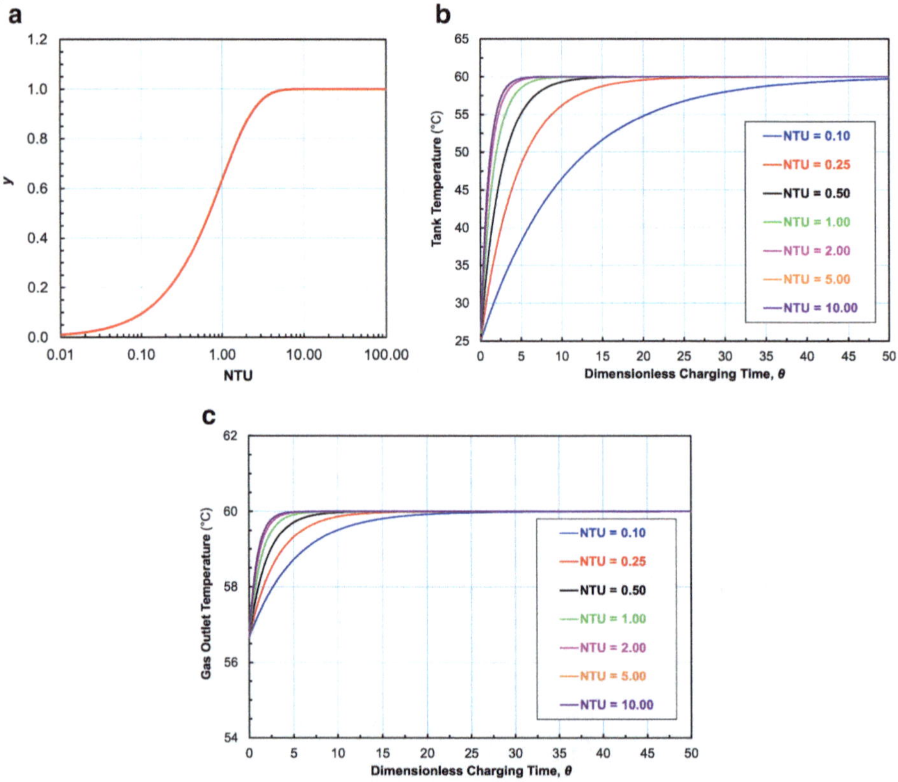


**Fig. 6.2** Influence of NTU on *y* parameter and timewise temperature variations. (**a**) The relation between *y* and NTU, (**b**) evolution of tank temperature, (**c**) evolution of outlet temperature of the gas

of stored energy inside the tank significantly increases at higher NTU values. Notice that beyond a critical NTU value, NTU > 2, no remarkable change is observed in the temperature variations. The amount of stored energy also increases with time. The time for complete charging reduces as increasing NTU. To sum up the ability to store energy increases as the charging time and NTU increase (Bejan 1978).

Bejan (1978) proposed a novel second law-based optimization approach to minimize the destruction of thermodynamic availability. For the current sensible heat TES tank, irreversibilities arise (*i*) due to heat transfer between the hot gas and the liquid storage medium, (*ii*) due to cooling of the hot exhaust gas to ambient temperature, and (*iii*) due to the friction inside the heat exchanger. In the current simple illustrative example, irreversibility arises from the friction that is neglected (Dincer and Rosen 2001). The irreversibilities

(continued)

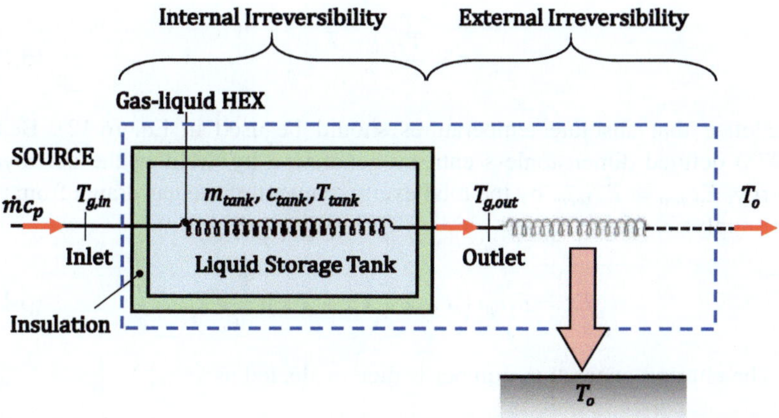


**Fig. 6.3** Sources of entropy generation in a liquid sensible heat TES unit. (Adapted from Bejan 1978)

are illustrated in Fig. 6.3. Eventually, due to the irreversibilities that arise from the finite temperature difference, only some portion of the exergy that is brought by the hot stream is stored in the storage tank (Bejan 1978).

The rate of irreversibility is defined as

$$\dot{I} = \dot{S}_{gen} T_o \quad (\text{in kW}) \tag{6.9}$$

where $\dot{S}_{gen}$ is the rate of entropy generation in the system that is defined by dashed lines in Fig. 6.3. The entropy generation due to heat transfer is evaluated by Bejan (1978) as

$$\dot{S}_{gen} = \dot{m}\,c_p \ln \frac{T_o}{T_{g,in}} + \frac{d}{dt}\left(mc \ln \frac{T}{T_o}\right)_{\text{tank}} + \frac{1}{T_o}\dot{m}\,c_p\left(T_{g,out} - T_o\right) \quad (\text{in kW/K}) \tag{6.10}$$

where the first term represents the entropy change of the ideal gas flow, the second term stands for the internal entropy variation of the storage, and the last term is entropy generation due to cool down of the exhaust gas through the ambient temperature. Bejan (1978) integrated Eq. (6.10) to evaluate the total amount of entropy production for a transient process

$$\frac{S_{gen}}{\dot{m}\,c_p t} = \tau - \ln(1+\tau) + \frac{1}{\theta}\{\ln[1 + \tau(1 - \exp(-y\theta))] - \tau(1 - \exp(-y\theta))\} \tag{6.11}$$

where $\tau$ is the characteristic temperature difference and defined as follows:

(continued)

$$\tau = \frac{T_{g,in} - T_o}{T_o} \tag{6.12}$$

Notice that absolute temperatures should be used in Eq. (6.12). Bejan (1978) defined dimensionless entropy generation by dividing the destroyed exergy, $Ex_{dest} = T_o S_{gen}$, by the total exergy content of the gas drawn from the hot supply:

$$Ex = \dot{m}\, c_{pt}\left(T_{g,in} - T_o - T_o \ln\frac{T_{g,in}}{T_o}\right) \tag{6.13}$$

The entropy generation number is then evaluated as

$$N_S = \frac{T_o S_{gen}}{Ex} = 1 - \frac{\tau[1 - \exp(-y\theta)] - \ln\left[1 + \tau(1 - \exp(-y\theta))\right]}{\theta[\tau - \ln(1+\tau)]} \tag{6.14}$$

Figure 6.4 illustrates the variation entropy generation number ($N_s$) regarding the dimensionless time. Computations are conducted for nine different NTU values, which vary from 0.1 to 10, and three different dimensionless temperatures, $\tau$. Notice that increasing the NTU value reduces the fraction of accumulated irreversibility or $N_s$. The optimum charging time ($\theta$) for any given NTU value and $\tau$ corresponds to the time when the minimum entropy generation number is evaluated. Bejan (1978) defined the two extreme cases of this storage process: for $\theta$ goes to zero and $\theta$ goes to infinity. In the $\theta \to 0$ limit, the whole exergy content of the hot gas at the inlet section of the storage tank is destroyed by the heat transfer between the gas and liquid storage medium. In the $\theta \to \infty$ limit, irreversibility occurs outside the tank. The temperature of the hot flow remains unchanged throughout the tank and leaves the tank with the same temperature as it enters. That is, the exergy content of the gas is totally destroyed by the heat transfer to ambient (Bejan 1978).

Bejan (1978) stated that the optimum charging time could be calculated explicitly in the limit $\tau \to 0$. For this case, the entropy generation number reduces from Eq. (6.14) to the following form:

$$N_S = 1 - \frac{1}{\theta}[1 - \exp(-y\theta)]^2 \tag{6.15}$$

Solving $\partial N_s / \partial \theta = 0$, Bejan (Dincer and Rosen 2001) evaluated the following expression for optimum time, where minimum entropy generation exists:

$$\theta_{opt} = \frac{1.256}{y} = \frac{1.256}{1 - \exp(-\mathrm{NTU})} \tag{6.16}$$

(continued)

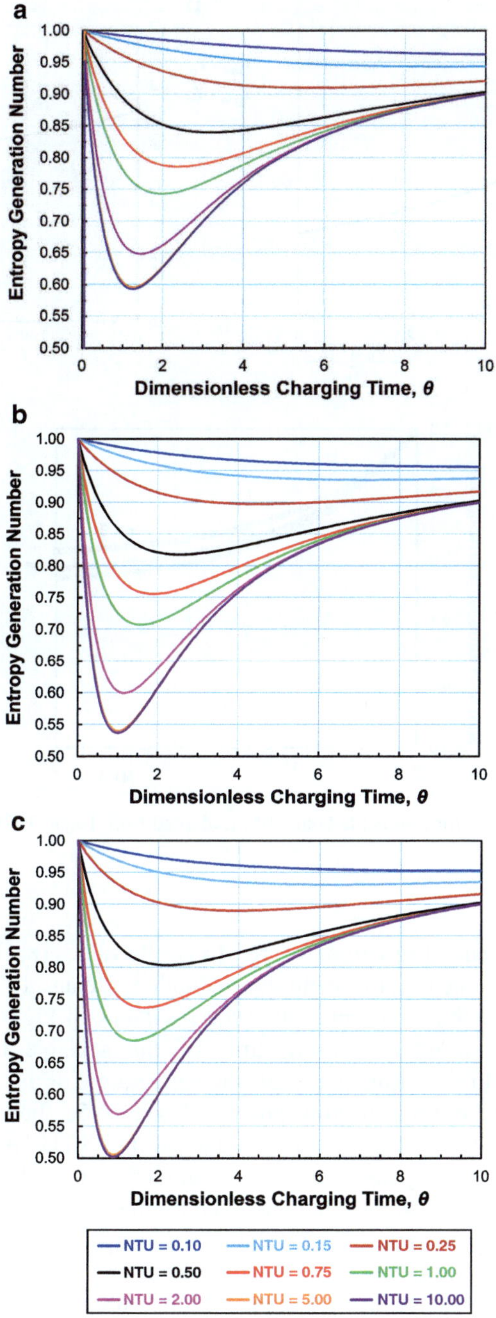


**Fig. 6.4** Variation of entropy generation number as a function of NTU. (**a**) $\tau \rightarrow 0$, (**b**) $\tau = 1$, (**c**) $\tau = 2$. (Adapted from Bejan 1978)

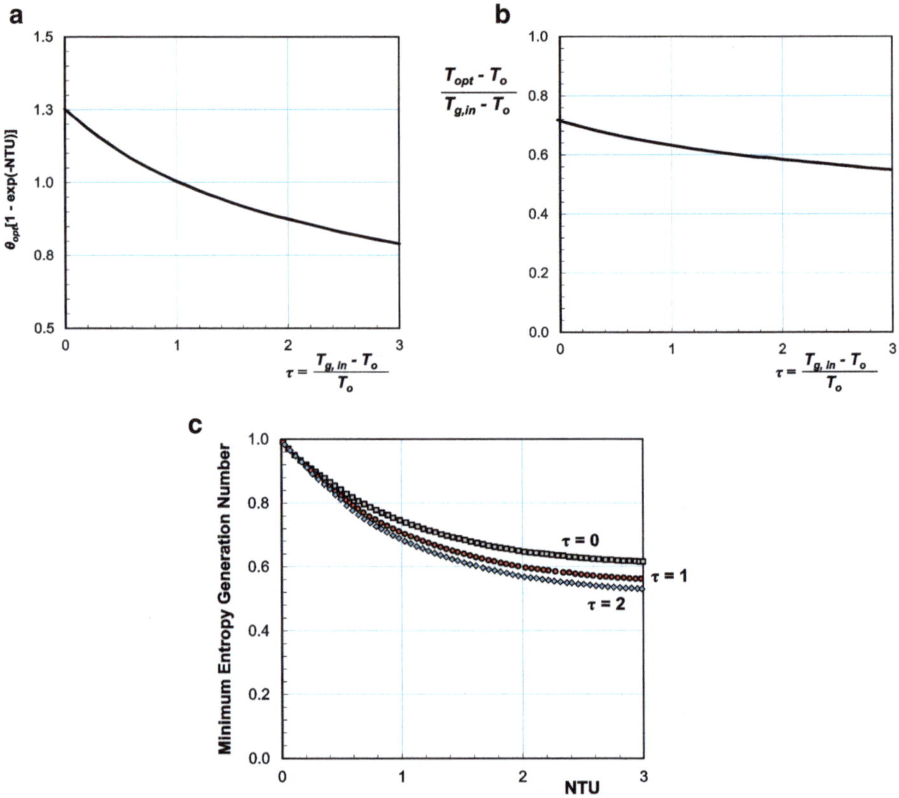


**Fig. 6.5** Optimal values for a sensible heat TES. (Adapted from Bejan 1978)

In the more general situation, where $\tau$ has finite values ($\tau > 0$), the optimum charging is evaluated from an implicit equation. Bejan (1978) numerically resolved the equation and represented the results as in Fig. 6.5. Figure 6.5a illustrates the optimum charging time of the sensible heat TES unit. Figure 6.5b, on the other hand, reveals the temperature of the storage medium at the end of the optimal heating process. Figure 6.5c shows the minimum entropy generation number variation regarding the NTU and $\tau$. Notice that regardless of the input parameters, the fraction of destroyed exergy is at least as high as 50%. It should be noted that even for the best design conditions, the exergy destruction is almost half of the stored exergy. That is, Bejan (Dincer and Rosen 2001) proposed the implementation of a series of sensible heat storage tanks to reduce entropy generation in sensible heat TES units further. The results of Taylor and Krane (1991) reveal that the entropy generation values are in the range 0.2–0.8.

**Illustrative Example 2: Cost Optimization of a Sensible TES System**
Badar et al. (1993) extended the second law-based optimization model of Bejan (1978) for sensible heat TES by including the economic concerns. The model includes the monetary values in addition to the irreversibilities that arise due to the finite temperature difference and the pressure loss. Badar et al. (1993) expressed the annualized total capital cost of owning the sensible heat TES system as

$$\dot{Z} = \dot{z}A + \dot{k}_o \quad (\text{in } \$/\text{year}) \tag{6.17}$$

where $\dot{z}$ ($/m$^2$year) represents the annualized capital cost of owning and maintaining the energy-storage system and $\dot{k}_o$ ($/year) is the sum of the fixed maintenance cost and any other annual costs that apply to the storage system. $A$ is the heat transfer area of the gas side. The total annual cost rate, on the other hand, includes owning and operating the sensible heat TES system. Badar et al. (1993) defined the total annual cost rate as

$$\dot{\Gamma} = \dot{Z} + \lambda_P T_o \dot{S}_{gen,P} + \lambda_T T_o \dot{S}_{gen,T} \quad (\text{in } \$/\text{year}) \tag{6.18}$$

where the last two terms are related to the entropy generation rates due to pressure drop and heat transfer. $\lambda$ is the unit cost of lost work ($/kW-hour). Notice that in the previous illustrative example, entropy generation due to pressure drop was neglected. Equation (6.10) defines entropy generation due to heat transfer. Entropy generation due to friction loss is expressed as

$$\dot{S}_{gen,P} = \dot{m}c_p\left(\frac{R}{c_p}\right)\ln\left(1+\frac{\Delta P}{P_o}\right) \quad (\text{in kW/K}) \tag{6.19}$$

Badar et al. (1993) expressed the total annual cost rate by introducing the cost per unit overall conductance ($\gamma_{UA}$) as

$$\dot{\Gamma} = \dot{m}c_p\left(\gamma_{UA}\text{NTU} + \lambda_T T_o\left[\tau - \ln(1+\tau) + \frac{1}{\theta}\{\ln[1+\tau(1-\exp(-y\theta))] - \tau(1-\exp(-y\theta))\}\right]\right) \tag{6.20}$$

where $\gamma_{UA}$ is defined as

$$\gamma_{UA} = \left[\dot{Z} + \lambda_P T_o \dot{S}_{gen,P}\right]/UA \quad (\text{in } \$^{\circ}\text{C/kWh}) \tag{6.21}$$

(continued)

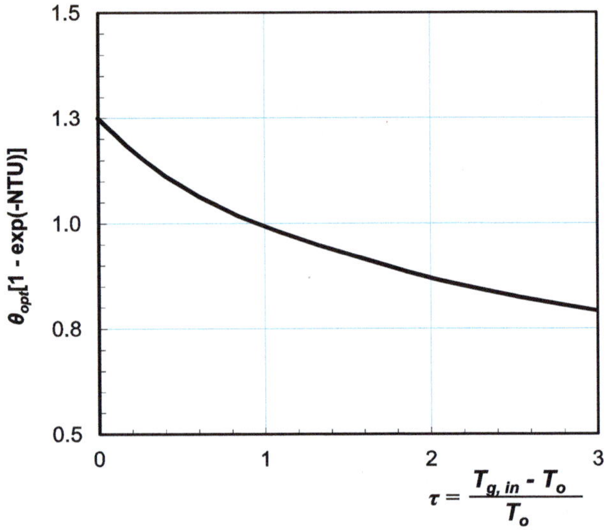


**Fig. 6.6** Optimum charging time for a sensible heat TES for minimum irreversibility cost. (Adapted from Badar et al. 1993)

Consequently, the optimum charging time of the hot gas is evaluated by numerically resolving Eq. (6.20). Figure 6.6 represents the optimum charging time to minimize the irreversibility cost. Badar et al. (1993) noted that the optimum charging time is strongly influenced by NTU and the effect of dimensionless temperature difference is quite weak on $\theta_{opt}$.

**Illustrative Example 3: Optimization of the Discharging Period of a Sensible Heat TES System**

Dincer and Rosen (2001) presented a mathematical model to predict the discharge efficiency of a thoroughly mixed sensible heat TES system. The model was initially developed by Gunnewiek et al. (1993) and includes both energetic and exergetic aspects. The schematic of the sensible heat TES is illustrated in Fig. 6.7. The system involves a storage tank, heat exchanger, and pumps. During discharge, the storage medium circulates in the primary loop and transfers thermal energy to the working fluid. The inlet and outlet temperatures of the storage medium to the heat exchanger are represented by $T_{s,i}$ and $T_{s,o}$, respectively. For the working fluid, the inlet and outlet temperatures are represented by $T_{w,i}$ and $T_{w,o}$, respectively. It is assumed that the outlet temperature of the tank ($T_{s,i}$) is equal to the tank temperature, i.e., $T_{s,i} = T_s$.

(continued)

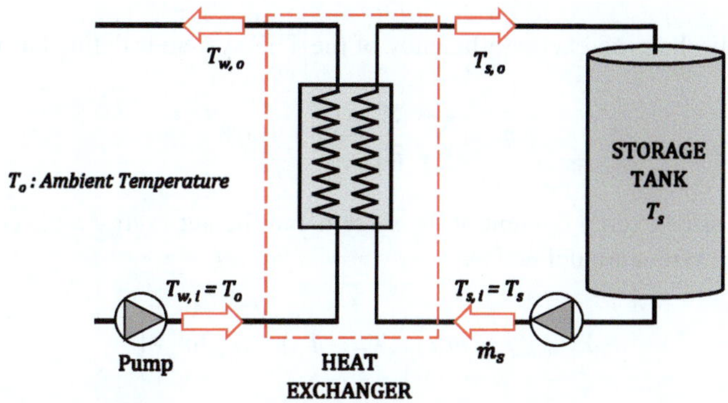


**Fig. 6.7** Sensible heat TES – discharging period. (Adapted from Dincer and Rosen 2001)

The balance equations for the storage tank and heat exchanger yield the following expression for the new fluid temperature within the storage tank:

$$T_s^{new} = T_s - \frac{C_{\min}\varepsilon\Delta t}{m_s c_s}(T_s - T_o) - \frac{UA\Delta t}{m_s c_s}(T_s - T_o) \tag{6.22}$$

where $T_o$ is the inlet temperature of the working fluid, which is also equal to the ambient temperature. $C_{min} = \dot{m}\, c$ is the minimum heat capacity rate. $\varepsilon$ is the effectiveness of the heat exchanger. $\Delta t$ is the time step size. $m_s$ and $c_s$, on the other hand, represent the mass and specific heat of the storage medium. The last term on the right-hand side corresponds to the heat exchange between the tank and the ambient. $U$ is the overall heat transfer coefficient, and $A$ is the outer-surface area of the tank. For an adiabatic storage tank, $UA = 0$.

The discharging energy efficiency of the TES system is defined as follows:

$$\eta = \sum_{i=1}^{n} \frac{E_{net}(i)}{E_{s,initial}} \times 100\% \tag{6.23}$$

The initial energy content of the system and the net energy recovered from the TES system are defined as

$$E_{s,initial} = m_s c_s (T_s - T_o) \tag{6.24}$$

$$E_{net}(i) = Q(i) - \dot{W}\,\Delta t \tag{6.25}$$

where $Q(i)$ is the recovered heat from TES and can be written as

$$Q(i) = \dot{m}_s c_s \Delta t(i)\left(T_s^{new}(i) - T_{s,o}(i)\right) \tag{6.26}$$

(continued)

The discharging exergy efficiency of the TES system is defined as follows:

$$\psi = \sum_{i=1}^{n} \frac{\Xi_{net}(i)}{\Xi_{s,initial}} \times 100\% \tag{6.27}$$

The initial exergy content of the system and the net exergy recovered from the TES system are defined as

$$\Xi_{s,initial} = m_s c_s \left[ (T_s - T_o) - T_o \ln \frac{T_s}{T_o} \right] \tag{6.28}$$

$$\Xi_{net}(i) = \Xi(i) - \dot{W}\,\Delta t \tag{6.29}$$

where $\Xi(i)$ is the recovered heat from TES and can be written as

$$\Xi(i) = \dot{m}_s c_s \Delta t(i) \left[ \left( T_s^{new}(i) - T_{s,o}(i) \right) - T_o \ln \frac{T_s^{new}(i)}{T_{s,o}(i)} \right] \tag{6.30}$$

Dincer and Rosen (2011) considered a fully mixed sensible heat TES. Water is used as the storage medium. The mass of the water is $m = 10{,}000$ kg. The specific heat of water is assumed to be constant as $c_s = 4.18$ kJ/kgK. At the beginning of the discharge period, the water temperature is at $T_s = 353$ K. Air flows within the heat exchanger (air, with $c_w = 1.007$ kJ/kg K) with a flow rate of $\dot{m}_w = 1.2$ kg/s. The mass flow rate of water, on the other hand, is $\dot{m}_s = 0.22$ kg/s. The NTU and effectiveness of the heat exchanger are $NTU = 2.5$ and $\varepsilon = 0.7$, respectively. The inlet temperature of the air to the heat exchanger is equal to the ambient. The reference ambient temperature is $T_o = 293$ K. Dincer and Rosen (2001) defined the time step size as $\Delta t = 600$ s.

Timewise variations of energy and exergy efficiencies are evaluated for ideal and real working conditions. Ideally, the pump work and heat loss through the ambient are neglected. Besides, to simulate the real working conditions, both the pump work and heat losses are considered in the computations. Figures 6.8 and 6.9 compare the influences of pumping work and heat loss from the tank on the energy and exergy efficiencies of the sensible heat TES system during the discharging period. Two points of significance are noted in the differences between the energy and exergy efficiency curves. First, the maximum exergy and energy discharge efficiencies differ and occur at different times; for example, for the nonadiabatic TES, the maximum exergy efficiency (28.7%) occurs at 13.5 h, while the maximum energy efficiency (72.1%) occurs at 57.3 h. Secondly, the net exergy recovered from a TES becomes negative (and consequently, the exergy discharge efficiency becomes

(continued)

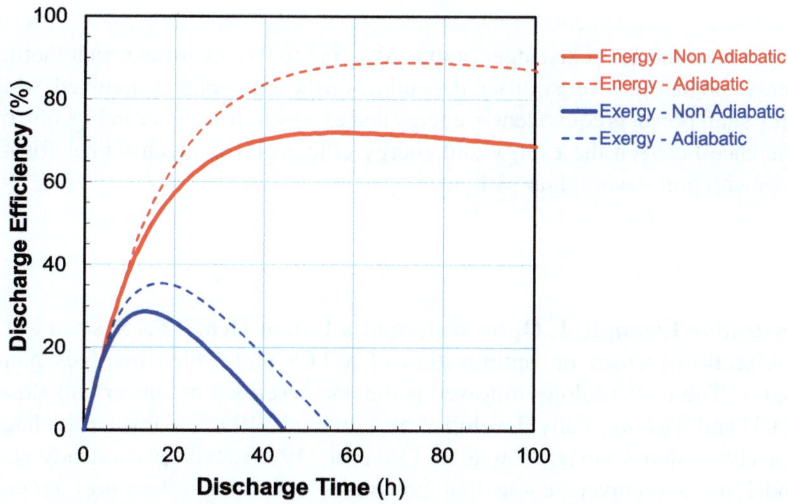


**Fig. 6.8** Evolution of energy and exergy efficiencies for adiabatic and nonadiabatic storage tanks (Adapted from Dincer and Rosen 2001)

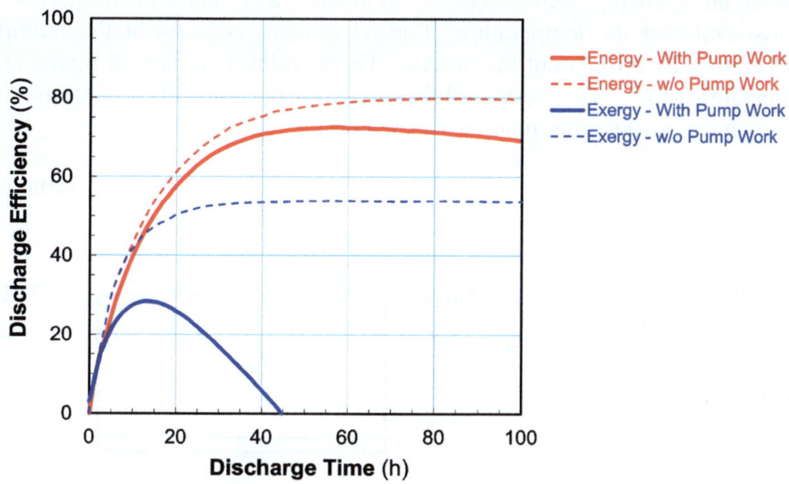


**Fig. 6.9** Evolution of energy and exergy efficiencies with and without pump work (Adapted from Dincer and Rosen 2001)

negative) before the maximum energy discharge efficiency is attained. If pump shaft power is considered negligible, the maximum discharge efficiencies are higher for both energy and exergy analyses, relative to the case in which the pump shaft power is considered nonzero. Also, with negligible pump shaft power, the maximum net energy recovery is not diminished by continued

(continued)

operation of the heat-exchanger pump. Also, Fig. 6.9 demonstrates that thermal energy and thermal exergy differ, depending on the temperatures involved, while pump shaft power is equivalent in energy and exergy terms. Hence in Fig. 6.9, the difference between the exergy and energy efficiencies is much higher for the cases with pump work than without.

**Illustrative Example 4: Optimization of a Latent Heat TES System**
This section focuses on optimization of a TES unit which involves phase change. The methodology followed is the one described in Dincer and Rosen (2001) and was originally developed by Lim et al. (1992). Consider the latent heat TES system shown in Fig. 6.10. Lim et al. (1992) developed a steady-state model for a complete cycle that comprises a melting (*charging*) process followed by a solidification (*discharging*) process. The hot working fluid enters the storage tank with an inlet temperature of $T_{in}$. The heat transfer surface area and overall heat transfer coefficient between the working fluid and the phase change material (PCM) are defined as $U$ and $A_s$, respectively. It is assumed that the temperature of PCM remains constant at the melting temperature $T_m$ throughout the process. The working fluid is well mixed at a temperature of $T_{out}$. The working fluid also leaves the tank at $T_{out}$. The ambient temperature, on the other hand, is defined as $T_o$.

(continued)

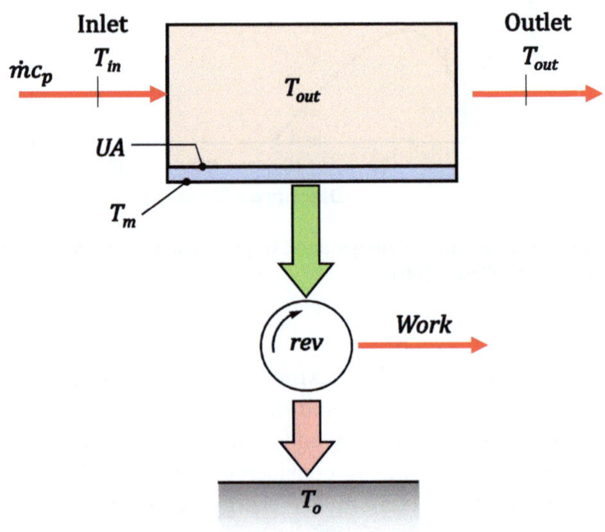


**Fig. 6.10** Power production using steady phase change and the mixed stream. (Adapted from Lim et al. 1992)

Suppose that a heat engine is considered working between the $T_m$ and $T_o$ as illustrated in Fig. 6.10. The heat transfer from the system is defined as

$$\dot{Q}_m = UA(T_{out} - T_m) \tag{6.31}$$

$$\dot{Q}_m = \dot{m}\, c_p(T_{in} - T_{out}) \tag{6.32}$$

NTU is then defined to eliminate the outlet temperature of the working fluid by combining Eqs. (6.31) and (6.32) to obtain

$$\dot{Q}_m = \dot{m}\, c_p \frac{\text{NTU}}{1 + \text{NTU}}(T_{in} - T_{out}) \tag{6.33}$$

The model of Lim et al. (1992) aims to maximize the rate of exergy or the useful work. That is, for the steady cycle, which works between $T_m$ and $T_o$, as described in Fig. 6.10, the useful work is defined as follows:

$$\dot{W} = \dot{Q}_m\left(1 - \frac{T_o}{T_m}\right) \tag{6.34}$$

Combining Eqs. (6.33) and (6.34) yields

$$\dot{W} = \dot{m}\, c_p \frac{\text{NTU}}{1 + \text{NTU}}(T_{in} - T_{out})\left(1 - \frac{T_o}{T_m}\right) \tag{6.35}$$

Lim et al. (1992) evaluated the optimal phase change temperature of the PCM (*melting/solidification*) to achieve the useful work output maximum as defined below:

$$T_{m,opt} = \sqrt{T_{in}T_o} \tag{6.36}$$

The maximum power output that could be achieved from the latent heat TES unit is then evaluated as follows:

$$\dot{W}_{\max} = \dot{m}\, c_p T_{in} \frac{\text{NTU}}{1 + \text{NTU}}\left[1 - \left(\frac{T_o}{T_m}\right)^{1/2}\right]^{-2} \tag{6.37}$$

Bejan and his colleagues further extended this basic model to include temperature distribution within the liquid phase and the superheating. De Lucia and Bejan (1991) investigated the superheating of liquid in an actual melting heat transfer problem. The dimensionless Stefan number defines the degree of liquid superheating:

(continued)

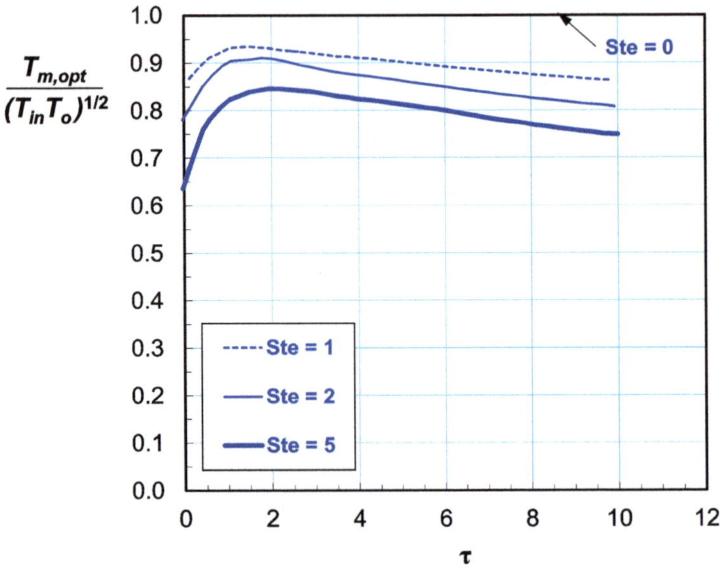


**Fig. 6.11** Influence of Ste number and melting duration on the optimum melting temperature (NTU = 1 and $T_{in}/T_o = 2$). (Adapted from De Lucia and Bejan 1991)

$$\text{Ste} = \frac{c_p(T_{in} - T_m)}{h_{sf}} \tag{6.38}$$

Notice that for small Ste numbers, Ste <1, the optimal melting temperature is the same as in Eq. (6.36). However, for Ste numbers in the order of unity, the optimum melting temperature should be defined regarding the Ste number. In Fig. 6.11, the variation in the optimum melting temperature is given depending on the Ste number and the dimensionless duration of melting ($\tau$). Here $\tau$ is defined as

$$\tau = \frac{tU^2(T_{in} - T_m)}{k\rho h_{sf}} \tag{6.39}$$

At Ste = 0, the optimum melting temperature approaches that in Eq. (6.36).

## 6.4 Optimization of a Multigeneration System with TES

The optimization techniques described so far are for individual storage systems. However, in a real-world application, TES units are mostly integrated with some other components to achieve power generation and heat supply. Such a system is called cogeneration. Recently, researchers are dealing with more complex energy conversion systems to meet several demands with a single input. Such systems are known as multigeneration (or poly-generation) systems. Multigeneration systems are designed to cover not only the electrical power but also the demands of heating, cooling, drying, and freshwater. More advanced multigeneration systems also include subsystems for hydrogen production and storage. Multigeneration systems are mostly designed as renewable energy source based and commonly integrated with solar, wind, and geothermal sources. In the case of solar-power-driven multigeneration systems, one of the most critical facts is that solar energy is not available throughout the day. Even during daytime, there may be significant variations and fluctuations. TES techniques allow to store excess energy during daytime, which maintains continuous energy supply to the system. There are several different control scenarios for multigeneration systems with TES units to achieve a better performance considering the energetic, exergetic, economic, and environmental factors. Evolutionary optimization algorithms are mostly used to identify the optimum design and working conditions for multigeneration systems. In the following illustrative example, the optimization of a multigeneration system with energy storage system is discussed.

**Illustrative Example 5: Optimization of a Multigeneration System with TES**

The current section follows the recent work of El-Emam and Dincer (2018). A novel solar-driven multigeneration system, which involves parabolic trough solar collector system, Rankine cycle, absorption cooling system, desalination unit, and electrolyzer, is proposed and investigated by considering the energetic and exergetic aspects. The schematic of the proposed system is illustrated in Fig. 6.12. Along with electrical power production, the multigeneration system produces hydrogen, provides cooling, and supplies freshwater and domestic hot water. A comprehensive thermodynamic model is built to simulate the energetic and exergetic characteristics of each component and the overall system. The operating parameters are given in Table 6.1. The energy and exergy efficiencies of the overall system are as follows:

$$\eta = \frac{\dot{W} + \dot{Q}_{cooling} + \dot{m}_{H_2} LHV_{H_2} + \left(\dot{m}\,h\right)_{DHW} + \left(\dot{m}\,h\right)_{freshwater}}{\left(\dot{Q}_u\right)_{solar}} \tag{6.40}$$

$$\Psi = \frac{\dot{W} + \dot{Q}_{cooling}\left(\frac{T_o}{T_c} - 1\right) + \dot{E}x_{H_2} + \dot{E}x_{DHW} + \dot{E}x_{freshwater}}{\dot{E}x_{solar}} \tag{6.41}$$

where $\dot{m}_{H_2} LHV_{H_2}$, $\left(\dot{m}\,h\right)_{DHW}$, and $\left(\dot{m}\,h\right)_{freshwater}$ are the rate of energy transported by the streams of hydrogen production, domestic hot water, and

(continued)

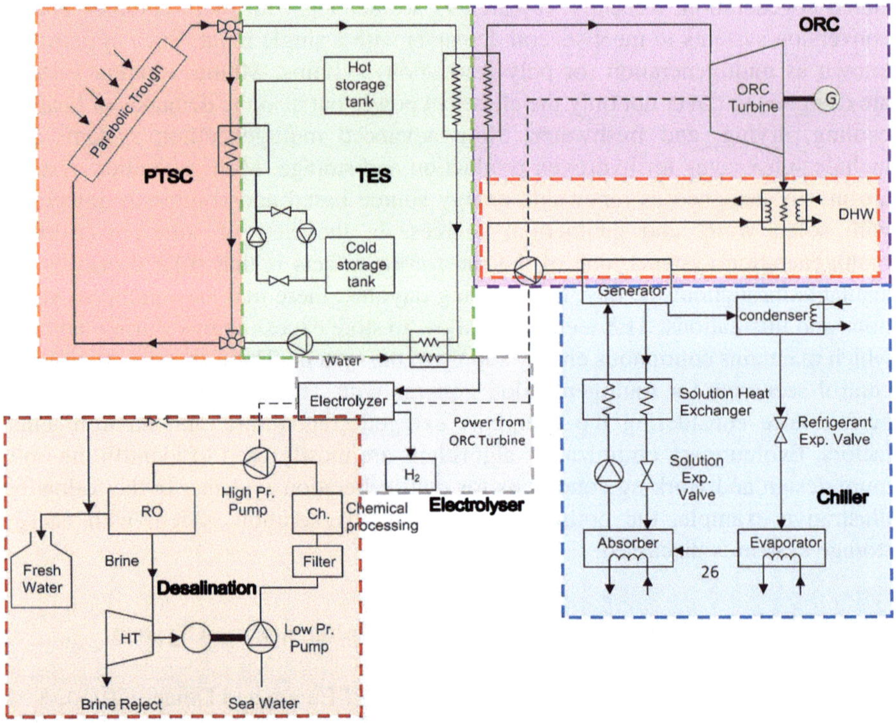


**Fig. 6.12** Schematic overview of a multigeneration system with TES units. (Adapted from El-Emam and Dincer 2018)

freshwater of the desalination system. The exergy rates of these systems are similarly defined in the exergy efficiency.

Thermo-economic-based optimization analyses conducted by El-Emam and Dincer (2018) evaluate the best operating efficiency at the condition of minimum cost. Before going through the details of the optimization algorithm, the steps of the procedure that is followed could be overviewed:

- Step 1: Identification of the system boundaries and the useful operating parameters. For each subsystem, the crucial parameters should be selected.
- Step 2: Identification of the optimization criteria. For a thermal system, these criteria can cover the energetic, exergetic, economic, and environmental factors.
- Step 3: Selection of the decision variables. The decision variables that affect the system performance and cost should be selected wisely.
- Step 4: Selection of the mathematical optimization model.

(continued)

**Table 6.1** Design and operating conditions of the system

| Parameter | Value |
|---|---|
| *Parabolic solar collector* | |
| Heat transfer fluid (HTF) | Therminol-66 |
| Width of the collector | 5.76 m |
| Length of the collector | 12.27 m |
| Solar radiation during high sun radiation | 850 W/m$^2$ |
| Solar radiation during low sun radiation | 500 W/m$^2$ |
| Storage medium | Hitec XL |
| *Organic Rankine cycle* | |
| Working fluid | n-octane |
| Turbine inlet pressure | 22 bar |
| Turbine isentropic efficiency | 85% |
| Evaporator effectiveness | 88% |
| Turbine exit pressure | 35 kPa |
| Pump inlet temperature | 85 °C |
| Pinch temperature of heat exchangers | 6 °C |
| Pump isentropic efficiency | 85% |
| *PEM electrolyzer* | |
| Operating temperature | 80 °C |
| Operating current density | 5000 A/m$^2$ |
| Anode activation energy, $E_{act,a}$ | 76 kJ/mol |
| Cathode activation energy, $E_{act,c}$ | 18 kJ/mol |
| Water content at anode membrane, $\lambda_a$ | 14 |
| Water content at cathode membrane, $\lambda_c$ | 10 |
| Anode exchange current density, $J_{o,a}$ | $1 \times 10^5$ A/m$^2$ |
| Cathode exchange current density, $J_{o,c}$ | 10 A/m$^2$ |
| *Absorption chiller system* | |
| Solution circulation ratio | 11.05 |
| Heat exchanger effectiveness | 0.7 |
| Pump isentropic efficiency | 85% |
| Weak solution concentration | 0.565 |
| Strong solution concentration | 0.62 |
| *Reverse osmosis desalination system* | |
| Salinity of product water | 450 ppm |
| Seawater salinity | 35,000 ppm |

Adapted from El-Emam and Dincer (2018)

The current multigeneration system requires multiple objectives with several constraints for its subsystems and components, that is, genetic-evolutionary algorithm is applied in the current model. Evolutionary algorithms are inspired by biological evolution mechanisms (reproduction,

(continued)

mutation, recombination, and selection). The generated optimal solution is evaluated considering evolutionary techniques of inheritance, learning, natural selection, and mutation. The process starts with a population of candidate solutions and then progresses through generations. As the process proceeds, each of the population is evaluated concerning its fitness. The predefined objective functions govern these. Multiple individuals are selected based on their fitness, and they get modified using mutation technique, generating a new population which is employed in the new iteration approaching an optimal solution.

The genetic evolutionary algorithm is applied by El-Emam and Dincer (2018) with a generation number of 150, a population of 67, and a mutation rate of 0.03. The objective functions are selected in a way to maximize the exergy efficiency and minimize the capital cost considering the capital cost of the system components based on the data provided in Table 6.2 and the correlations provided by Turton et al. (2012). The total cost rate is defined as

$$\dot{C}_{total} = \sum \dot{C}_k \tag{6.42}$$

(continued)

**Table 6.2** Cost parameters of the system components

| Component | Value |
|---|---|
| *Parabolic solar collector* | |
| HTF cost | 3 \$/kg |
| Concentrators cost rate | 320 $\$/m^2$ |
| Storage-specific cost | 22 $\$/kWh_{th}$ |
| *Organic Rankine cycle* | |
| Turbine purchase cost | $4750\ \dot{W}_T(kW)^{3/4}\$$ |
| Evaporator purchase cost | $309.14\ A_{ev}(m^2)^{0.85}\$$ |
| Pump purchase cost | $200\ \dot{W}_P(kW)^{0.65}\$$ |
| Domestic hot water heater cost | $516.62\ A_h(m^2)^{0.6}\$$ |
| *PEM electrolyzer* | |
| Specific capital cost | 1000 \$/kW |
| *Absorption chiller system* | |
| Purchase equipment cost | $1154\ \dot{Q}_{Evap}(kW)^{0.66}\$$ |
| *Reverse osmosis desalination system* | |
| Specific capital cost | $1700\$/m^3/day$ |
| Seawater salinity | 35,000 ppm |
| *Financial parameters* | |
| Interest rate, $i$ | 6% |
| Maintenance factor, $\phi$ | 1.09 |
| Equipment life span, $n$ | 15 year |
| Annual operating hours, $N$ | 7000 h |

Adapted from El-Emam and Dincer (2018)

The total cost rate is considered as a good indicator of the economic impact of operating the system. The cost rate of each of the system components, $\dot{C}_k$, is estimated based on the purchase cost of the component using the following relation:

$$\dot{C}_k = \frac{CRF \times \phi}{N} P_k \tag{6.43}$$

where $P_k$ is the purchase cost of the component $k$, $\phi$ is the maintenance factor, and $N$ is the total number of operating hours of the unit annually. $CRF$ is the capital recovery factor, which is defined as the ratio of a constant annuity to the present value of receiving that annuity for a given period. It is formulated as a function of the interest rate, $i$, and the system's total operating life in years, $n$,

$$CRF = \frac{i(1+i)^n}{(1+i)^n - 1} \tag{6.44}$$

The decision variables set for this analysis are turbine isentropic efficiency, 80–90%; turbine inlet temperature, 175–340 °C; evaporator temperature, 2–7 °C; evaporator pinch temperature, 5–30 °C; effectiveness of the solar heat exchanger, 80–90%; and pump inlet temperature set higher than 80 °C.

In Figs. 6.13, 6.14, and 6.15, the selected results of El-Emam and Dincer (2018) are represented. Figure 6.13 illustrates the influences of the turbine inlet

(continued)

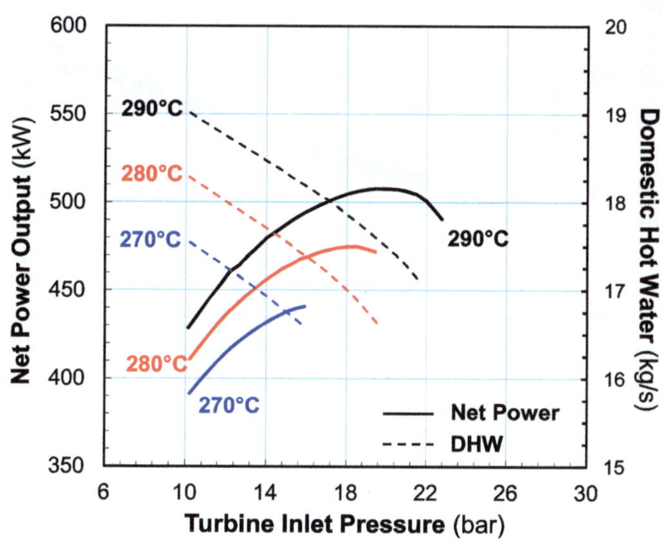


**Fig. 6.13** Influences of turbine inlet pressure and temperature on the energy and exergy efficiencies. (Adapted from El-Emam and Dincer 2018)

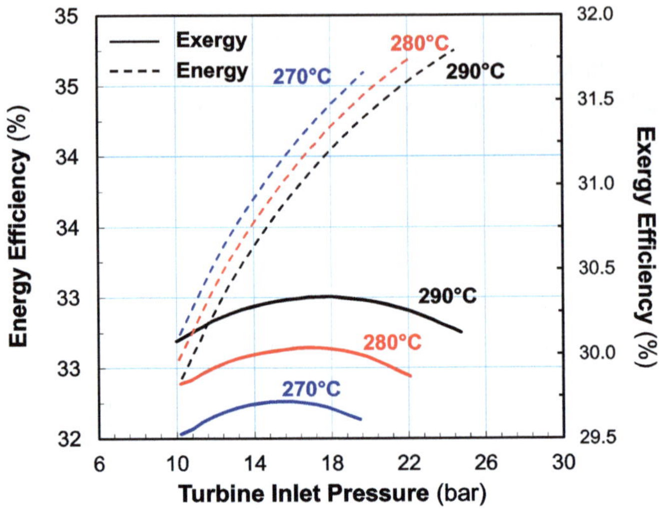


Fig. 6.14 Influences of turbine inlet pressure and temperature on the net power output and DHW. (Adapted from El-Emam and Dincer 2018)

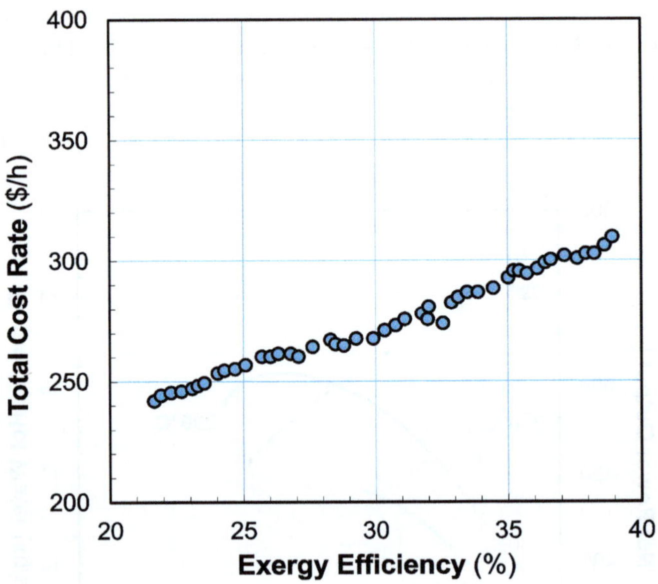


Fig. 6.15 The Pareto frontier curve for best value of the optimization results. (Adapted from El-Emam and Dincer 2018)

pressure and the superheating value of the working fluid in the ORC on the energetic and exergetic efficiencies of the overall system. Energy efficiency increases at higher turbine inlet pressures and lower inlet temperatures. The response of the exergy efficiency to the variation of the inlet pressure is nonlinear. Increasing pressure tends to improve the exergy efficiency until a critical value, and beyond this maximum, efficiency starts to reduce slightly. Also, note that increasing the superheating temperature of the working fluid improves the exergy efficiency. Figure 6.14 shows the variations of the net power output and the mass flow rate of generated domestic hot water (DHW). It is clear that increasing the pressure and temperature improves the DHW flow rate and power output. It is interesting to note that especially at higher superheating temperatures, increasing the turbine pressure reduces the power output beyond a critical pressure. The nonlinear behavior of the inlet pressure and the power output is clearly seen at 290 °C. The Pareto frontier that is obtained from the optimization analysis of the integrated multigeneration system is illustrated in Fig. 6.15. The figure depicts the relation of two objectives considered in the work of El-Emam and Dincer (2018). The relationship between the two objectives has an almost linear tendency. This curve represents all best points that resulted from the analysis. The two extreme points of the curve represent the highest and lowest limits. The highest gained efficiency is 39% with the highest cost rate of 309.56 $/h; on the other hand, the lowest exergy efficiency point is 21.7% associated with a cost rate of 241.7 $/h.

## 6.5 Optimization of a Thermal Management System with PCM

Thermal management of hybrid electrical vehicles (HEVs) has great importance to maintain adequate power output from the batteries and prevent overheating. The battery is one of the most significant components of HEVs, since any improvement or reduction in the performance of the battery will affect the range, cost, and safety of HEVs. More heat must be extracted from the cell to access higher power in the HEVs. However, this means that higher current has to be supplied to the battery, which, in turn, generates more heat (according to Ohm's law). Among the rest of the complications, high temperature has the most destructive influence on the performance of battery cells and modules, as it can lead to thermal runaway, capacity fading, limited driving range, and other safety issues (Javani et al. 2014). That is, the design of an effective thermal management system for the battery module of HEVs is of vital importance. In the following illustrative example, the optimization procedure of a PCM-integrated active cooling system of an HEV is introduced.

**Illustrative Example 6: Optimization of an Active Cooling System of an HEV with PCM**

The current illustrative example summarizes the recent work of Javani et al. (2014). The schematic of the thermal management system of an HEV is illustrated in Fig. 6.16. The system consists of two loops. One loop provides heat rejection from the battery pack to maintain the cell temperature below the critical limit, and the other loop is for air-conditioning the vehicle cabin. The thermal energy that is rejected from the battery and vehicle cabin is then extracted through ambient with a cooling system. Three working modes are studied by Javani et al. (2014) to cool down the hot working fluid that is received from the battery. Among the three alternatives, the type of cooling is determined considering the outlet temperature of the coolant. For the lower loads, the bypass loop and radiator are capable of managing the cooling. However, as the outlet temperature of the working fluid increases, the four-way valve switches the coolant to pass through the chiller and to use the compressor-driven cooling system. These two options are readily in use with a controller switch. Javani et al. (2014) proposed a new idea to decrease the coolant flow rate passing through the chiller by sending a fraction of the coolant (which is called PCM fraction by Javani et al. 2014) through the latent heat thermal energy storage system (LHTES) consisting of a shell and tube heat exchanger, filled with octadecane as the phase change material (PCM).

(continued)

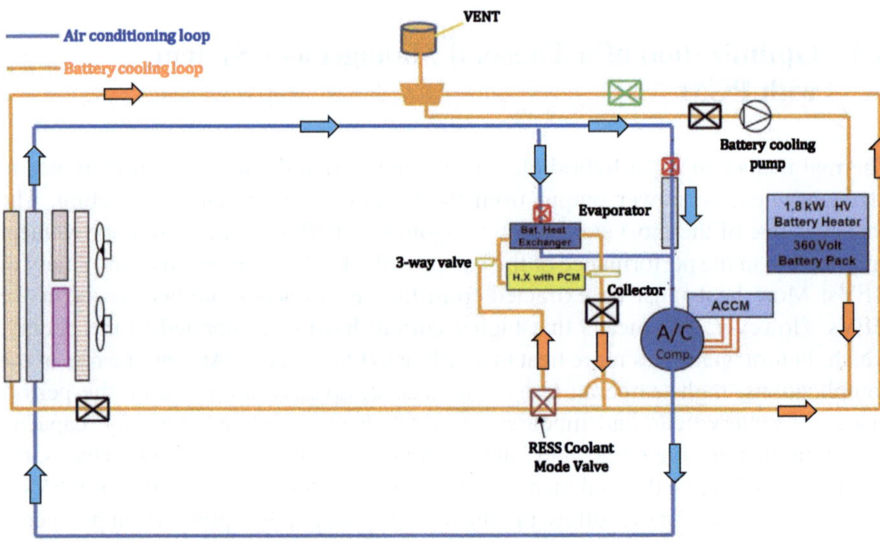


**Fig. 6.16** Schematic of a thermal management system of an HEV. (Adapted from Javani et al. 2014)

Javani et al. (2014) resolved the energy and exergy balance equations for each component to evaluate the performance of the system under various design and working conditions. Exergoeconomic analyses are also carried out to determine the optimum working conditions for the thermal management system. The primary goal of applying the thermodynamic analyses is to improve the efficiencies of the different components which are used in the cycle, where exergy analysis can be a more effective and crucial tool to achieve the goal. In any combining approach, the variation range should be considered for the variables, in a way that the resultant set of optimum solutions (which is called Pareto front in case of multi-objective optimization) are in the defined borders. Production cost and adverse environmental impact should be regulated in the variable definition domain. Specific exergy costing (SPECO) is a methodology introduced by Lazzaretto and Tsatsaronis (2006). In this way, a link between the fuel and product terms of the component and corresponding costs is formed. The cost flow rate $\dot{C}$ (in \$/h) and a cost balance equation should be written for each flow in the system. The resulting balance equation, which is required for exergoeconomic analysis, has the following general form:

$$\dot{C}_{q,k} + \sum_{i} \dot{C}_{i,k} + \dot{Z}_k = \sum_{e} \dot{C}_{e,k} + \dot{C}_{w,k} \tag{6.45}$$

where $\dot{C}_{j,k} = c_j \dot{E}x_j$. For each component, the cost balances are required to be solved. When the number of unknown cost parameters exceeds the number of equations, there should be auxiliary equations to solve the matrix of unknowns. The required auxiliary exergoeconomic equations are created using F (fuel) and P (product) in a way that the number of equations is equal to the number of unknowns. More details can be found elsewhere (Bejan 1996). The exergetic and exergoeconomic objectives concerning the decision variables and constraints are first defined before going through the optimization. A Pareto frontier will be obtained for the system, and based on a linear programming technique, a single desirable optimal solution will be selected for the multidimensional analysis of the preference decision-making processes (LINMAP) as given in Fig. 6.17.

There can be different variables in a problem. A relevant set of the parameters should be selected as the decision variables to obtain a desirable multi-objective optimized set of solutions. Javani et al. (2014) considered a thermal management system for a hybrid electric vehicle with a vapor-compression cycle. Hence, the essential system variables that are selected in the optimization survey are evaporator saturation temperature and air mass flow rate, the condenser saturation temperature, the magnitude of superheating in the evaporator, and the magnitude of subcooling in the condenser. Also, considering the importance of the compressor, as the main component to consume the energy, compressor efficiency is also considered as another

(continued)

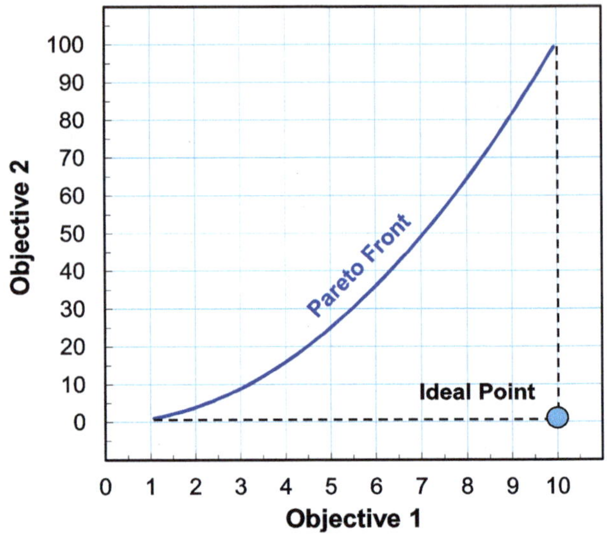


**Fig. 6.17** Pareto frontier for maximizing objective 1 and minimizing objective 2. (Adapted from Javani et al. 2014)

**Table 6.3** Ranges of each parameter

| Parameter | Range of variation |
|---|---|
| Compressor speed (rev/min) | 1500–5000 |
| Compression ratio | 1–5 |
| Evaporating air temperature (°C) | 0–15 |
| Superheating temperature (°C) | 0–12 |
| Evaporator air mass flow rate (kg/s) | 0.1–0.5 |
| Cooling capacity (kW) | 1–5 |
| Condensing air temperature (°C) | 40–55 |
| Condenser air mass flow rate (kg/s) | 0.1–0.5 |
| Subcooling temperature (°C) | 0–12 |

Adapted from Javani et al. (2014)

variable. Once the critical variables are selected, the appropriate ranges of their change should be considered in a way that feasibility, commercial availability, and engineering constraints of the thermal management system are satisfied. The lower and upper limits for the considered variables are listed in Table 6.3, where each interval for the decision-making variables satisfies the limitations mentioned above, such as the feasibility of the practical aspects of a system's operating conditions.

At first, balance equations are coded in Engineering Equation Solver (EES) to carry out a baseline analysis. Figure 6.18 illustrates the exergy efficiencies of each component and the overall system. Notice that the heat exchangers have the lowest exergy efficiencies. This is due to the phase change that takes place within the evaporator and the condenser. Entropy generation associated

(continued)

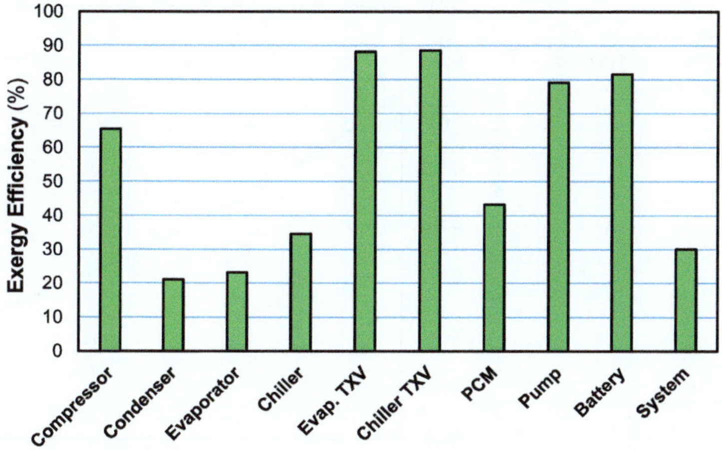


**Fig. 6.18** Exergy efficiencies of each component and the system. (Adapted from Javani et al. 2014)

with the phase change at a constant temperature causes irreversibilities and significantly reduces the exergy efficiency. An in-depth discussion on entropy generation minimization options is introduced by Javani et al. (2014).

An option to reduce the entropy generation associated with the compressor is to reduce the compressor work. Javani et al. (2014) varied the PCM mass fraction, which is the fraction of the coolant passing through the LHTES tank, to evaluate the influence on the compressor work. Figure 6.19 depicts that the PCM mass fraction significantly reduces the compressor work. Increasing the PCM mass fraction from 0.6 to 0.9 reduces the compressor work almost by 9%.

The genetic algorithm optimization technique is conducted for 500 generations, using a search population size of M = 100 individuals, crossover probability of pc = 0.90, gene mutation probability of pm = 0.03, and controlled elitism value c = 0.60. To solve the multi-objective optimization, a fast non-dominating sorting genetic algorithm (NSGA-II) is implemented. The results of optimum exergy efficiency and total cost rate for all points evaluated over 500 generations are depicted in Fig. 6.20. The Pareto-optimal front (best rank) is visible in the lower part, which is separately shown in Fig. 6.21.

A precise approximation of the Pareto front is visible on the lower part of the curve. The Pareto optimum results reveal the conflict between two objectives, the exergy efficiency, and the total cost rate. Any change that increases the exergy efficiency leads to an increase in the total cost rate and vice versa. This shows the need for multi-objective optimization technique in optimal design of cycle. The maximum exergy efficiency exists at design point A (0.3479), while the total cost rate is the biggest at this point. On the other hand,

(continued)

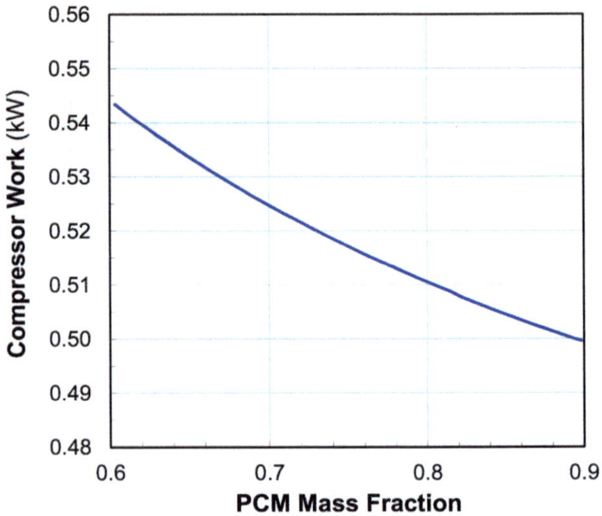


**Fig. 6.19** Influence of PCM mass fraction on the compressor work. (Adapted from Javani et al. 2014)

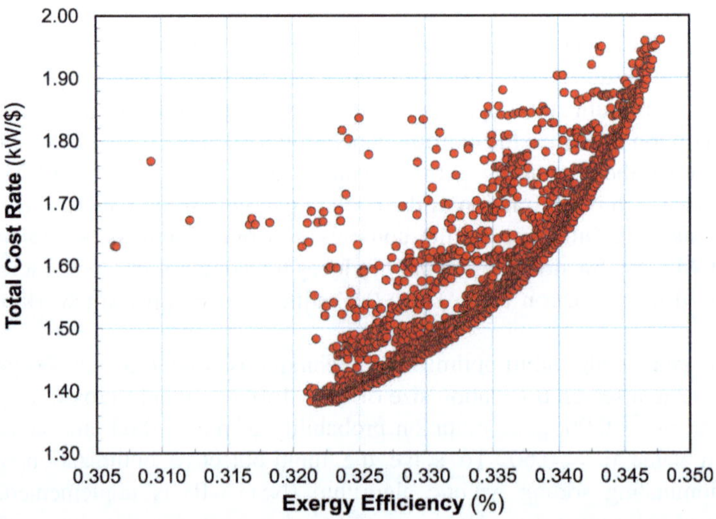


**Fig. 6.20** Results of all evaluations during 500 generations using NSGA-II. (Adapted from Javani et al. 2014)

the minimum total cost rate occurs at design point C (1.377 $/h), with the smallest exergy efficiency value (0.3217) at that point. Design point A is the optimal situation in which exergy efficiency is a single-objective function, while design point C is the optimum condition at which the total cost rate is a single-objective function. Based on the procedure presented in the final optimum solution, optimum design point B is selected as the final optimum design using the LINMAP method.

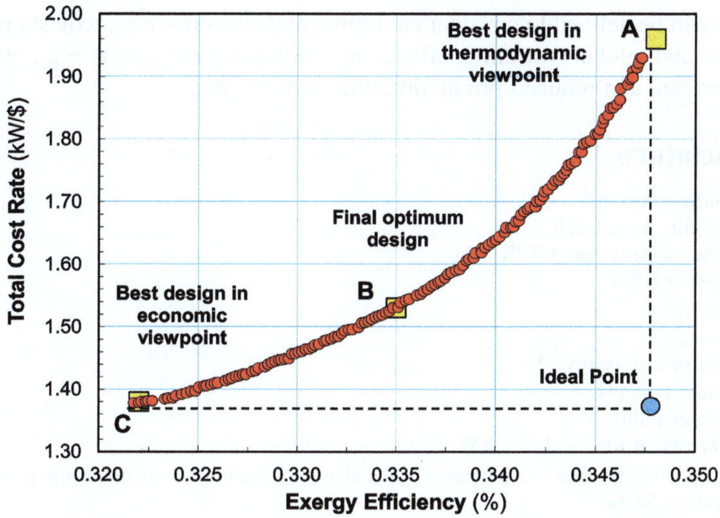


**Fig. 6.21** Distribution of Pareto-optimal points solutions using NSGA-II. (Adapted from Javani et al. 2014)

## 6.6 Closing Remarks

The current chapter presents the optimization studies that are conducted for stand-alone and integrated TES systems. The chapter includes a total of six illustrative examples. The first three examples deal with the second law-based optimization of sensible and latent heat TES systems. One should note that before going through an optimization study, a mathematical model of the system should be built by considering the fundamental balance equations. For stand-alone TES systems, first, reduced heat transfer models are developed to predict timewise variation in the temperatures of the system and at the inlet/outlet sections. Balance equations and efficiencies (either energy or exergy) are then defined to reduce the outputs of the system and end up with objective functions. Objective functions may be determined by considering energetic, exergetic, environmental, or economic aspects. More complex algorithms, such as evolutionary, should be followed to conduct a multi-objective optimization. Notice that the second-law-based optimization of stand-alone TES unit provides a better understanding of the thermal behavior of the system, constraints, and available working ranges in which the system performs efficiently. However, in engineering applications, TES units are mostly integrated into a system as a component to work as thermal accumulators. In the last two illustrative examples, optimization analyses of TES-integrated systems are introduced. Results show that the second-law-based multi-objective optimization technique enables selection of the optimum design condition for a thermal system to achieve both better thermodynamic and economic performance. Further objective functions can be defined to consider environmental concerns, e.g., reduced pollutant emissions and global warming affect, and technological goals, e.g., improved production rate and reduced production time and weight.

## Nomenclature

| | |
|---|---|
| $A$ | Surface area, $m^2$ |
| $c$ | Specific heat, J/kgK |
| $C$ | Heat capacity rate, kW/K |
| $\dot{C}$ | Cost rate, $/h |
| $E$ | Energy, J |
| $Ex$ | Exergy, J |
| $\dot{E}x$ | The rate of exergy, W |
| $i$ | Interest rate, % |
| $I$ | Irreversibility, J |
| $\dot{I}$ | The rate of irreversibility, kW |
| $\dot{k}_o$ | The sum of the fixed maintenance cost and any other annual costs that apply to the storage system, $/year |
| $L$ | Length of the tube, m |
| $m$ | Mass, kg |
| $\dot{m}$ | Mass flow rate, kg/s |
| $n$ | Number of years, years |
| $N$ | Total number of operating hours, hours |
| $N_s$ | Dimensionless entropy generation number |
| NTU | Number of transfer units |
| $P$ | Pressure (Pa) or perimeter (m) |
| $q$ | Heat transfer rate, W |
| $Q$ | Total heat transfer, J |
| $\dot{Q}$ | Heat transfer rate, W |
| $R$ | Ideal gas constant, J/kgK |
| $S$ | Total entropy, J/K |
| $S_{gen}$ | Total entropy generation, J/K |
| $\dot{S}_{gen}$ | The rate of entropy generation, kW/K |
| Ste | Stefan number |
| $T$ | Time, s |
| $T$ | Temperature, K or °C |
| $U$ | Overall heat transfer coefficient, $W/m^2K$ |
| $\dot{W}$ | Work, kW |
| $x$ | Cartesian coordinates, m |
| $y$ | Dimensionless parameter $=1 - \exp(-\text{NTU})$ |
| $\dot{Z}$ | The annualized total capital cost of owning the sensible heat TES system ($/year) |

## Greek Letters

| | |
|---|---|
| $\varepsilon$ | The effectiveness of the heat exchanger |
| $\Delta$ | Difference |
| $\eta$ | Energy efficiency |
| $\phi$ | Maintenance factor |
| $\Xi$ | Exergy, kJ |
| $\lambda$ | The unit cost of lost work ($/kW-hour) |
| $\gamma_{UA}$ | Cost per unit overall heat conductance ($°C/kW-hour) |

| | |
|---|---|
| $\dot{\Gamma}$ | Total annual cost of owning and operating the storage system ($/year) |
| $\rho$ | Density, kg/m$^3$ |
| $\theta$ | Dimensionless charging time |
| $\tau$ | Characteristic temperature difference |
| $\psi$ | Exergy efficiency |

## Subscripts

| | |
|---|---|
| *g* | Gas |
| *i* | Initial or inner |
| *in* | Inlet |
| *m* | Melting |
| *min* | Minimum |
| *o* | Dead state |
| *opt* | Optimum |
| *out* | Outlet |
| *P* | Pressure |
| *s* | Surface or storage |
| *sf* | Solid to liquid |
| *T* | Temperature |
| *u* | Useful |
| *w* | Working fluid |

## Abbreviations

| | |
|---|---|
| *ANN* | Artificial neural network |
| *CRF* | Capital recovery factor |
| *GA* | Genetic algorithm |
| *HEV* | Hybrid electric vehicle |
| *LINMAP* | Linear programming technique for multidimensional analysis of preference |
| *LHV* | Lower heating value |
| *TXV* | Thermal expansion valve |

## References

Badar, M. A., Zubair, S. M., & Al-Farayedhi, A. A. (1993). Second-law-based thermoeconomic optimization of a sensible heat thermal energy storage system. *Energy, 18*(6), 641–649.

Bejan, A. (1978). Two thermodynamic optima in the design of sensible heat units for energy storage. *Journal of Heat Transfer, 100*(4), 708–712.

Bejan, A. (1996). Entropy generation minimization: The new thermodynamics of finite-size devices and finite-time processes. *Journal of Applied Physics, 79*(3), 1191–1218.

De Lucia, M., & Bejan, A. (1991). Thermodynamics of phase-change energy storage: The effects of liquid superheating during melting, and irreversibility during solidification. *Journal of Solar Energy Engineering, 113*(1), 2–10.

Dincer, I., & Rosen, M. (2001). *Thermal energy storage: Systems and applications*. Chichester: Wiley.

Dincer, I., & Rosen, M. (2011). *Thermal energy storage: Systems and applications* (2nd ed.). Hoboken: Wiley.

Dincer, I., Rosen, M., & Ahmadi, P. (2017). *Optimization of energy systems*. Chichester: Wiley.

El-Emam, R. S., & Dincer, I. (2018). Investigation and assessment of a novel solar-driven integrated energy system. *Energy Conversion and Management, 158*, 246–255.

Gunnewiek, L. H., Nguyen, S., & Rosen, M. A. (1993). Evaluation of the optimum discharge period for closed thermal energy storages using energy and exergy analyses. *Solar Energy, 51*(1), 39–43.

Javani, N., Dincer, I., Naterer, G. F., & Yilbas, B. S. (2014). Exergy analysis and optimization of a thermal management system with phase change material for hybrid electric vehicles. *Applied Thermal Engineering, 64*(1–2), 471–482.

Lazzaretto, A., & Tsatsaronis, G. (2006). SPECO: A systematic and general methodology for calculating efficiencies and costs in thermal systems. *Energy, 31*(8–9), 1257–1289.

Lim, J. S., Bejan, A., & Kim, J. H. (1992). Thermodynamic optimization of phase-change energy storage using two or more materials. *Journal of Energy Resources Technology, 114*(1), 84–90.

Taylor, M. J., & Krane, R. J. (1991). Second law optimization of a sensible heat thermal energy storage system with a distributed storage element – Part 1: Development of the analytical model. *Journal of Energy Resources Technology, 113*, 20–26.

Turton, R., Bailie, R., Whiting, W. B., Shaeiwitz, J., & Bhattacharyya, D. (2012). *Analysis, synthesis, and design of chemical processes* (4th ed.). Upper Saddle River: Pearson.

# Chapter 7
# System Characterization and Case Studies

## 7.1 Introduction

This chapter introduces case studies on TES applications. The selected cases cover the characterization of storage materials and application of heat storage techniques to a wide range of TES systems in J to GJ scale regarding energetic content. Numerous case studies on the following specific subjects, but not limited to them, are illustrated in this chapter:

- Characterization of heat storage materials
- Clathrates of refrigerants as phase change material
- Heat storage materials in building elements
- Natural convection-driven phase change
- Aquifers with TES
- Greenhouse with TES
- High-temperature TES for solar thermal energy
- Passive thermal control of battery cells
- Borehole thermal energy storage

For each subject, a brief overview of the technical background of the problem is discussed, and then the steps of manufacturing, characterization, or numerical/thermodynamic modeling steps of the cases are represented with significant outcomes.

## 7.2 Characterization of Heat Storage Materials

One of the most crucial steps in the design process of a TES unit is the selection of suitable storage medium. Here, it should be noted that to determine a suitable material for a particular TES application, one should conduct in-depth analyses to



I. Dincer, M. A. Ezan, *Heat Storage: A Unique Solution For Energy Systems*, Green Energy and Technology, https://doi.org/10.1007/978-3-319-91893-8_7

achieve the thermal, mechanical, and chemical properties. Either in a sensible or a latent heat storage unit, researchers conduct a set of characterization experiments for candidate materials before building up a test setup for long-term energetic/exergetic considerations. Several sample preparation and characterization routines should be repeated to reveal influences of the amount of ingredients and methods of sample preparation and characterization, on the thermal properties of candidate materials.

In a sensible heat thermal energy storage (SHTES) unit with liquid storage material, the following thermophysical properties should be measured to assess whether the candidate material is suitable for an application:

- Density, $\rho$ (in $kg/m^3$)
- Viscosity, $\mu$ (in Pas)
- Thermal conductivity, $k$ (in W/mK)
- Specific heat capacity, $c$ (in J/kgK)

Possessing a higher *volumetric heat capacity* ($C = \rho c$ in $J/m^3K$) for a candidate material is a significant advantage since the required mass in a storage tank may be significantly reduced. Reduction in the tank volume not only provides an advantage on the cost of the storage material but also enables the usage of smaller tanks with low costs that can be used in reduced spaces in various applications. Enhanced thermal mass (or *volumetric heat capacity*) is not the only parameter that is taken into account while selecting a proper storage material in a SHTES unit. The thermal conductivity of the material ($k$) of the storage material is another characteristic that determines the dynamic response of the storage unit. Higher thermal conductivity values enhance the thermal diffusion within the storage material and reduce the time for complete charging or discharging. One should note that thermal properties may vary significantly in the working temperature range. In some instances, the density variation of the solid materials causes volume expansion and leads to physical damages for the storage tanks. It is desired to have a limited thermal expansion due to temperature variation of storage materials in the solid phase. In the case of the liquid storage medium, the variation of density with respect to the temperature initiates the natural convection within the storage tank. That is, density measurements for the liquid storage materials should be conducted in a wide temperature range to evaluate the density as a function of the temperature, otherwise it is not possible to predict the heat transfer mechanism within the storage tank adequately. For liquid storage materials, the viscosity is also another significant property that affects the thermal response of the storage tank. For instance, it may not be possible to conduct a CFD simulation for a SHTES system with a liquid material to evaluate the thermal stratification phenomena in the absence of the temperature-dependent variations of the viscosity and density.

In a latent heat thermal energy storage (LHTES) unit, on the other hand, in addition to the properties that are listed above, the following characteristics also have vital importance:

- Latent heat of fusion, $h_{sf}$ (in J/kg)
- Phase change temperature, $T_m$ or $T_s$ (in °C)

As already mentioned in previous chapters, in an LHTES, storage material undergoes a phase change process (commonly *solid* to *liquid* or vice versa), so storage material that is used in LHTES units is called phase change materials (PCMs). Like the sensible heat storage, in an LHTES unit, possessing high volumetric heat capacity and latent heat will reduce the required mass of the storage medium and investment costs. In comparison to the sensible heat storage materials, PCMs are quite complicated since the storage medium should undergo reversible melting and solidification cycles for long-term usage. The cyclic behaviors of the candidate materials are systematically observed to reveal the long-term thermal stability of PCMs. Moreover, unlike the sensible storage materials, properties of a PCM should be determined both for the liquid and solid states. Since the thermal conductivity and specific heat values vary significantly for solid and liquid phases of material, measurement of the properties as a function of temperature becomes very important for characterization of PCMs. Viscosity plays a significant role in the case of natural convection-driven phase change. For instance, consider a cylindrical capsule that is filled with solid PCM. Heating up the outer shell of the cylinder initiates inward melting. At the beginning of the process, the dominant heat transfer mechanism is conduction. Increasing the liquid volume within the capsule, however, will initiate a temperature gradient within the liquid phase of the PCM. In such a case, the buoyancy forces lead upward and downward motions for the liquid that is adjacent to the hot and cold surfaces, respectively. The influence of the natural convection on the phase change process (either solidification or melting) can only be predicted if the temperature-dependent properties are evaluated accurately.

Mechanical and chemical characteristics, on the other hand, identify stability of the material. A thorough understanding of a candidate material regarding the thermal, mechanical, or chemical could only be achieved by carrying out long-term thermal cycles. As a typical approach, thermophysical and mechanical properties are measured at the end of long-term charging and discharging cycles and compared with the baseline measurements. That is, a systematic approach should be followed to evaluate the variation of each property after the thermal cycling experiments.

In the following, commonly used characterization methods to evaluate thermophysical properties that are used in TES systems are discussed.

### 7.2.1 Density Measurement

The densities of solid samples are mostly measured by applying the classical Archimedes methods. In the case of liquids, there are several alternative approaches. There are some standard measurement methods for liquids in a specific range of viscosity. For instance, Gschwander et al. (2011) suggest the following test standards to measure the density of liquid-state PCMs as a function of temperature:

- **ASTM D1481-17**(2017): Standard Test Method for Density and Relative Density (Specific Gravity) of Viscous Materials by Lipkin Bicapillary Pycnometer

- **ISO 3675:**1998: Crude petroleum and liquid petroleum products – Laboratory determination of density – Hydrometer method
- **ISO 12185:**1996: Crude petroleum and petroleum products – Determination of density – Oscillating U-tube method

In some instances, researchers build up dedicated test setups to measure the temperature-dependent liquid densities. In the following, two case studies for temperature-dependent density measurement of liquids are reported.

***Case Study 1:* Temperature-Dependent Density Measurements of PCMs**
A common approach to measure the density of liquid substances is to use pycnometers and hydrometers. The schematic of a simple pycnometer is given in Fig. 7.1. The exact volume of the flask is determined by conducting a calibration measurement with reference material (i.e., water). The pycnometer is immersed in a constant temperature liquid bath with high accuracy to determine the temperature-dependent density of a substance. The volume of pycnometer is obtained at various temperatures in the calibration runs before measuring the density of the unknown liquid. Density is then evaluated merely from the measurements of the mass and volume of the liquid at different temperatures.

Bauer et al. (2010) measured the density of the eutectic mixture of $NaNO_3$-$KNO_3$ in solid phase via pycnometer. A commercial helium pycnometer (AccuPyc 1330, Micromeritics) was used to determine the densities of the samples at 25 °C. Salt pellets were prepared by compressing the samples with 50 N/mm$^2$. The measurements were repeated five times to ensure the accuracy of the evaluated data. The measurements revealed that the variations in the volume of the samples were less than ±0.01%. In Fig. 7.2, the measured densities are compared with the experimental data from the literature. Here, DLR represents the results of Bauer et al. (2010).

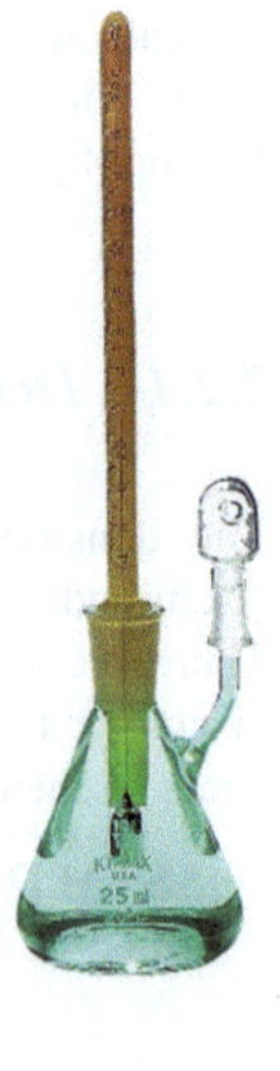

**Fig. 7.1** A schematic for the pycnometer. (Coleparmer 2017)

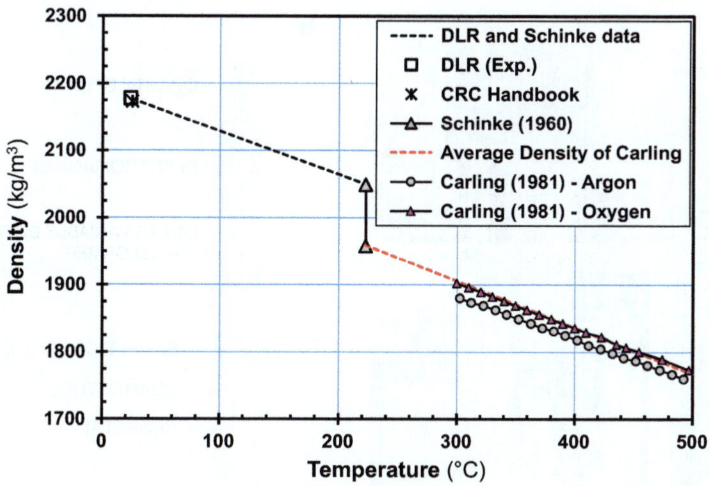


**Fig. 7.2** Density measurements of a eutectic mixture of $NaNO_3$-$KNO_3$. (Adapted from Bauer et al. 2010)

Besides, in the comprehensive report of Carling et al. (1981), the thermophysical properties of the molten salts are reported. In that report, Nissen (1982) introduces a dedicated experimental setup to measure the temperature-dependent viscosity, surface tension, and density. The photograph and cutaway view of the apparatus are shown in Fig. 7.3. It is reported that the apparatus has a function of the Archimedean densitometer. The density of the liquid sample is evaluated from the weight loss of a solid volume. Nissen (1982) evaluated variation of the density of the eutectic molten salt in the range of 300–600 °C. The results of Nissen (1982) are indicated as Carling et al. (1981) in Fig. 7.2.

Ukrainczyk et al. (2010) developed a simple measurement unit which is based on the Archimedes law to determine the density of the substance in the liquid phase as a function of the temperature. The system includes a scale, well-insulated temperature-controlled chamber, sinker, and data acquisition unit. The schematic of the setup is illustrated in Fig. 7.4. A sinker, with a known volume, is immersed within the liquid paraffin wax with a thin wire, and the chamber is slowly cooled down throughout a day. Density is then obtained as a function of the temperature from the temperature and mass data.

Ukrainczyk et al. (2010) investigated five paraffin wax samples with different melting ranges of 46–50 °C, 52–54 °C, 58–62 °C, 62–70 °C, and 70–75 °C. Figure 7.5 represents the variations of the densities of paraffin. The corresponding density functions ($\rho(T)$) are given in Table 7.1. For each sample density, the variation of density with respect to the temperature is linear, and the density of liquid paraffin decreases with increasing temperature. It is interesting to note that higher density values are obtained for paraffins with high melting temperatures. Ukrainczyk et al. (2010) stated that this is a reasonable finding since both the melting point and the density of paraffins increase as the length of the hydrocarbon chain increases.

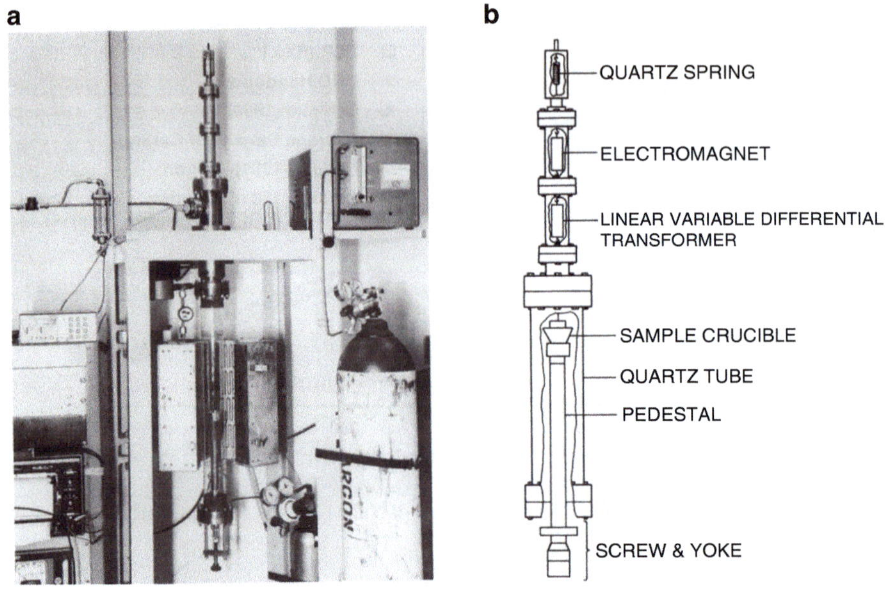


**Fig. 7.3** Experimental apparatus for measurement of viscosity, surface tension, and density. (Carling et al. 1981)

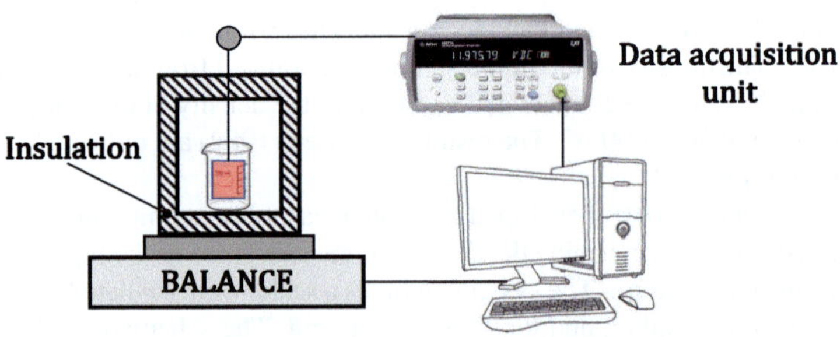


**Fig. 7.4** Experimental setup of Ukrainczyk et al. (2010). (Reproduced from Ukrainczyk et al. 2010)

Once the density is evaluated as a function of the temperature, it is possible to predict the natural convection within the liquid PCM. To do so, there are two commonly followed approaches to incorporate the buoyancy term in the momentum equations. In the first, the density of the material is defined as a function of temperature and determined for each computational node to evaluate the buoyancy forces. In the second, instead of using the density function, the slope of the density-temperature curve is simply used in the buoyancy force term. For each approach, the buoyancy forces are defined as follows:

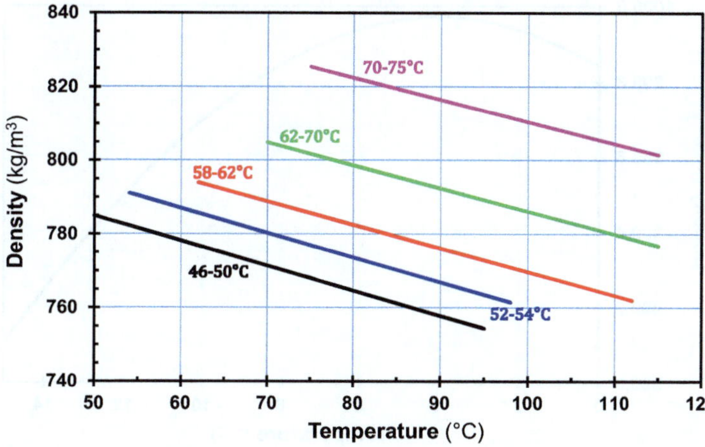


**Fig. 7.5** Density variations of paraffin waxes with respect to temperature. (Data from Ukrainczyk et al. 2010)

**Table 7.1** Density functions for paraffin waxes with different melting ranges [$\rho(T) = aT + b$]

| Melting range (°C) | $a$ | $b$ |
|---|---|---|
| 46–50 | 819 | −0.68 |
| 52–54 | 827.3 | −0.6721 |
| 58–62 | 833.36 | −0.6398 |
| 62–70 | 848.4 | −0.6231 |
| 70–75 | 870.2 | −0.5979 |

Adapted from Ukrainczyk et al. (2010)

$$F_B = -(\rho - \rho_{ref})g \quad (\mathrm{N/m^3}) \tag{7.1}$$

or

$$F_B = \rho g \beta (T - T_{ref}) \quad (\mathrm{N/m^3}) \tag{7.2}$$

where $\beta$ (in $K^{-1}$) is the *volume expansion coefficient* and defined as follows:

$$\beta = -\frac{1}{\rho}\left(\frac{\partial \rho}{\partial T}\right)_P \quad (1/\mathrm{K}) \tag{7.3}$$

For most of the liquids, the density-temperature variations have a constant slope. However, some liquids have a density maximum point with an inversion. Water, molten bismuth, antimony, and gallium are some of the widely known liquids that have density extremum in density-temperature curves. In such a case, the thermal expansion coefficient changes its sign, and it is not adequate to define a constant

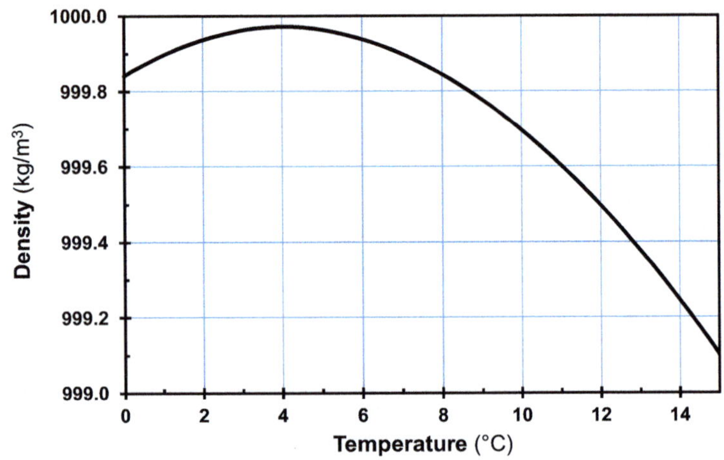


**Fig. 7.6** Density inversion of water

$\beta$ value of around the density inversion temperature. In the case of water, the density inversion point is 4 °C, and the nonlinear density variation is given in Fig. 7.6. The density of water reaches a maximum value at 4 °C. Before or after this temperature, density of water decreases. Because of this nonlinearity, convective motion in water behaves in a rather peculiar manner when the temperature encompasses the 4 °C. Several researchers have proposed different correlations for the density of water as a function of temperature. Lin and Nansteel (1987) indicated that the correlation of Gebhart and Mollendorf (1977) is widely used because of its simple form. Gebhart and Mollendorf (1977) defined a relationship between the temperature and the density as

$$\rho(T) = \rho_m(1 - \omega|T - T_m|^q) \tag{7.4}$$

where $\rho_m$ is the maximum density of water ($\rho_m = 999.972$ kg/m$^3$), $\omega = 9.297173\text{E}{-}6°\text{C}^{-q}$, $T_m = 4.0293$ °C, and $q = 1.894816$.

## 7.2.2 *Viscosity Measurement*

The viscosity of storage material, either for sensible or latent heat storage systems, is a crucial thermophysical property as it significantly influences the flow field both in natural or forced convection applications. According to a recent report by Gschwander et al. (2011), the following test standards are followed to measure the viscosity of liquid-state PCMs as a function of temperature:

- **ASTM D2534-88**(2017): Standard Test Method for Coefficient of Kinetic Friction for Wax Coatings
- **ASTM D2669-16**: Standard Test Method for Apparent Viscosity of Petroleum Waxes Compounded with Additives (Hot Melts)
- **ASTM D1986-14-(2014)**: Standard Test Method for Determining the Apparent Viscosity of Polyethylene Wax

PCMs are not only used in storage tanks to increase the thermal storage capacity of the TES units but also incorporated within the heat transfer fluids (HTFs) to improve the apparent specific heat values of HTFs. Such fluids are commonly known as "latent functional thermal fluids" (LFTF) or "PCM slurries." Dispersion of encapsulated PCM into the base fluid improves the heat storage capability of the HTF significantly. However, the viscosity, or the fluidity, of the LFTF is adversely affected by this treatment. In the following, case studies on PCM slurries are briefly reported.

***Case Study 2:*** **Viscosity Measurements of PCM Slurry**
Yamagishi et al. (1999) experimentally investigated the hydrodynamic and thermal characteristics of the slurry with *microencapsulated phase change materials* (MCPCMs). The novel slurry includes octadecane ($C_{18}H_{38}$), which is encapsulated within microcapsules with diameters in the range of 2–10 μm. Water is selected as the base liquid, and the volume fraction of the microcapsules vary from $\varphi = 0.07$ to 0.30. A lab-scale experimental unit was designed to evaluate the viscosity of the slurry. In the system, the flow rate of the slurry was set by a pump, and pressure taps monitored the pressure drop across the test section. The viscosities of slurries with different particle loadings were then evaluated from the pressure-drop data. On the other hand, the bulk properties of the slurries were evaluated from the following correlations for mixtures:

$$Cp_{bulk} = c_{PCM}Cp_{PCM} + c_{cw}Cp_{cw} + (1 - c_{cw} - c_{PCM})Cp_f \tag{7.5}$$

where $c$ is the weight fraction of each material in the slurry, and it was determined from the mass balance. Subscripts $cw$ and $f$ refer to the properties of the coating wall and the base fluid. The thermal conductivity of the slurry is calculated from Maxwell's relation:

$$k_{bulk} = k_f \frac{2k_f + k_p + 2\varphi(k_p - k_f)}{2k_f + k_p - \varphi(k_p - k_f)} \tag{7.6}$$

where $k_f$ and $k_p$ correspond to the thermal conductivity of the base fluid and the MCPCM particle, respectively. The following relation is used by Yamagishi et al. (1999) to evaluate the thermal conductivity of the particles:

**Table 7.2** Thermophysical properties of slurry PCM

| MCPCM slurry particle volume fraction | Specific heat (J/kgK) | Thermal conductivity (W/mK) | Latent heat (kJ/kg) | Viscosity (mPa.s) |
|---|---|---|---|---|
| $\varphi = 0.07$ | 4009 | 0.597 | 10.2 | 1.17 |
| $\varphi = 0.12$ | 3887 | 0.571 | 17.4 | 1.19 |
| $\varphi = 0.15$ | 3824 | 0.566 | 21.8 | 2.26 |
| $\varphi = 0.25$ | 3572 | 0.531 | 36.3 | 6.26 |
| $\varphi = 0.30$ | 3489 | 0.483 | 43.6 | 10.4 |

Adapted from Yamagishi et al. (1999)

$$\frac{1}{k_p d_p} = \frac{1}{k_{PCM} d_{PCM}} + \frac{d_p - d_{PCM}}{k_w d_p d_{PCM}} \tag{7.7}$$

The bulk properties of the slurry HTF are given in Table 7.2. It is important to note that increasing the fraction of the particles in the slurry improves the latent heat of the HTF significantly. As an instance, increasing the fraction from 0.07 to 0.30 increases the latent heat more than four times. However, the viscosity of the slurry is also influenced by the fraction of particle loading. The viscosity of slurry gradually increases as the fraction of microcapsules increased. Increasing the particle volume fraction from 0.07 to 0.30 causes ten times increment in the viscosity of the slurry. It is clear that the fraction of loading should be decided both considering the heat transfer and hydrodynamic characteristics of the PCM slurries. Yamagishi et al. (1999) also noted that the influence of MCPCM loading inside water shows different characteristics depending on the flow structure. Local heat transfer coefficient increases as the MCPCM melts for the turbulent flow condition. In contrast, in the case of laminar flow, the heat transfer performance degrades in comparison with turbulent flow.

Recently Wang et al. (2008) followed the same path of Yamagishi et al. (1999) to obtain a Nusselt correlation for the MCPCM-loaded HTF. Industrial-grade 1-bromohexadecane, with $T_m = 14.3$ °C and $h_{sf} = 160$ kJ/kg, is used as the PCM. PCM is encapsulated inside amino plastic shell material. The capsule diameters vary in the range of 2.0–31.7 μm, and the average capsule diameter is 10 μm. The following Nusselt correlation is evaluated for the turbulent slurry flow in circular tubes:

$$\begin{aligned} \mathrm{Nu}_m &= 4.8527 \times 10^{-4} \mathrm{Re}_m^{0.7733} \mathrm{Pr}_m^{2.7941} \mathrm{Ste}^{0.3159} \\ &\quad \times [(L_1 + L_2)/D]^{-0.333} (\mu_m/\mu_w)^{-2.4349} \end{aligned} \tag{7.8}$$

where $\mathrm{Re_m}$ and $\mathrm{Pr_m}$ are the average Reynolds number and Prandtl number, respectively. On the other hand, $[(L_1 + L_2)/D]$ is the phase change region length, and $\mu_m/\mu_w$ is the correction factor to account for the viscosity values of the slurry in the phase change region and the viscosity of water near the tube wall. Stefan number (Ste) is defined as follows:

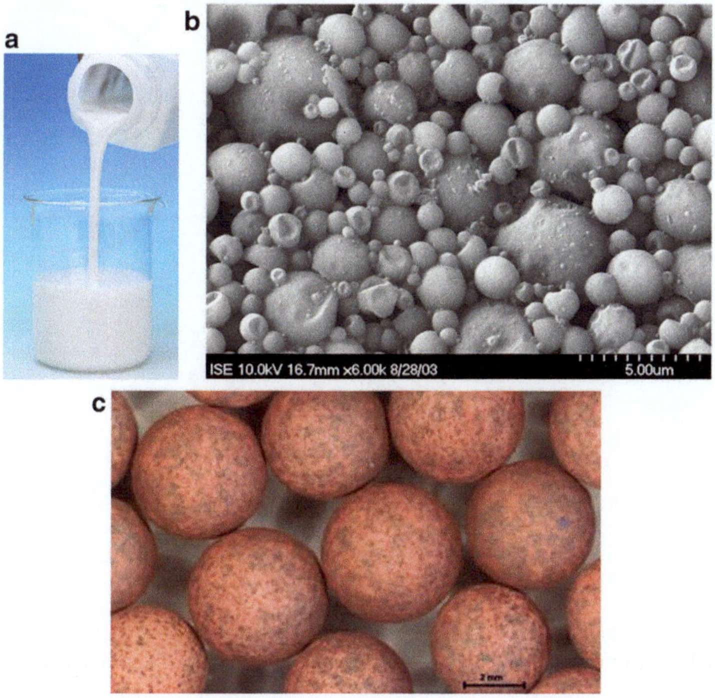


**Fig. 7.7** PCM slurries. (**a**) Photograph for PCM slurry produced by BASF (2018), (**b**) SEM image of a PCM slurry (Fraunhofer ISE 2018), (**c**) PCM-filled copper-coated spheres (Fraunhofer IFAM Dresden 2018)

$$\mathrm{Ste} = \frac{Cp_b |q_w R_d / k_b|}{w_p h_{sf,m} \rho_p / \rho_b} \tag{7.9}$$

where $R_d$ is the radius of the test tube, $q_w$ is the heat flux, and $w_p$ is the mass fraction of the particle. Figure 7.7a, b illustrate the photograph and SEM image of PCM slurries. The PCM-filled heat storage capsules are dispersed in the working fluid. Fraunhofer IFAM (2018) developed PCM-filled and copper-coated spheres, with 2–6 mm in diameter, to improve the heat transport capacities of the conventional heat transfer fluids.

### 7.2.3 *Thermal Conductivity Measurement*

One of the main drawbacks of PCMs that are commercially available with high storage capacity, i.e., fatty acids or paraffins, is that they mostly have lower thermal conductivities. Possessing lower conductivities increases the thermal resistance due to conduction and significantly extends the melting and solidification periods of

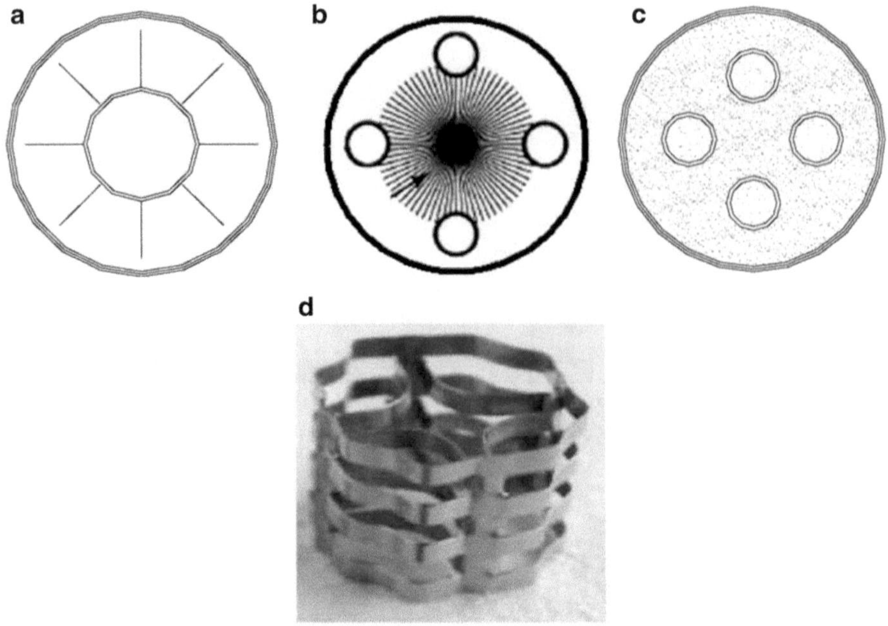


**Fig. 7.8** Alternative methods to improve heat transfer inside PCMs. (Agyenim et al. 2010)

PCMs. In many TES applications, either high or low temperature, there is a limited duration for charging or discharging periods, that is, dynamic thermal response of the PCM should be higher in such systems. In the last decades, numerous approaches have been followed by many researchers to improve the thermal conductivity of PCMs without any reduction in latent heat and cyclic thermal stability.

In the early works, the heat diffusion within the PCM domain is improved with the addition of macro-level high-conductive metals, such as copper fins or pieces. In Fig. 7.8, some macro-level heat transfer enhancement methods are overviewed. The most common approach is the use of extended surfaces (*fins*) inside the PCM domain to increase the rate of solidification or melting. Immersing a metal matrix within the PCM is also an economical and readily applicable approach. Dispersing nano-size highly conductive particles within PCMs is a recent trend to enhance thermal conductivity. Khodadadi et al. (2013) reviewed the usage of nanostructure-based additives to improve the thermal conductivities of traditional PCMs. They stated that implementation of nano-sized materials in PCM had been studied since 2005. It is reported that carbon-based nanostructures (nanofibers, nano-platelets, and graphene flakes), carbon nanotubes, metallic (Ag, Al, C/Cu, and Cu) and metal-oxide ($Al_2O_3$, CuO, MgO, and $TiO_2$) nanoparticles, and single nanowires have been studied as additives.

There is an increasing trend in the research that is investigating the effects of nano-sized additives into PCMs. Due to the high cost of nanomaterials, the maximum content of the additives is commonly dispersed inside the PCM with a low

**Table 7.3** Thermal conductivity enhancement of PCMs with nanostructures

| Base PCM | Nanomaterial | Mass/volume fraction of nanomaterial | Characterization technique | Thermal conductivity enhancements |
|---|---|---|---|---|
| Palmitic acid | Carbon nanotubes | 1 *wt%* | Short hot-wire method | 46% at 25 °C, 38% at 65 °C |
| Palmitic-stearic acid binary eutectic mixture | Carbon nanotubes | 5–8 *wt%* | Hot-disk method | In the range of 20–30% |
| Paraffin wax | Expanded graphite | 3 *wt%* | Commercial instrument | In the range of 14–24% |
| Paraffin wax | Carbon nanotubes | 2 *wt%* | Transient short hot-wire method | Solid, 35%; liquid, 45% |
| Paraffin wax | $Al_2O_3$ and carbon black | 1 *wt%* | Commercial instrument | $Al_2O_3$, NA carbon black, 24–35% |
| $BaCl_2$ | $TiO_2$ | 1.13 *vol%* | Transient hot-wire method | 12.76% |
| Paraffin wax | $Al_2O_3$ | 0–38 *wt%* | Hot-disk thermal conductivity meter | 27.1% for 16 *wt* % loading |

Adapted from Reddy et al. (2018)

fraction which is not more than 10% (Şahan et al. 2015). In the recent work of Reddy et al. (2018), the thermal conductivity enhancements that are obtained with nanostructure loadings are overviewed, and a condensed list is provided in Table 7.3. Significant improvements are achieved with the nano-enhanced PCMs (nePCMs). The increments in the conductivities reach as much as 45% for the case of carbon nanotubes. There are various methods to measure the conductivity of nePCMs. Şahan et al. (2015) listed the commonly used methods such as hot-wire, hot-strip, laser flash, thermo-reflectance, simple inverse, and 3-$\omega$. In the following, a case study on the measurement of thermal conductivities of nePCMs with 3-$\omega$ technique is given.

***Case Study 3:* Thermal Conductivity Measurement of nePCMs with 3-$\omega$ Technique**

Ezan et al. (2014) proposed graphene-loaded myristic acid (MA/Gr) as a nePCM. Myristic acid was purchased from Alfa Aesar that had a purity of 98%, and graphene was obtained from Grafen Chemical Industries Company (Ankara, Turkey). The particle size and the thickness of the graphene were 5 μm and 5–8 nm, respectively. The main steps of sample preparation are illustrated in Fig. 7.9. As a first step, myristic acid is heated inside an incubator, which is set to 60 °C, to ensure that MA ultimately turns into the liquid phase. Graphene is then added to the molten MA. In the last step, the MA/Gr mixture is subjected to an ultrasonic homogenizer (Misonix, Sonicator 3000) for approximately 2 min to procure homogeneous distribution. In order to determine the effect of graphene additive on the thermal conductivity of MA, the mass ratio of the graphene inside the nePCM is varied as 0.5%, 1%, and 2%.

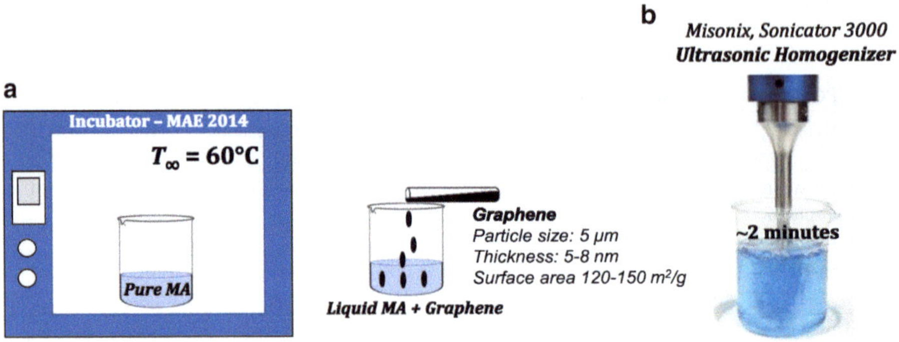


**Fig. 7.9** Preparation of MA/Gr composite nePCMs

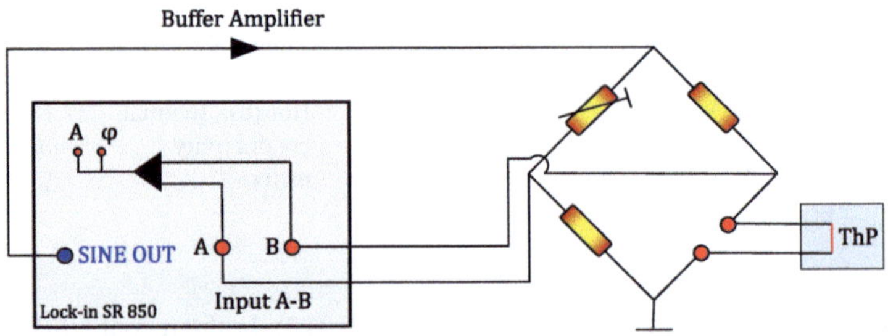


**Fig. 7.10** Schematic diagram of 3-$\omega$ experimental setup. (Ezan et al. 2014)

Thermal conductivity values of PCMs are measured for the liquid phase. Instead of using a commercial instrument, a dedicated experimental setup is designed on the basis of the 3-$\omega$ thermal conductivity measurement method. The method originates from the well-known hot-wire technique. A thin metal wire submerged into a liquid sample acts as a heater and thermometer. In the current setup, Ni wire with dimensions of 40 μm in diameter and $2\ l = 19.0$ mm long is used as a thermal probe (ThP). An alternating input current with a frequency of $\omega$ heats the wire, consequently the sample receives $2\omega$ frequency. Since the electrical conductivity of metal varies linearly with temperature, output voltage includes a 3-$\omega$ component which depends on the thermal conductivity of the sample. The first harmonic in the voltage signal is dominant and must be canceled by a Wheatstone bridge arrangement. The selection of the third harmonic from the differential signal across the bridge is performed by a Stanford SR850 lock-in amplifier tuned to this frequency (Turgut et al. 2008). In Fig. 7.10, the schematic of the 3-$\omega$ measurement system is illustrated. The method was validated with pure fluids, such as water, methanol, ethanol, and ethylene glycol, yielding accurate thermal conductivity ratios within $\pm 2\%$. The main components of the experimental setup are illustrated in Fig. 7.11. Further details about the setup and the theory of the 3-$\omega$ thermal conductivity measurement method can be found elsewhere (Turgut et al. 2008; Tavman and Turgut 2010).

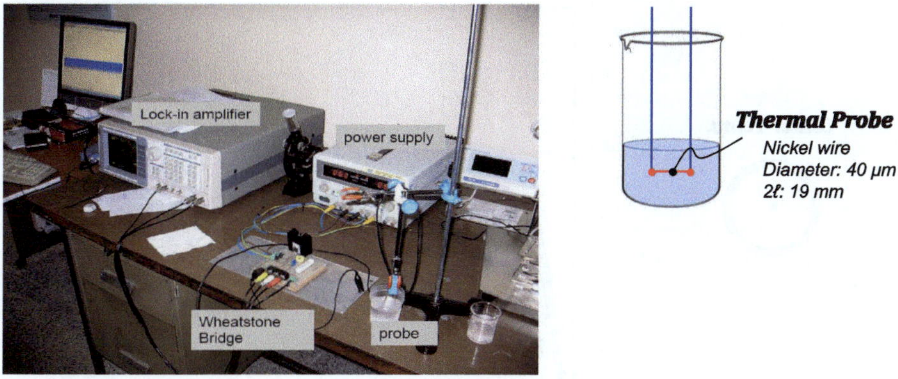


**Fig. 7.11** Main components of the 3-$\omega$ experimental setup. (Ezan et al. 2014)

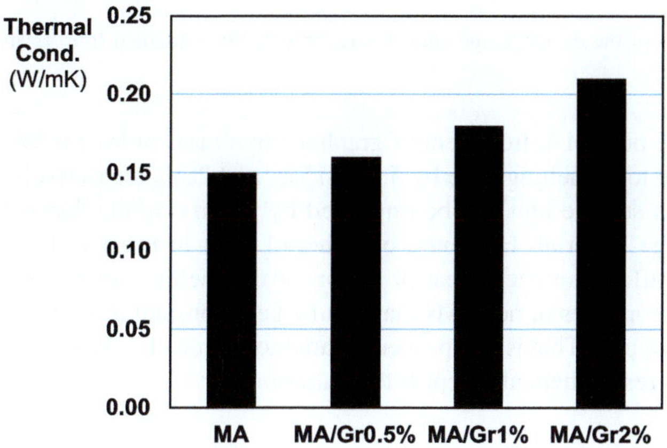


| Sample | Thermal conductivity (W/mK) |
|---|---|
| MA | 0.15 |
| MA/Gr0.5% | 0.16 |
| MA/Gr1% | 0.18 |
| MA/Gr2% | 0.21 |

**Fig. 7.12** Thermal conductivities of pure MA and MA/Gr nePCMs. (Adapted from Ezan et al. 2014)

For each sample, five sets of measurements are carried out to determine the thermal conductivity values of the pure and composite PCMs. Thermal conductivity values of pure MA, MA/Gr0.5%, MA/Gr1%, and MA/Gr2% are given in Fig. 7.12. As expected, increasing the fraction of graphene inside the composite PCM improves the thermal conductivity. As the mass fraction of graphene increases from 0.5% to 2%, the increment in the thermal conductivity value rises from 8% to 38%.

Ezan et al. (2014) also conducted a numerical simulation to reveal the influence of enhancement in the thermal conductivity on the charging performance of a storage unit. A spherical capsule that is filled with pure and composite PCMs is subjected to a specific temperature during the melting process. The evaluation of the center temperature of the sphere is then compared as in Fig. 7.13. Here, it should be noted that increasing the thermal conductivity considerably reduces the total melting

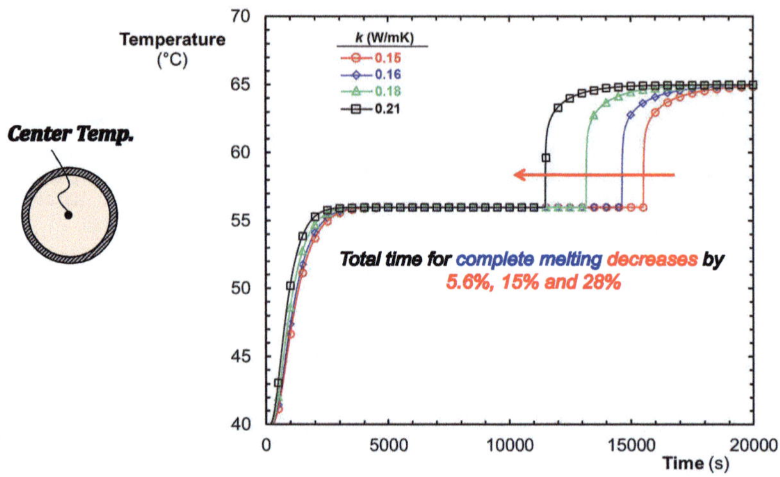


**Fig. 7.13** Timewise variation of the center temperature for various PCMs. (Adapted from Ezan et al. 2014)

time. In comparison to the pure MA, treatment of graphene in a mass ratio of 0.5%, 1%, and 2% decreases the total melting time by 5.6%, 15%, and 26%, respectively. The thermal response of a storage unit can be improved by modifying the thermal conductivity of the storage material. However, one should keep in mind that the nano-additives may also affect the latent heat of fusion or the melting time. Most importantly, the thermal properties of nePCMs may not be kept constant throughout the melting/solidification cycles. That is, the properties are monitored after long-term reversible cycles, and the researchers also report the variations.

### *7.2.4 Measurement of Specific Heat and Latent Heat of Fusion*

TES systems may involve only temperature variation (sensible heat TES) or phase change with temperature change (latent heat TES). In a sensible heat TES system, the amount of energy that is stored within the system is related to the specific heat of the matter. On the other hand, for a latent heat TES unit, in addition to the specific heat of the storage material, the latent heat of the storage medium is another critical parameter to determine the exact amount of energy that can be stored within the system. Hence, determination of the specific heat and latent heat values of the storage materials have vital importance to conduct proper analyses during the design period of a storage unit. Researchers mostly carry out one of the following analyses to determine the heat storage characteristics of the candidate materials:

- Differential scanning calorimetry (DSC)
- Temperature history methods (T-History)

#### 7.2.4.1 Differential Scanning Calorimetry (DSC)

DSC is a widely used technique to determine temperature-dependent specific heat, phase change temperature (both *melting* and *solidification*), and phase change enthalpy. DSC is a standard measurement method in thermal analysis, and there are several ready-to-use instruments from various manufacturers. DSC includes two identical furnaces as shown in Fig. 7.14. As a commonly followed procedure, an empty pan is placed into the reference furnace, and the sample is put inside the sample furnace. The sample and reference furnaces are heated (or *cooled*) with a constant rate throughout the analyses, and the variations of the heat fluxes are measured as a function of the temperature. To characterize the thermal properties of a sample, the difference in the amount of heat required to increase (or reduce) the temperature of the sample and the reference material is then directly evaluated. The reference material should not undergo any thermal or chemical transformation during the analyses. The temperatures of two furnaces are kept identical by adjusting the heat fluxes. If the sample undergoes an endothermic reaction, i.e., melting, additional energy is required to provide the same rate of change in the temperature. During the endothermic reaction, the heat flux that is supplied to the sample furnace will be higher than the reference furnace. On the contrary, in the case of an exothermic reaction, i.e., solidification, since heat is released from the sample, extra heat should be supplied to the reference pan to keep the same rate of heating or cooling.

In Fig. 7.15, a sample thermogram of DSC is illustrated. Exothermic and endothermic peaks correspond to the solidification (*crystallization*) and melting (*thawing*) of a material, respectively. Once the variation of the heat flux is evaluated as a function of temperature (or time), the area underneath the heat flux peaks are calculated to obtain the heat of fusion values. As an instance, in Fig. 7.16, an endothermic curve is represented. The shaded area corresponds to the total energy (or enthalpy, $\Delta H$) that is required to melt down the sample at a constant heating rate ($dT/dt$). In DSC analysis, the weight of the sample is recorded precisely before the experiments to evaluate the specific quantities such as specific heat or latent heat of

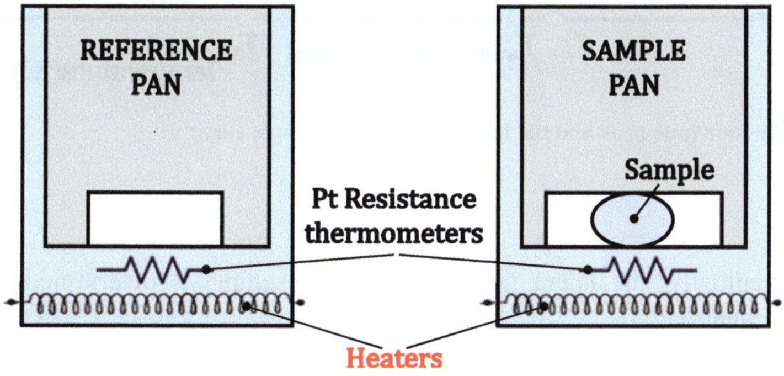


**Fig. 7.14** Section view of a DSC furnace

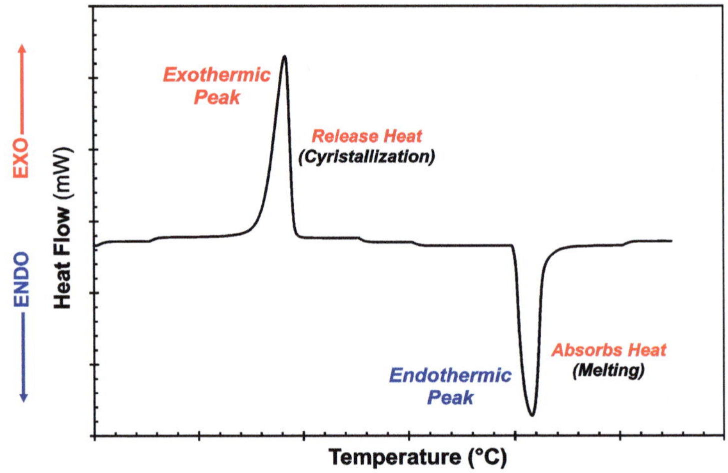


Fig. 7.15 A characteristic heat flow curve for endothermic and exothermic reactions

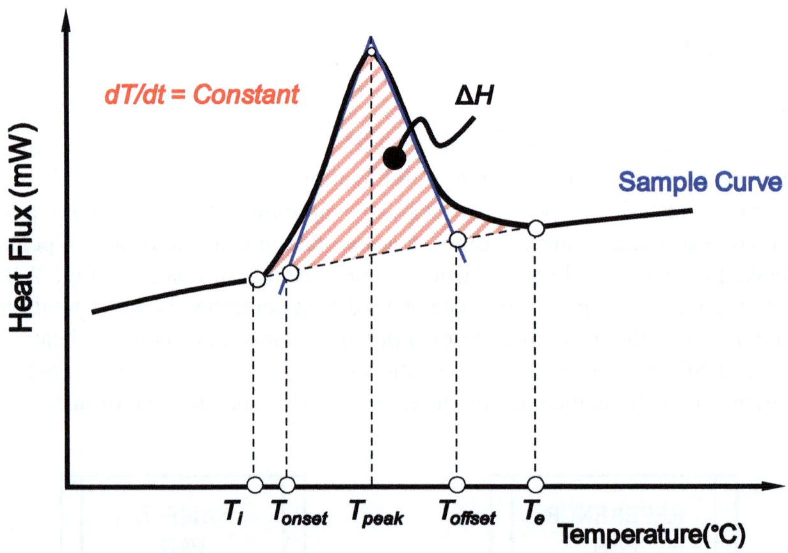


Fig. 7.16 Endothermic peak and the temperatures on a sample curve

fusion. Eventually, the latent heat value of the sample is determined from the following equation:

$$h_{sf} = \Delta H/m \quad (\text{kJ/kg}) \tag{7.10}$$

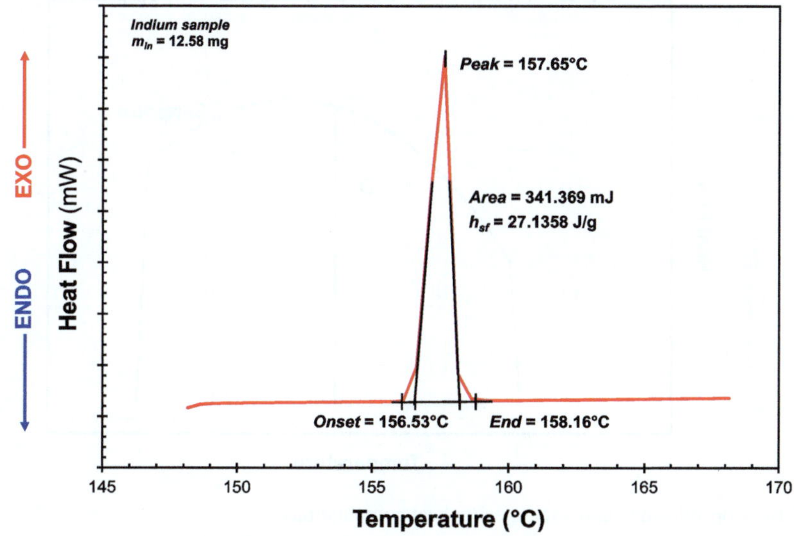


**Fig. 7.17** A DSC thermogram for calibration material (indium)

Some specific terms that are used in the characterization of a DSC analysis are also indicated in Fig. 7.16. $T_i$ is the initial peak temperature. At the initial temperature, the sample curve starts to deviate from the baseline. $T_{onset}$ is the extrapolated onset temperature. It is merely the intersection of the slopes of the ascending peak and the baseline. $T_{peak}$ is the peak temperature. At the peak temperature, the difference between the sample and the reference heat flux values is the highest. $T_{offset}$ is the extrapolated offset temperature. Offset temperature indicates the intersection of the descending peak and baseline. $T_e$ is the final peak temperature at which the sample curve intersects the baseline. The sample curve for the indium is provided in Fig. 7.17. Indium is a widely used calibration material to evaluate calibration curves for both temperature and heat flow measurements. It should be noted that the procedure that is followed to calibrate the DSC significantly affects the accuracy of the measurements. Following a standard test method is essential to achieve a reasonable result from a measurement. As an instance, ASTM E967-08 (2014) (standard test method for temperature calibration of differential scanning calorimeters and differential thermal analyzers) describes a whole routine to obtain a reliable calibration curve for a DSC.

DSC is also used to determine the specific heat of the substances as a function of the temperature for both liquid and solid states. According to the ASTM E1269-11 (2011) (standard test method for determining specific heat capacity by differential scanning calorimetry), the measurement of specific heat includes three steps: *empty pans*, *sapphire*, and the *sample*. In the first step, empty pans are placed in both furnaces to evaluate a baseline curve. In the second run, sapphire is placed into one furnace to obtain a calibration curve. The sample is then placed in the furnace and the sample curve is evaluated. Notice that in each step, the scan rate and the temperature range should be identical. Moreover, it is desired to use pans with similar weights to

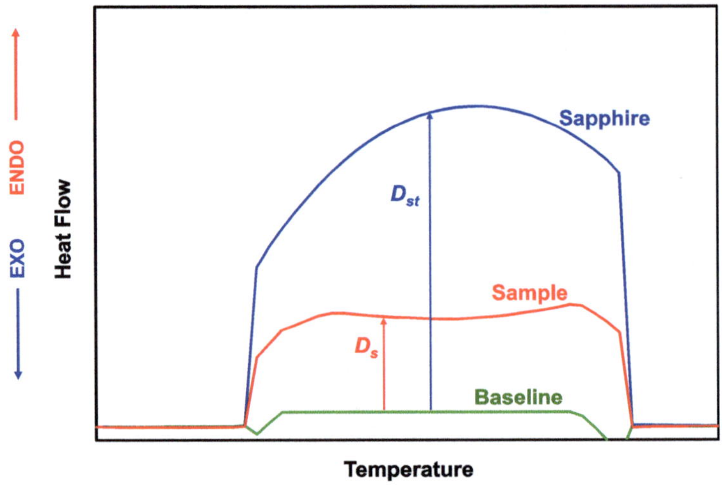


**Fig. 7.18** Consolidated curves for specific heat computations

reduce errors. Once the baseline, sapphire, and sample curves are evaluated, they are consolidated into a single chart as shown in Fig. 7.18. The calibration coefficient is then evaluated by using the following equation:

$$E(T) = [b/(60D_{st}(T))][W_{st}Cp_{st}(T) + \Delta WCp_c(T)] \tag{7.11}$$

where

$b$: Scan rate (°C/min)
$D_{st}$: Heat flow difference between the sapphire and baseline (mW)
$W_{st}$: The mass of the sapphire (mg)
$Cp_{st}$: Specific heat of sapphire (J/g°C)
$Cp_c$: Specific heat of the pan (J/g°C)
$\Delta W$: The difference between the mass of the empty pans that are used in reference furnace and sapphire

The specific heat of the sample as a function of temperature is calculated by using the calibration coefficient as follows:

$$Cp(T) = \frac{60E(T)D_s(T)}{W_s b} - \frac{\Delta WCp_c(T)}{W_s} \tag{7.12}$$

where

$D_s$: Heat flow difference between the sapphire and baseline (mW)
$W_s$: Mass of the sample (mg)
$\Delta W$: The difference between the mass of the empty pans that are used in reference furnace and sapphire

In Fig. 7.19, DSC results that are obtained for a substance are given. The results reveal that the specific heat measurements give almost identical variations before the

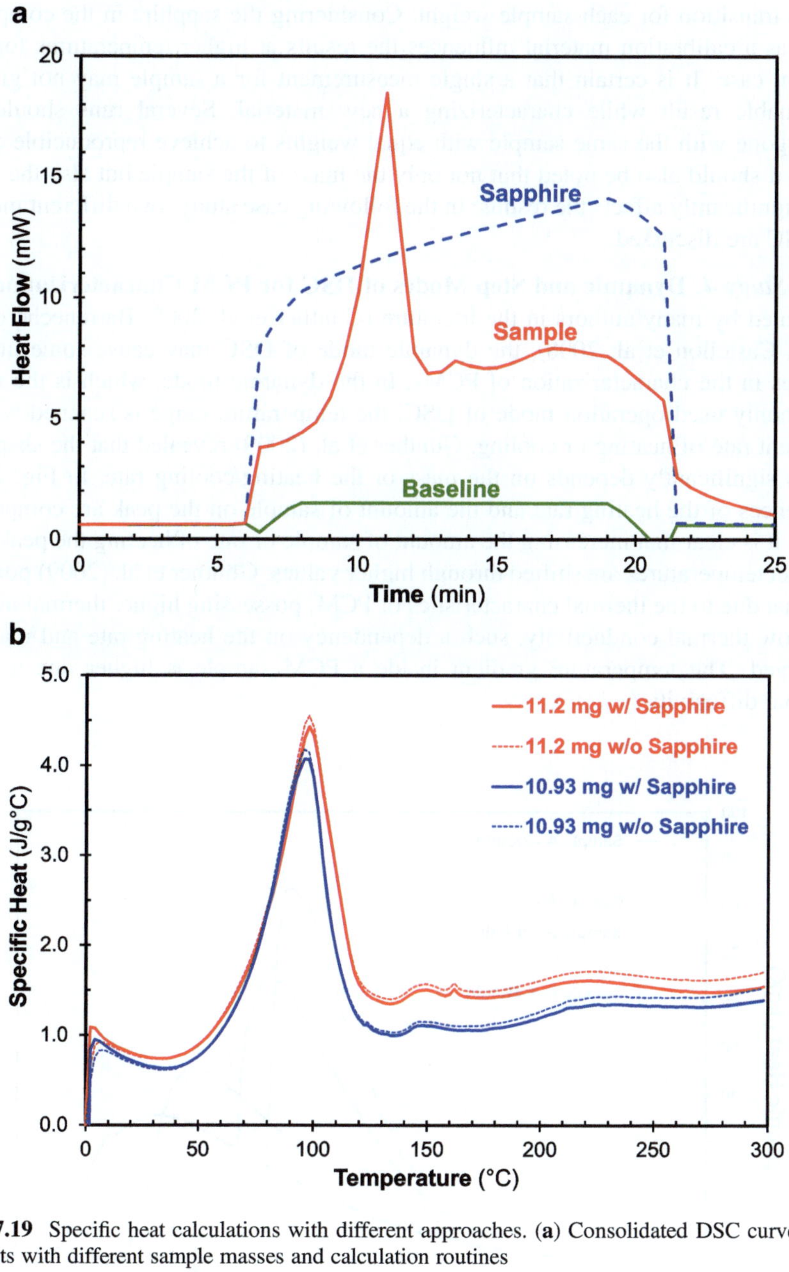


**Fig. 7.19** Specific heat calculations with different approaches. (**a**) Consolidated DSC curves (**b**) Results with different sample masses and calculation routines

phase transition for each sample weight. Considering the sapphire in the computations as a calibration material influences the results at higher temperatures for the current case. It is certain that a single measurement for a sample may not give a reasonable result while characterizing a new material. Several runs should be undergone with the same sample with equal weights to achieve reproducible data. Here, it should also be noted that not only the mass of the sample but also the scan rate significantly affects the results. In the following case study, two different modes of DSC are discussed.

***Case Study 4:* Dynamic and Step Modes of DSC for PCM Characterization**
As stated by many authors in the literature (Günther et al. 2009; Barreneche et al. 2013; Castellon et al. 2008), the dynamic mode of DSC may cause some uncertainties in the characterization of PCMs. In the dynamic mode, which is the most commonly used operation mode of DSC, the temperature range is scanned with a constant rate of heating or cooling. Günther et al. (2009) revealed that the shape of peaks significantly depends on the mass or the heating/cooling rate. In Fig. 7.20, influences of the heating rate and the amount of sample on the peak are compared. Here, it is clear that increasing the amount of sample or rate of heating the peak and end-set temperatures are shifted through higher values. Günther et al. (2009) pointed out that due to the thermal characteristics of PCM, possessing higher thermal inertia and low thermal conductivity, such a dependency on the heating rate and mass is observed. The temperature gradient inside a PCM sample is higher due to low thermal diffusivity ($=k/\rho c$).

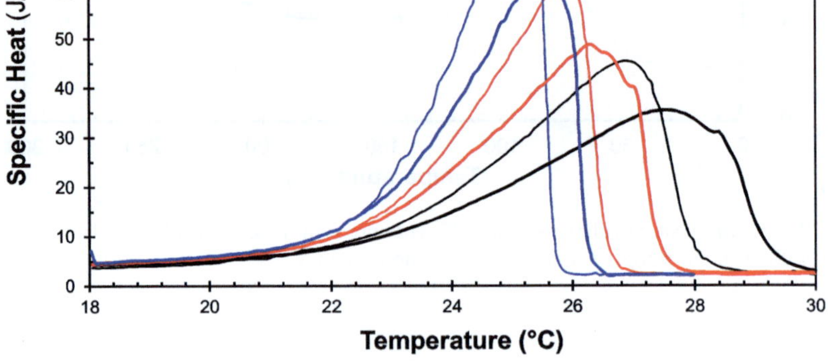


**Fig. 7.20** Influences of the scan rate and sample mass on the DSC peak for a PCM. (Adapted from Günther et al. 2009)

Günther et al. (2009) also investigated the isothermal step mode to regulate the uncertainties that may arise due to thermal nonuniformities during the transient heat transfer inside the PCM sample. In the isothermal step mode, instead of a constant rate of heating/cooling, the sample is heated/cooled stepwise within predefined temperature ranges. Timewise variations of the sample and the signal are given in Fig. 7.21. Step-like temperature increments are observed. Within a temperature range, the signal increases suddenly at the early stages to achieve ramp increment in temperature and gradually reduces to zero. By this way, temperature gradients

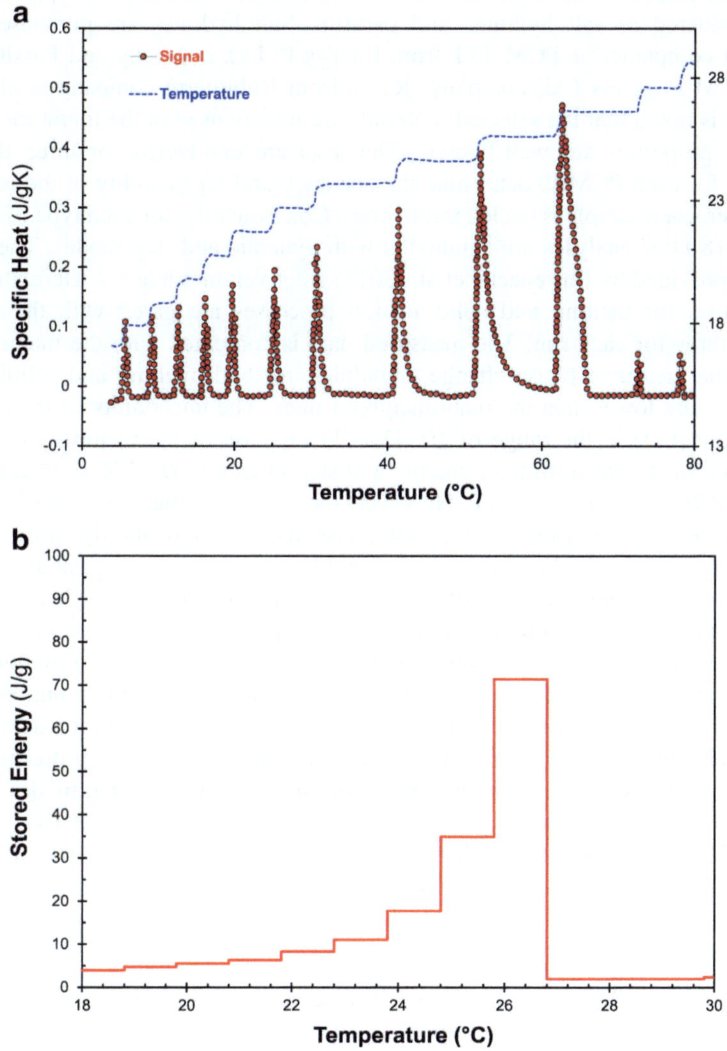


**Fig. 7.21** Variation of the DSC signal and stored energy in step mode. (**a**) Variation of signal (**b**) Variation of the stored energy. (Adapted from Günther et al. 2009)

within the sample and reference could be minimized. At the end of each step, a uniform temperature distribution is achieved. The amount of energy that is stored within the sample is then evaluated by integrating the signal for each temperature range. As the temperature ranges are defined precisely, the uncertainty due to internal temperature variation could be minimized in the step mode. It is also noted that dividing the whole range within small temperature ranges may reduce the signal output. In such a case, uncertainty may arise due to inaccurate measurement of the signal.

Barreneche et al. (2013) evaluated the melting temperature and latent heat values of three different PCMs for solidification and melting processes. Two types of PCMs are considered as salt hydrates and paraffin. Salt hydrates are purchased from different companies as PCM 17T from Energy P. Ltd. company and PlusICE S27 from PCM Products Ltd. company. RT27 from Rubitherm company is also analyzed. It is noted that the selected materials are widely used in the literature and the material properties are well-known. The tests are conducted for three different samples for each PCM to determine the accuracy and repeatability of the methods. Moreover, each sample is cycled three times. Consequently, for each type of PCM, a total of 18 DSC analyses are conducted with dynamic and step modes. The results that are provided by Barreneche et al. (2013) are given in Table 7.4. Here, the latent heat values for melting and solidification processes are given with the melting temperatures for each run. The measured data is compared with the manufacturer value. The measured phase change enthalpies, for both melting and solidification processes, are lower than the manufacturer values. The uncertainty of the enthalpy measurements is in the range of 10–47%. In the case of the melting process, the results are evaluated from the dynamic and step modes of DSC close to each other for PlusICE S27 and PCM 17T. However, the melting enthalpies of RT27 significantly depend on the mode of the DSC. The uncertainty of the dynamic mode is nearly 2.5 times that of the step mode. It should be noted that the step mode could not be successful to determine the solidification enthalpy of PlusICE S27. The solidification enthalpy could be obtained only in one sample. It is a well-known fact that the eutectic mixtures mostly show a subcooling. Subcooling is simply initiation of the crystallization while cooling at lower temperatures than the melting point. For PCM 17T and RT27, both methods give similar results, and the standard deviation is around 12% and 1%, for two different materials, respectively. No significant change is observed regarding the melting temperature for dynamic and step modes. Consequently, the dynamic and step mode of DSC has some advantages and disadvantages and could be successful depending on the type of the material. One should keep in mind that there is no specific method to measure the properties of all types of PCMs since the sample weight, scan rate, and the type of the heating/cooling (dynamic or step) significantly affect the results.

***Case Study 5:* Influence of Thermal Cycling on the Properties of PCM**

In Sect. 7.2.4, the methods for improving the thermal conductivity of materials are discussed in detailed. Development of nanoparticle-loaded novel PCMs has commonly followed practice to enhance the thermal response of the conventional

**Table 7.4** Thermal conductivity enhancement of PCMs with nanostructures

| | Melting enthalpy (kJ/kg) | | | | | | Solidification enthalpy (kJ/kg) | | | | | | Melting temperature (°C) | | | | | |
|---|---|---|---|---|---|---|---|---|---|---|---|---|---|---|---|---|---|---|
| | PlusICE S27 | | PCM 17T | | RT27 | | PlusICE S27 | | PCM 17T | | RT27 | | PlusICE S27 | | PCM 17T | | RT27 | |
| Run | Dynamic | Step | Dynamic | Step | Dynamic | Step | Dynamic | Step | Dynamic | Step | Dynamic | Step | Dynamic | Step | Dynamic | Step | Dynamic | Step |
| R1.1 | 163 | 173 | 93 | 105 | 134 | 163 | 170 | – | 87 | 66 | 140 | 139 | 28 | 27 | 21 | 21 | 26 | 26 |
| R1,2 | 161 | 177 | 93 | 88 | 134 | 162 | 169 | 162 | 87 | 58 | 139 | 139 | 28 | 27 | 21 | 21 | 26 | 26 |
| R1.3 | 160 | 178 | 93 | 95 | 133 | 159 | 165 | – | 87 | 64 | 139 | 137 | 28 | 27 | 21 | 23 | 26 | 26 |
| R2.1 | 168 | 155 | 73 | 93 | 134 | 162 | 153 | – | 69 | 57 | 138 | 140 | 28 | 27 | 21 | 21 | 26 | 26 |
| R2.2 | 151 | 115 | 79 | 93 | 134 | 161 | 156 | – | 71 | 57 | 139 | 138 | 28 | 27 | 21 | 23 | 26 | 26 |
| R2.3 | 129 | 129 | 79 | 99 | 135 | 161 | 171 | – | 72 | 58 | 140 | 138 | 28 | 27 | 20 | 21 | 26 | 26 |
| R3.1 | 165 | 127 | 99 | 113 | 134 | 162 | 167 | – | 97 | 81 | 140 | 138 | 28 | 27 | 21 | 21 | 26 | 26 |
| R3.2 | 158 | 124 | 107 | 126 | 134 | 161 | 178 | – | 102 | 85 | 140 | 137 | 28 | 27 | 21 | 21 | 26 | 26 |
| R3.3 | 157 | 176 | 111 | 100 | 133 | 163 | 160 | – | 105 | 77 | 139 | 136 | 28 | 27 | 21 | 21 | 26 | 26 |
| Mean | 157 | 150 | 92 | 101 | 134 | 162 | 165 | 162 | 86 | 67 | 139 | 138 | 28 | 27 | 21 | 21 | 26 | 26 |
| Standard deviation | 11 | 26 | 13 | 12 | 0 | 1 | 8 | – | 13 | 11 | 0 | 1 | 0 | 0 | 0 | 1 | 0 | 0 |
| Manufacturer value | 183 | 183 | 175 | 175 | 179 | 179 | 183 | 183 | 175 | 175 | 179 | 179 | 27 | 27 | 17 | 17 | 28 | 28 |
| % Uncertainty | 14 | 18 | 47 | 42 | 25 | 10 | 10 | 12 | 51 | 62 | 22 | 23 | 2 | 0 | 23 | 26 | 6 | 7 |

Adapted from Barreneche et al. (2013)

materials. Loading of the nano-sized particle into the materials may affect the cyclic stability of PCMs. To examine the influence of nanoparticle loading into the pure PCM, thermal cycling experiments should be conducted.

In a recent work of Seki et al. (2015), a novel PCM which is composed of arachidic acid and graphite nanoplates was proposed. Pure arachidic acid (AA), $C_{20}H_{40}O_2$, was purchased from Alfa Aesar. Graphite nanoplates (GNPs) have a purity of 98% with thickness and particle size of 5–8 nm and 5 μm, respectively. The GNPs were loaded into the molten AA with three different weight fractions: 0.5%, 1.0%, and 2.0%. The substances were mixed with an ultrasonic homogenizer for 3 min to obtain a uniform GNP distribution within the AA. The thermal cycling experiments were conducted with DSC. The samples underwent continuous heating/cooling processes from 60 °C to 90 °C with a scan rate of 5 °C/min. The variations in the phase change temperatures and the latent heat values were evaluated for both melting and solidification processes. In Fig. 7.22, the thermograms of pure AA and 1% GNP-loaded AA are given at the end of the $10^{th}$, $40^{th}$, $70^{th}$, and $100^{th}$ cycles. Even though the continuous heating/cooling cycles distort the shape of curves, the onset of the melting does not vary significantly. As the melting or solidification enthalpy depends on the area under the peak, the variations could only be seen by evaluating the numerical values as given in Table 7.5. For the pure AA, the melting temperature reduces from 76.4 °C to 75.8 °C by 0.7% at the end of 100 thermal cycles. The variation in the solidification temperature, on the other hand, is less than 1 °C. The melting and solidification enthalpies remain almost unchanged throughout the thermal cycles. The difference between the melting and solidification temperature, or supercooling, for pure AA, is around 2 °C. Loading 1% of GNP into the pure AA reduces the degree of supercooling. At the end of the first cycle, the degree of supercooling is 1 °C for AA with 1% GNP, and the supercooling drops less than 1 °C cycles. One can deduce that GNP loading in AA has no adverse effect on the energy storage properties of the PCM, moreover, it also reduces the gap between the solidification and melting temperatures. GNP particles that are dispersed inside the pure AA may act as nucleation agent. Such an observation is also reported by Yu et al. (2014).

#### 7.2.4.2 T-History (Temperature History) Method

The DSC method ensures precise control in the temperatures of the sample and reference furnaces since the size of each furnace is small. A uniform temperature distribution could also be achieved by using tiny samples, mostly varying in the range of 1–50 mg. However, for the materials that do not possess homogenous distribution, such as eutectic PCMs, using a sample with small size may not reflect the behavior of the bulk material. Yinping et al. (1999) stated that characterization of composite PCMs within a small pan might raise some uncertainties due to the size of the samples. Especially for the salt hydrates, the DSC measurements may not give the effects of subcooling and phase separation. It is noted that as the sample size reduces, the effect of subcooling increases in DSC. Moreover, it may not be possible

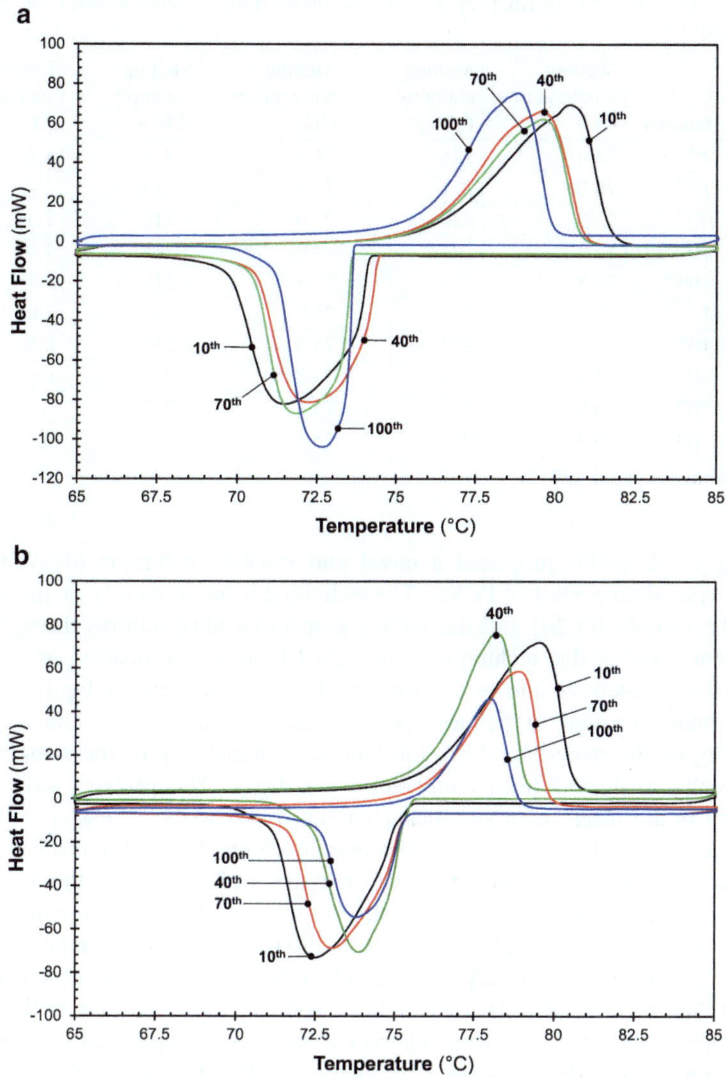


**Fig. 7.22** Thermograms for (**a**) pure AA and (**b**) 1% GNP-loaded AA at the end of thermal cycles (Data from Seki et al. 2015)

to observe the phase separation when a small amount of sampling is used in DSC. Besides, the scan rate and mass of the sample have a significant influence on the thermograms in the DSC method, which means that the results hardly depend on the selected parameters (Solé et al. 2013). For pure substances with homogenous structures, DSC can give highly accurate and reliable results. However, in the case of composite PCMs, care must be taken as uncertainties may arise due to the mass of the sample, scan rate, and type of crucible. In 1999, Yinping and his colleagues

**Table 7.5** Influence of thermal cycles on the heat storage characteristics of pure and GNP-loaded AA

| Sample | Cycle number | Melting temperature (°C) | Melting enthalpy (kJ/kg) | Melting temperature (°C) | Melting enthalpy (kJ/kg) | Degree of supercooling (°C) |
|---|---|---|---|---|---|---|
| Pure AA | 1st | 76.4 | 223 | 74.6 | −218 | 1.8 |
| | 10th | 76.0 | 221 | 73.6 | −218 | 2.4 |
| | 40th | 75.9 | 220 | 73.6 | −218 | 2.3 |
| | 70th | 75.9 | 221 | 74.0 | −218 | 1.9 |
| | 100th | 75.8 | 223 | 73.7 | −218 | 2.1 |
| AA +1% GNP | 1st | 76.2 | 223 | 75.2 | −218 | 1.0 |
| | 10th | 75.9 | 224 | 75.3 | −218 | 0.6 |
| | 40th | 75.9 | 224 | 75.3 | −217 | 0.6 |
| | 70th | 75.9 | 223 | 75.3 | −217 | 0.6 |
| | 100th | 75.9 | 223 | 75.3 | −218 | 0.6 |

Adapted from Seki et al. (2015)

(Yinping et al. 1999) proposed a novel and reliable technique to evaluate the thermophysical properties of PCMs. The technique is based merely on the monitoring of the sample (PCM) and the reference material temperatures throughout the experiment, leading this technique to be called temperature history, or T-History. The T-History method allows determining the specific heat in liquid and solid phases, phase change enthalpies, phase change temperatures, and degree of subcooling of the materials. Also, the thermal conductivity of the samples could be indirectly evaluated from temperature variations. The primary advantage of T-History is the usage of a significant amount of PCMs, i.e., $m_{PCM} = 15–20$ g (Solé et al. 2013). That is, in this technique, more reliable and realistic properties could be obtained. One of the most critical issues in T-History is the accuracy of temperature measurements. Yinping et al. (1999) stated that improving the accuracy of temperature measurement from ±0.5 °C to ±0.01 °C, the uncertainties of specific heat and phase change enthalpy reduce from 30% to 0.6%. Some early works revealed that the original T-History includes some uncertainties due to the computational algorithm. That is, several alternative methods are proposed to improve the computation of the thermophysical properties of PCMs with T-History method (Marín et al. 2003; Hong et al. 2004; Sandnes and Rekstad 2006; Peck et al. 2006).

Unlike the DSC, there is no commercial device for the T-History method in the market. Researchers developed their own setup to evaluate the properties of PCMs with T-History. In the following, a measurement setup for the T-History method is illustrated with some preliminary results.

***Case Study 6:*** **Evaluation of Properties of PCM with T-History Method**

The T-History technique is based on monitoring the temperature variations of the samples and the reference material throughout the experiments. The materials, both samples, and the reference material are placed inside tubes. Let the tube diameter be

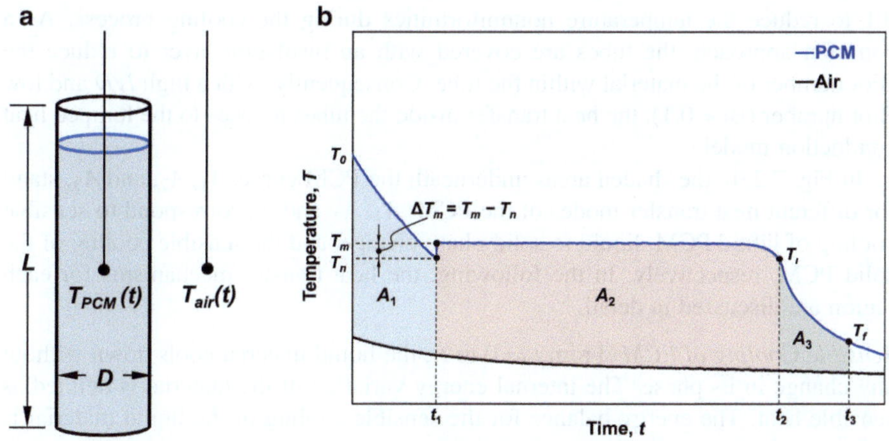


**Fig. 7.23** PCM tube and T-History curve for PCM. (**a**) PCM tube immersed in the air, (**b**) timewise variation of PCM and air temperature

*D* and the height of the tube *H*. A thermocouple is located in the middle of the tubes, both sample, and reference tubes, to measure the temperature of the material. A schematic of the PCM-filled tube is given in Fig. 7.23a. Initially, the temperature of the reference and the samples (PCMs) are at a temperature that is higher than the melting temperature of the PCMs ($T_o > T_m$). The tubes are then put into a constant temperature environment, which is lower than the melting temperature of the PCM, to cool down the reference and samples under the same thermal conditions. The temperature of the surrounding fluid, here air ($T_{air}$), is also recorded during the experiments. For instance, the evolution of the temperatures is represented in Fig. 7.23b. Here, the PCM undergoes subcooling. The temperature of the PCM reduces below the melting temperature ($T_m$) through the nucleation temperature ($T_n$). Here, the degree of subcooling is defined as the difference between the melting temperature and the nucleation temperature as $\Delta T_m = T_m - T_n$. At the nucleation temperature, the solidification initiates, and then the temperature of the PCM suddenly rises through the phase change temperature.

The heat transfer mechanism inside the tubes should be considered before writing an energy balance equation of the system. Notice that the reference material does not undergo a phase change process during the experiment, and it is in liquid phase. PCM, on the other hand, is initially in liquid phase and then turns into solid phase. Depending on the temperature difference between the material and the tube surface, natural convection may take place within the liquid substance. In such a case, the heat balance will become quite complicated. That is, tall and thin tubes are preferred in the T-history method to restrain natural convection. Yinping et al. (1999) state that tubes should have an aspect ratio of $L/D \geq 10$ to eliminate natural convection. Unlike DSC, in the T-history, significant amount of materials, order of gram, are used in the measurements. Increasing the mass of the samples may cause local temperature variations within the tubes. The Biot number should be kept below

0.1 to reduce the temperature nonuniformities during the cooling process. As a common approach, the tubes are covered with an insulation layer to reduce the Biot number of the material within the tube. Consequently, with a high *L/D* and low Biot number ($\mathrm{Bi} < 0.1$), the heat transfer inside the tubes reduces to the lumped heat conduction model.

In Fig. 7.23b, the shaded areas underneath the PCM curve, $A_1$, $A_2$, and $A_3$, stand for different heat transfer modes of the PCM. $A_1$, $A_2$, and $A_3$ correspond to sensible cooling of liquid PCM, liquid to solid phase change, and the sensible cooling of the solid PCM, respectively. In the following, the heat transfer mechanisms for each region are discussed in detail.

*Sensible Cooling of PCM* From $t = 0$ to $t_1$, the liquid material cools down without any change in its phase. The internal energy variation of the material is defined as sensible heat. The energy balance for the sensible cooling of the liquid material is defined as follows:

$$\left[(mc)_{tube} + (mc)_{PCM,\,liq}\right](T_0 - T_n) = h_1 A_{lateral} A_1 \tag{7.13}$$

where the subscript *tube* denotes the properties of the tube material. $A_{lateral}$ is the heat transfer surface area between the tube and the air. $A_1$ is the integral of the temperature difference between the PCM and air. $A_1$ is defined as

$$A_1 = \int_{t=0}^{t_1} [T_{PCM}(t) - T_{air}(t)]dt \tag{7.14}$$

Solidification takes place between $t = t_1$ and $t = t_2$. The energy balance for the phase change period can be written as

$$m_{PCM} h_{sf} = h_2 A_{lateral} A_2 \tag{7.15}$$

At $t = t_2$, the PCM completely turns into the solid phase. Further cooling reduces the temperature, and the energy balance of the system simply yields the following equation:

$$\left[(mc)_{tube} + (mc)_{PCM,\,solid}\right](T_i - T_f) = h_3 A_{lateral} A_3 \tag{7.16}$$

where $A_2$ and $A_3$ are defined as $A_2 = \int_{t=t_1}^{t_2} [T_{PCM}(t) - T_{air}(t)]dt$ and $A_3 = \int_{t=t_2}^{t_3} [T_{PCM}(t) - T_{air}(t)]dt$, respectively. In Eq. (7.16), $T_i$ corresponds to the temperature at which the phase change is completed. There are different approaches in the literature to determine the inflection temperature, $T_i$. Yinping et al. (1999)

suggest using nucleation temperature as $T_i$. Hong et al. (2004) proposed a systematic approach to determine the temperature for complete solidification. They suggest evaluating the first derivative of the timewise variation of PCM temperature. The inversion points in the first derivative correspond to the initial and final temperatures.

$h_1$, $h_2$, and $h_3$ in Eqs. (7.13), (7.15), and (7.16) stand for the convective heat transfer coefficients around the tube during the sensible cooling of liquid PCM, solidification, and sensible cooling of solid PCM, respectively. One could simply evaluate the specific heat values of solid and liquid phases and the latent heat of fusion as soon as the convective heat transfer coefficients around the tube are known. However, it should be noted that the procedure that is followed to evaluate the convective heat transfer coefficients significantly determines the accuracy of the T-History method. Yinping et al. (1999) proposed to use two identical tubes in the experiments, one filled with PCM and the other one containing reference material. Please note that the reference material does not undergo phase change within the working temperature range and the thermophysical properties should be known as a function of the temperature. The timewise variation of the reference material is also monitored under the same conditions as shown in Fig. 7.24. Yinping et al. (1999) suggest the following equations to determine the convective heat transfer coefficients:

$$\left[(mc)_{tube} + (mc)_{reference}\right](T_0 - T_n) = h_1 A_{lateral} A'_1 \tag{7.17}$$

$$\left[(mc)_{tube} + (mc)_{reference}\right]\left(T_n - T_f\right) = h_3 A_{lateral} A'_2 \tag{7.18}$$

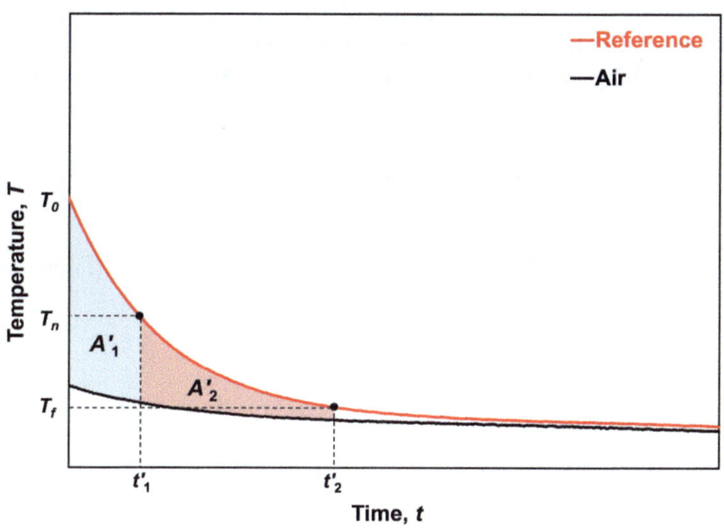


**Fig. 7.24** Timewise variation of the reference temperature

Please note that instead of three equations for three regions in Fig. 7.23, Yinping et al. (1999) proposed two equations to evaluate the convective heat transfer coefficients. Combining Eqs. (7.17) and (7.18) into (7.13), (7.15), and (7.16) yields

$$c_{PCM,solid} = \frac{(mc)_{tube} + (mc)_{reference}}{m_{PCM}} \frac{A_3}{A_2'} - \frac{m_{tube}}{m_{PCM}} c_{tube} \tag{7.19}$$

$$c_{PCM,liquid} = \frac{(mc)_{tube} + (mc)_{reference}}{m_{PCM}} \frac{A_1}{A_1'} - \frac{m_{tube}}{m_{PCM}} c_{tube} \tag{7.20}$$

$$h_{sf} = \frac{(mc)_{tube} + (mc)_{reference}}{m_{PCM}} \frac{A_2}{A_1'} (T_0 - T_n) \tag{7.21}$$

The procedure mentioned above is applicable to PCMs that show subcooling effect. If subcooling is not observed for a PCM during the solidification process, the equations are modified as follows:

$$h_{sf} = \frac{(mc)_{tube} + (mc)_{reference}}{m_{PCM}} \frac{A_2}{A_1'} (T_0 - T_{m,1}) - \frac{(mc)_{tube}(T_{m,1} - T_{m,2})}{m_{PCM}} \tag{7.22}$$

where $T_{m,1}$ and $T_{m,2}$ are the initial and final temperature of phase change. The method that is proposed by Yinping et al. (1999) is an excellent alternative to determine the significant properties of PCMs. However, the original version of the T-History method does not include a systematical approach to determine phase change temperature. As pointed out by Marín et al. (2003), determination of the solid-liquid phase change temperature has significant importance for the commercial PCMs that are used in TES applications. Marín et al. (2003) proposed a procedure to determine the enthalpy-temperature curve from the timewise variation of the temperature. Unlike Yinping's approach, in the modified T-History method, the energy balance is defined for a finite temperature range ($\Delta T_i$) as illustrated in Fig. 7.25.

The energy balance equations for the PCM and reference material are written as

$$[mc(T_i)]_{tube}(T_i - T_{i+1}) + [m\Delta h(T_i)]_{PCM} = hA_{lateral} \int_{t_i}^{t_i+\Delta t_i} [T(t) - T_{air}(t)]dt = hA_{lateral} I_i \tag{7.23}$$

$$\left\{[mc(T_i)]_{tube} + [mc(T_i)]_{reference}\right\}(T_i - T_{i+1}) = hA_{lateral} \int_{t_i'}^{t_i'+\Delta t_i'} [T(t) - T_{air}(t)]dt = hA_{lateral} I_i' \tag{7.24}$$

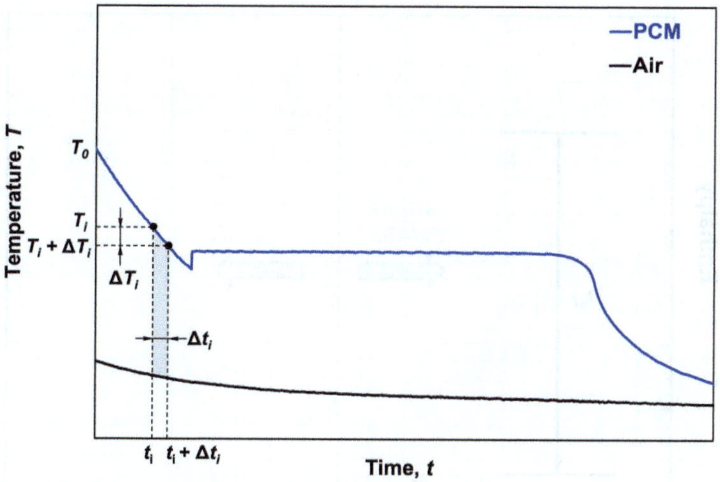


**Fig. 7.25** Finite temperature method of T-History

where $\Delta h_{PCM}(T_i)$ indicates the change of the enthalpy of PCM corresponding to an increment of $\Delta T_i$ in temperature. Combining Eqs. (7.23) and (7.24) yields the following enthalpy-temperature relation:

$$\Delta h_{PCM}(T_i) = \left\{ \frac{[mc(T_i)]_{tube} + [mc(T_i)]_{reference}}{m_{PCM}} \right\} \frac{I_i}{I_i'} \Delta T_i - \left[ \frac{m_{tube}}{m_{PCM}} c_{tube}(T_i) \right] \Delta T_i \tag{7.25}$$

Eventually, the variation of enthalpy is achieved as a function of temperature as in Fig. 7.26 by applying the following summation:

$$h_{PCM}(T_i) = \sum_{i=1}^{N} \Delta h_{PCM}(T_i) + h_{PCM,0} \tag{7.26}$$

where $h_{PCM,0}$ is the reference enthalpy of the material. The method of Marín et al. (2003) has some advantages over the original method of Yinping et al. (1999). Application of the energy equation for a finite time step allows taking into account the properties of tube and reference material as a function of temperature. Moreover, the specific heat values of PCM for the solid and liquid phases could be achieved from the derivative of the enthalpy-temperature curve (Fig. 7.26) as follows:

$$c_{PCM}(T_i) = \left. \frac{\partial h_{PCM}}{\partial T} \right|_{T=T_i} \tag{7.27}$$

An experimental setup is developed by Erek et al. (2015) to determine the properties of GNP-loaded eutectic PCMs. The setup is composed of PCM-filled tubes, a reference tube, refrigerated space, data acquisition system, and a computer.

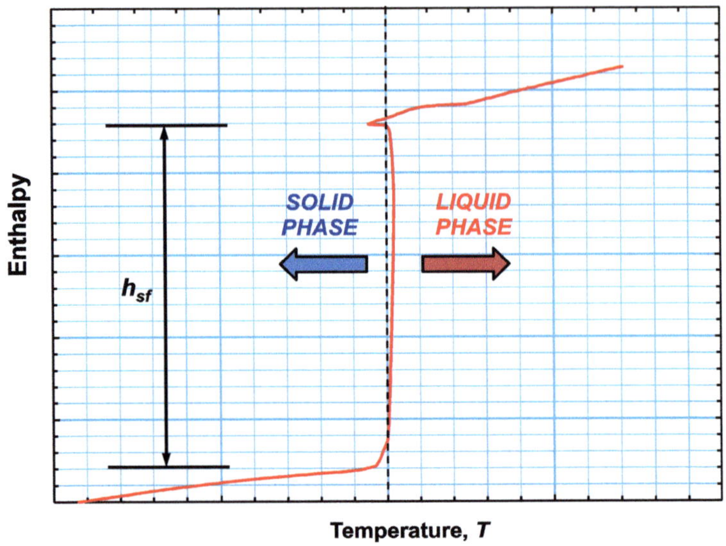


**Fig. 7.26** Enthalpy-temperature curve

The experimental setup is shown in Fig. 7.27. The procedure mentioned above of T-History method could only be followed if the heat transfer mechanism inside the tube is reduced into lumped heat conduction. Hence, the Biot number should be kept below 0.1 throughout the experiments. Erek et al. (2015) followed the advice of Lázaro et al. (2006) and covered the outer surfaces of the tubes with 12 mm of insulation. Application of an insulation layer reduces the rate of heat transfer by slowing down temperature variations and allows to capture the timewise variation of temperature more accurately. In the preliminary experiments, Erek et al. (2015) used water as PCM and revealed the uncertainties that may arise due to the measurement of temperature and the computational approach. Ethanol is used as reference material. The dimensions of the tubes and the thermophysical properties of the materials are given in Table 7.6. Note that the specific heats of the tube and insulation material are assumed to be constant. Temperature measurements are conducted by using T-type thermocouples. A total of three thermocouples are used in each tube. Thermocouples are placed at the center of the tube and the inner and outer surfaces of the insulation.

Erek et al. (2015) evaluated the timewise variations of thermocouples as shown in Fig. 7.28. Thermocouples 101 and 105 are placed inside water and ethanol, respectively. 102 (*or* 106) and 103 (*or* 107), on the other hand, are used to monitor the temperature variations at the inner and outer surfaces of the insulation. Figure 7.28 also includes the timewise variations of the ambient temperature and the first derivative of the water temperature. The vertical dashed lines indicate the inflection points for the temperature variation of water. It is clear that the phase change initiates at around $t = 4000$ s and finalizes at $t = 13{,}000$ s.

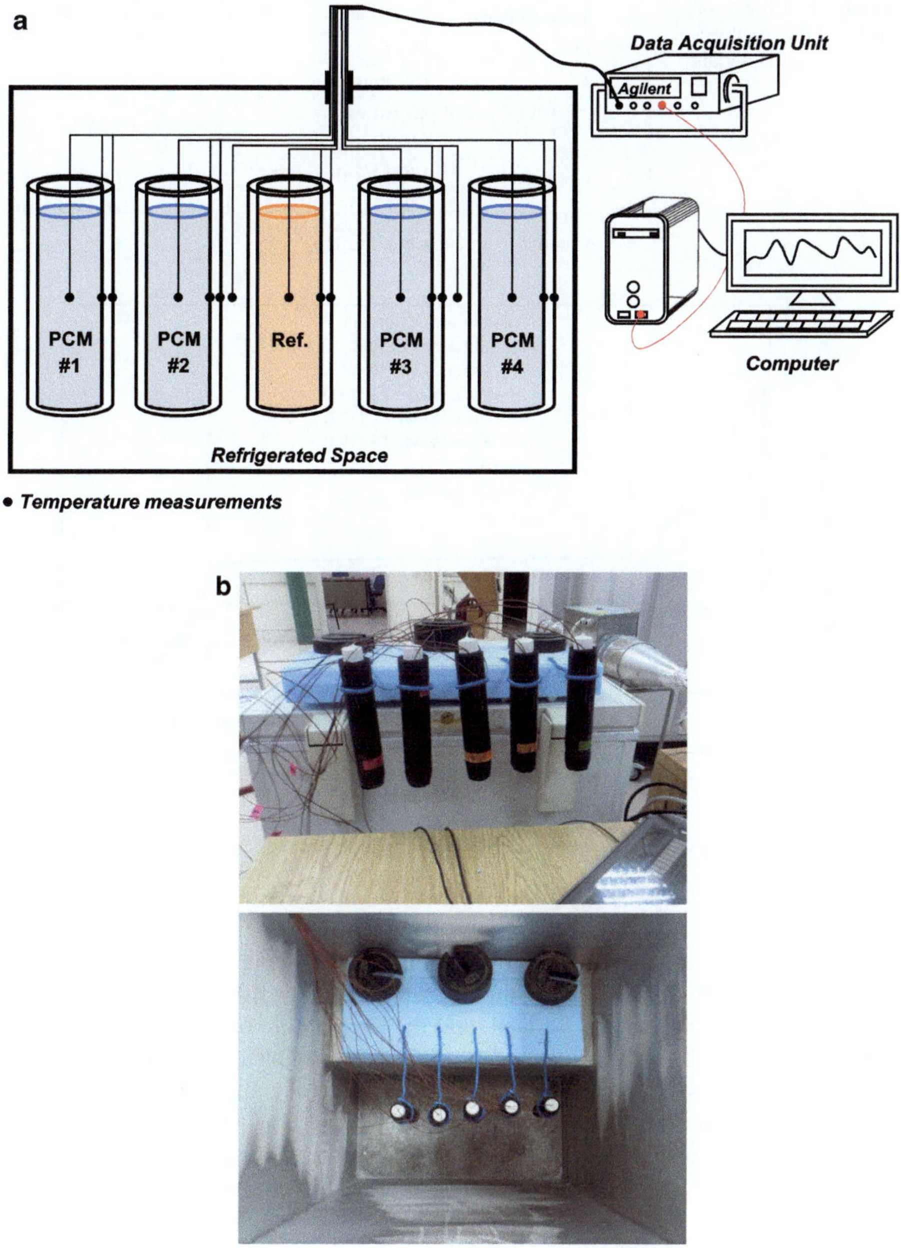


**Fig. 7.27** Experimental setup for *T-History* method. (**a**) Schematic (**b**) The insulated tubes. (Erek et al. 2015)

**Table 7.6** Dimensions of tube and thermophysical properties of materials

| Tube dimensions | |
|---|---|
| $D$ | 15 mm |
| $L$ | 150 mm |
| Thermophysical properties | |
| $c_{tube}$ | 1100 J/kgK |
| $c_{insulation}$ | 1300 J/kgK |
| $c_{reference}$ | $-3 \times 10^{-5}$ T$^3$ + 5 $\times$ 10$^{-2}$ T$^2$ + 2268.8 J/kgK |

Adapted from Erek et al. (2015)

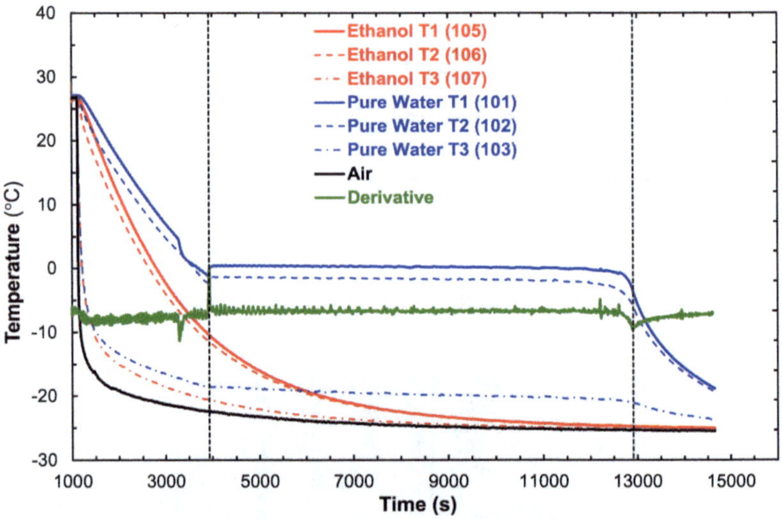


**Fig. 7.28** Evolution of sample (water), reference (ethanol), and air temperatures in T-History method

In Fig. 7.29, a representative enthalpy-temperature curve is given. Enthalpy of water reduces with almost a constant slope from 20 °C to 0 °C. The slope of the curve then approaches nearly vertical, and the temperature of the water varies in a narrow range near 0 °C. The latent heat of fusion is evaluated merely by subtracting the enthalpy values at top and bottom points of the vertical line. For the given case, the latent heat of solidification is evaluated as 348 kJ/kg. When it is compared with the literature, the uncertainty of the current method is found to be 4%. As noted above, the accuracy of the temperature measurement significantly determines the uncertainty of the T-History method. Recently Stanković and Kyriacou (2012) conducted a detailed experimental work to reveal the influence of temperature sensors on the measurement uncertainties of the T-History method. It is stated that the thermocouples have an accuracy of ±0.5 °C and such a high error band increases the uncertainty of the T-History method. Pt-100, on the other hand, has improved accuracy, but the dynamic response times of such sensors are high. Stanković and Kyriacou (2012) conducted a comprehensive study using NTC-type thermistor with

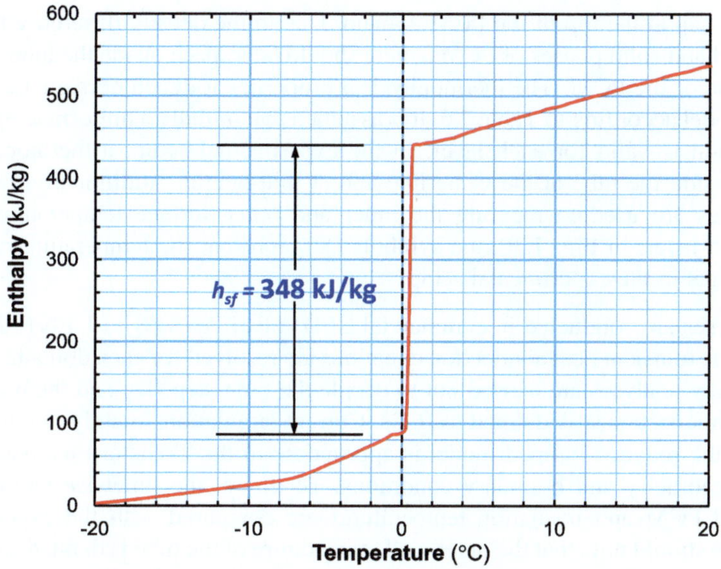


Fig. 7.29 A sample enthalpy-temperature curve for water obtained from the T-History method

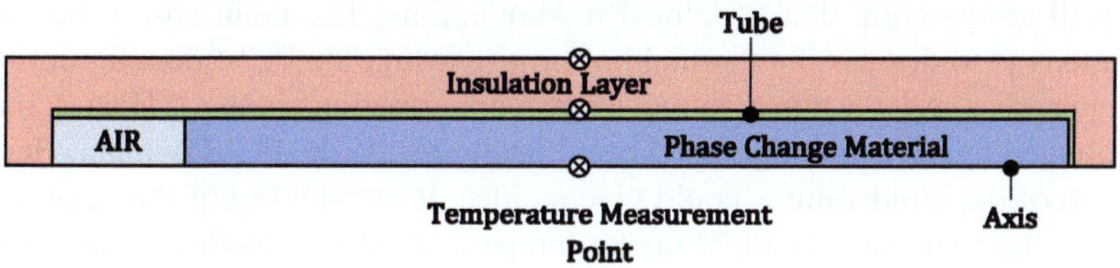


Fig. 7.30 2D computational domain

two different linearization algorithms. Results revealed that the serial-parallel-resistor (SPR) technique has an uncertainty less than ±0.1 °C. On the other hand, application of Wheatstone bridge (WB) increases the uncertainty to ±1.5 °C.

***Case Study 7:* A CFD Model to Simulate the T-History Method**

As discussed in the previous case, the uncertainty of the T-History methods depends on many factors such as the measurement accuracy of the temperature sensors, the shape of the tube, the thermophysical properties of the reference material, and the insulation. In this case study, we have demonstrated a CFD model to understand the heat transfer mechanism within a PCM-filled tube that is used in T-History experiments. The thermophysical properties that are evaluated from the T-History method, i.e., specific heat for liquid and solid phases, solidification temperature, and the latent heat of fusion, are used in the mathematical model. A two-dimensional axisymmetric geometry is considered as illustrated in Fig. 7.30. The computational model includes the PCM, air domain, insulation material, and the tube material. Notice that there is a small air gap at the top of the PCM domain. Due to the density difference between the liquid and solid phases of PCMs, there should be a space inside the tube to allow expansion of the PCM. The thermophysical properties for each domain are defined in the model according to Table 7.6. It is assumed that initially a uniform temperature distribution is valid for each layer. In the experimental setup, a thermocouple is placed inside the tube at the centerline and at mid-height. Additionally, two thermocouples are used to measure the inner and outer surface temperature of the insulation layer. In Fig. 7.30, the symbols (X) represent the temperature measurement points in the experimental setup.

Analyses are conducted in commercial CFD solver ANSYS-FLUENT. 2D axis-symmetric transient heat conduction equations are resolved for each domain. A set of preliminary analyses are carried out to decide the time-step size and the number of mesh. Time-step size is defined as 0.1 s. Current simulations mainly aim to understand if the reduced lumped model is appropriate in the T-History method for the current geometry and boundary conditions. To do so, the timewise variations of predicted PCM and insulation temperatures are compared with the experimental data. One should note that the outer wall temperature of the tube is defined according to the experimental measurements. In Fig. 7.31, the variations of the measured and predicted temperatures are compared for two different experiments. The solid red lines indicate the experimental measurements. The blue curves with circles are the outer wall temperature that is defined in simulations. The main aim of the current simulations is to discuss the thermal uniformity inside the PCM domain during the cooling process. As it is noted above, the balance equations in the T-History method are evaluated by assuming a lumped heat transfer inside the PCM domain. Here, the evolution of the temperatures could give an idea of the validity of this approach. In Fig. 7.31, the variations of PCM center temperature that is evaluated from experiments are compared with the one that is obtained in simulations. Moreover, the mean temperature of the PCM domain is also reported in simulations and is shown on the same figure with the black curve. The results reveal that the temperature variations that are obtained from experiments and simulations are close to each other, especially in the sensible cooling regions. The curves slightly differ at the end of phase

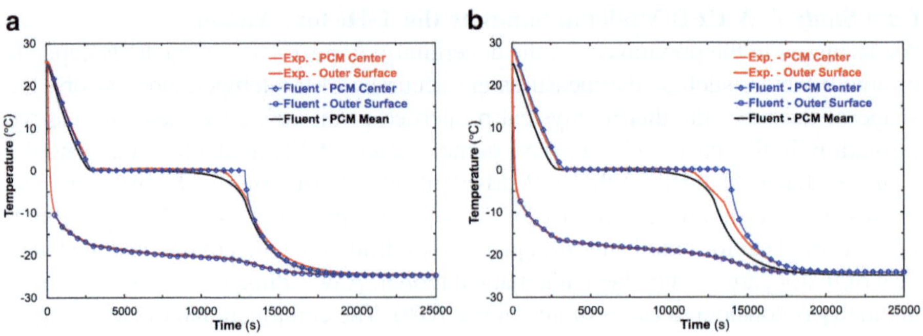


**Fig. 7.31** Comparison of the temperature history results. (**a**) Experiment #1 (**b**) Experiment #2

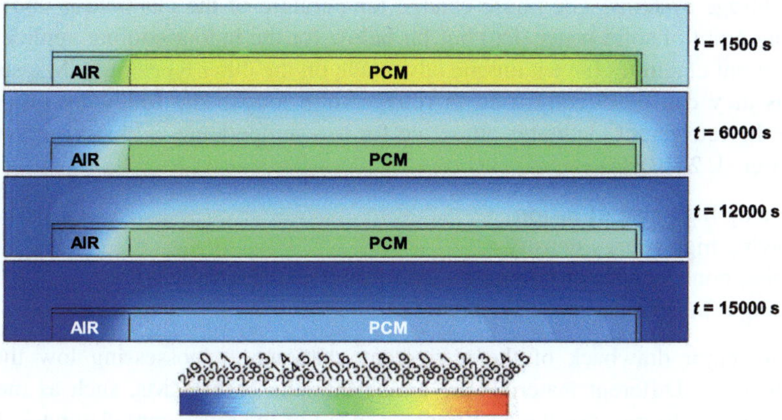


**Fig. 7.32** Predicted temperature contours

change process. Several facts may cause the discrepancies between the experimental measurements and the predicted results as the uncertainties that may arise due to the temperature readings, the position of the thermocouples, the thermophysical properties that are defined in the mathematical model, the thermal resistance between the layers, and the volume of air.

In Fig. 7.32, temperature contours are given at four selected instances. As the thermal conductivity of air is close to the insulation material, no significant temperature gradient along the vertical direction of the tube is observed. The thick insulation layer around the tube, on the other hand, slows down the heat transfer process and reduces the temperature gradient along the radial direction. For the current simulations, one can conclude that the cooling process of the PCM fits with the lumped model. The current case study shows that reducing the 3D convection and diffusion problem into a lumped model could be appropriate as soon as the suggested geometrical (aspect ratio of the tube) and design (covering the tube with insulation) concerns are followed adequately.

## 7.3 Clathrates of Refrigerants as Phase Change Materials

In a latent heat thermal energy storage system, the materials that are used to store the thermal energy undergo solid to liquid (or vice versa) phase change. Due to the massive volume change during the liquid to gas phase change, even the high latent heat of evaporation, it is not readily applicable for energy storage systems. One novel approach to form a phase change material (PCM) is introducing gas/liquid into water molecules which is called clathrate hydrates (Zafar et al. 2017). When a refrigerant is used as the supplied gas, the clathrate is called *clathrates of refrigerant*. Refrigerant clathrates are alternative storing materials which are suitable for low-temperature cold storage systems. The phase change temperature of the clathrate is above the melting point of solid water (ice) but far below for the indoor cooling applications. Refrigerant clathrates have a unique advantage on the other types of PCMs, such as ice, as they can be directly used in refrigeration loops. The following properties make the refrigerant clathrates attractive for low-temperature cold storage systems (Zafar et al. 2017):

- Having high heat of fusion
- Having high energy density
- Being noncorrosive or nontoxic
- Being efficient and cost-effective

One major drawback of the refrigerant clathrates is possessing low thermal conductivity. Different materials are added into the base PCMs, such as metallic nanoparticles, to improve the heat transfer speed of the refrigerant clathrates. In the following case study, the influence of additives on the charging and discharging performance of R134a + water refrigerant clathrate is represented.

***Case Study 8:* R134a Clathrate and Water as PCM for Cold Storage and Thermal Management**

Zafar (2015) conducted comprehensive experimental work to determine the potential improvements of the thermal properties of refrigerant clathrate and application to the thermal management of a battery block. The PCMs are formed using R134a clathrate and distilled water. The charging and discharging performances of the R134a clathrates are evaluated by varying the mass fraction of the refrigerant and dispersing different materials. Charging and discharging experiments are conducted in a constant temperature bath. In Fig. 7.33, the formation of R134a clathrate is illustrated for six different refrigerant mass fractions for tubes at 3 °C (276 K) and 5 °C (278 K). The pressure inside the tubes is 300 kPa. From (*i*) to (*vi*), the mass

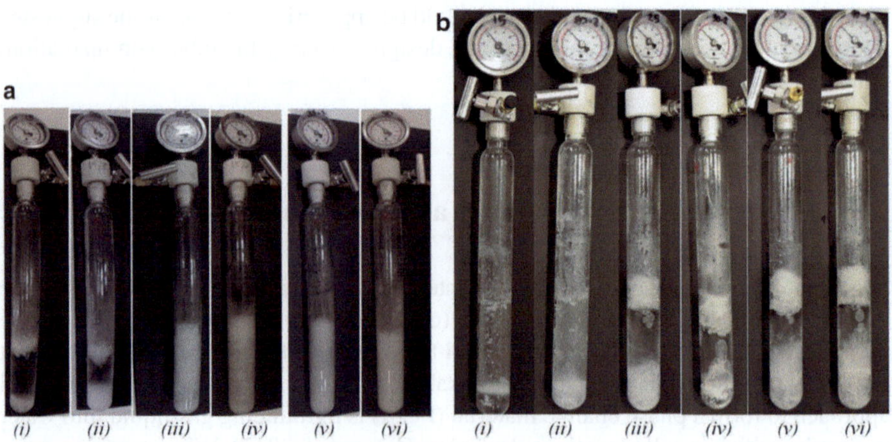


**Fig. 7.33** R134 clathrates in tubes with different refrigerant mass fractions (**a**) at 276 K (**b**) at 278 K. (Zafar 2015)

fractions of refrigerant are (*i*) 0.15, (*ii*) 0.2, (*iii*) 0.25, (*iv*) 0.3, (*v*) 0.35, and (*vi*) 0.4. The partial clathrate formation is observed for mass fractions of 0.15 and 0.2 at 276 K. At higher mass fractions of 0.35 and 0.4, a better formation of clathrate and more solidified clathrate is reported. On the other hand, at 278 K the clathrate formation takes longer time, and the solidified clathrate amounts are lower when it is compared with the results for 276 K. At higher mass fractions, i.e., 0.35 and 0.4, better clathrate formations are observed.

Moreover, to improve the heat transfer speed of the refrigerant clathrate, various additives are incorporated into the refrigerant, such as copper, $MgNO_3$, ethanol, aluminum, and NaCl, with different mass fractions. In Fig. 7.34, average onset durations of PCMs are compared for six additives with mass fractions of 0.01–0.05. PCMs with copper and aluminum additives turn into solid phase in around 10 min. It is reported that the Cu and Al additives reduce the onset time nearly by 25 min when compared to the refrigerant clathrate without additives. At low-additive concentrations, ethanol reduces the clathrate formation time, but at higher concentrations, the incorporation of ethanol adversely affects the onset time.

In Fig. 7.35, on the other hand, the time to liquify for each PCM is compared. In melting experiments, the tubes are exposed to hot air flow at 42 °C with a mass flow rate of 50 g/s. Results reveal that, except $MgNO_3$ and NaCl, the selected additives do not reduce the time for melting. On the contrary, for instance, the additive of ethanol almost doubles the time for melting. For Al- and Cu-based PCMs, the melting time also increases. Zafar (2015) stated that the structures of refrigerant clathrates significantly vary with the additives. Ethanol, Cu, and Al additives make the solid structure of the refrigerant harder with large crystals. Moreover, it is also stated that the additives settle at the bottom of the tube, that is, the dispersion of additives

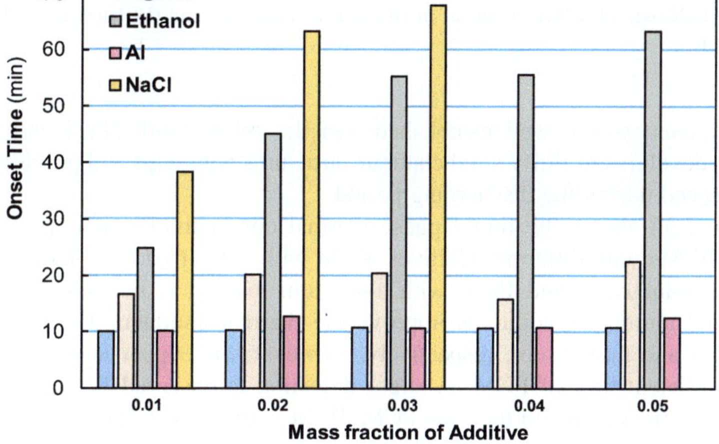


**Fig. 7.34** Influence of additive and the mass fraction on the onset time of PCM. (Data from Zafar 2015)

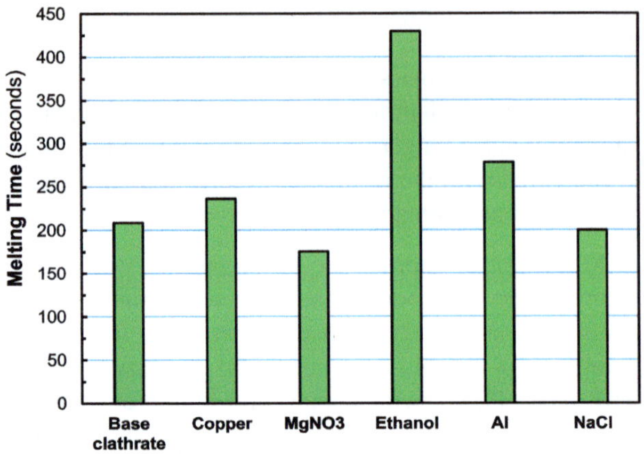


**Fig. 7.35** Influence of additives on the melting time. (Data from Zafar 2015)

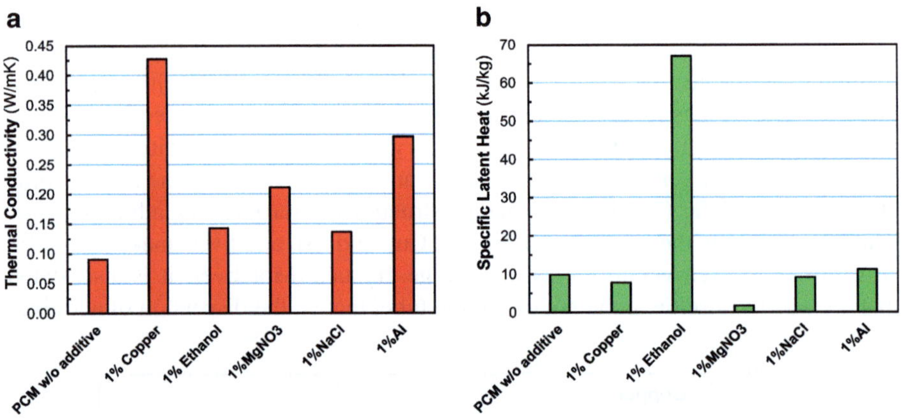


**Fig. 7.36** Influence of additives on (**a**) the thermal conductivity and (**b**) latent heat. (Data from Zafar 2015)

into refrigerant is not well established. On the other hand, NaCl and $MgNO_3$ additives develop soft fluffy solid clathrate structures with gaps and provide a better melting speed during the discharging period.

Figure 7.36 shows the liquid phase thermal conductivities and specific latent heats of PCMs with different additives. PCM with copper additive has the highest thermal conductivity, and PCM with aluminum loading is in second place. The copper and aluminum additives improve the thermal conductivity of base PCM nearly four and three times, respectively. However, the copper additive adversely affects the latent heat of PCM. A slight reduction is observed for the PCM with copper in comparison with the base PCM. PCM with $MgNO_3$ has the lowest latent heat. It is interesting to note that ethanol additives considerably improve the latent heat value of the PCM, more than six times. The solid structure of the PCM with ethanol is harder and releases a vast amount of heat during discharge. No notable change is observed for the Al and NaCl additives considering the latent heat values of the PCMs.

Zafar (2015) also built up a passive battery cooling pack with PCM to determine the influence of PCMs on the transient heat transfer inside the pack. In Fig. 7.37, the battery and the aluminum jacket are shown. 6s LiPo 5000 mAh 60C battery is used in the experiments. The PCM is filled inside the aluminum jacket. Figure 7.38 compares the cooling performances of PCMs with different additives. With each PCM, three experiments are conducted. The blue-shaded bars correspond to the

**Fig. 7.37** Battery with aluminum jacket. (Zafar 2015)

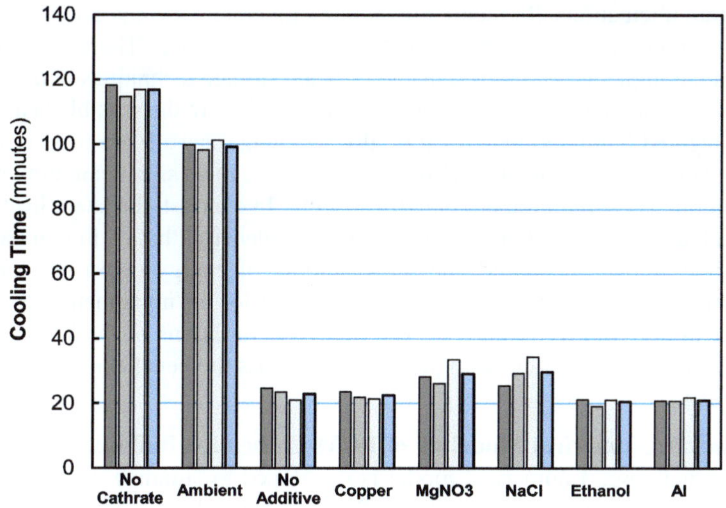


**Fig. 7.38** Influence of additives on the cooling time of battery. (Data from Zafar 2015)

average cooling times for each set of experiments. The required time for cooling significantly reduces with the application of PCM around the battery pack. Ethanol and Al additives provide slightly faster cooling when compared with the base PCM. It should also be noted that the performance of the base clathrate PCM is better when compared with the other PCMs.

This case briefly illustrated the characterization and performance experiments of refrigerant clathrates as PCM with different additives. Results of the case study depict that additives into the base PCM improve the thermal properties significantly and the proposed PCMs suitable for low-temperature cooling or passive thermal applications.

## 7.4 Heat Storage Materials in Building Elements

The total energy consumption that arises from the heating and cooling demands in a typical home corresponds to 66% of the overall energy demand. There is a challenge to reduce the demand for energy consumption and green gas emission that is related to space conditioning. A direct approach to reducing the heat transfer between a space and its environment is to increase the insulation thickness and/or reduce the thermal conductivity of the insulation material. However, the thermal conductivity of the building material is not the only parameter that controls the dynamic heat transfer. The heat storage capacity of the building elements has a significant role in the transient temperature variation within the conditioned space. Balaras (1996) states that the thermal mass of a building can reduce the temperature fluctuations within the conditioned space and peak cooling load. It is also stated that the improved thermal mass could be beneficial for the locations with significant diurnal temperature fluctuations. Reducing the indoor temperature swings also improves the thermal comfort of the occupants in the conditioned space. There are numerous works on the implementation of sensible or latent storage techniques in the building elements to improve the storage capacity of the building. In the sensible heat storage, heavy weighted building elements (i.e., floor, ceiling, or wall) are used to increase the thermal mass of the element. However, one of the most significant drawbacks of this technique is the requirement of extra spacing. In contrast, the latent heat storage has the advantage of excellent volumetric storage density; that is, in comparison to the sensible storage technique, the same amount of energy could be stored in a smaller mass. In the following, a case study is given for the implementation of PCM into the exterior wall of a building. The timewise variations of the heat fluxes are compared for different brick configurations to discuss the benefits of heat storage in building skin.

***Case Study 9:* Numerical Modeling of PCM-Embedded Building Wall**
In the current case, a numerical model is developed to simulate the transient heat transfer within a conventional wall and PCM-embedded walls with different configurations. The schematic of the mathematical model is given in Fig. 7.39. The

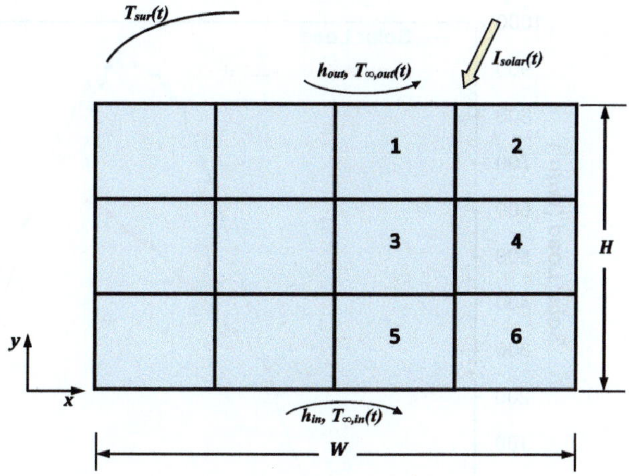


**Fig. 7.39** PCM-embedded brick geometry

height and width of the model are $H = 150$ mm and $W = 250$ mm. The top surface of the domain is exposed to the surroundings. A mixed thermal boundary condition is defined on the top surface to consider the convection, radiation, and incident solar radiation:

$$-k\frac{dT}{dy}\bigg|_{y=H} = I_{solar} - q''_{conv} - q''_{rad} \tag{7.28}$$

The ambient temperature and solar load are defined according to monthly average daily weather data of Izmir in July 2016. The variations of the solar load and ambient temperature are given in Fig. 7.40. A user-defined function (UDF) is coded in C++ and interpreted into the ANSYS-FLUENT to incorporate the transient boundary conditions. The surrounding temperature (or sky temperature), on the other hand, is evaluated from the following equation that is suggested by Hendricks and Sark (2013):

$$T_{sur}(t) = 0.037536T_{\infty}^{1.5} + 0.32T_{\infty} \tag{7.29}$$

The convective and radiative heat transfers that are defined in Eq. (7.28) are expressed as follows:

$$q''_{conv} = h_{out}\left(T_{surf} - T_{\infty}\right) \tag{7.30}$$

$$q''_{rad} = \varepsilon\sigma\left(T_{surf}^{4} - T_{sur}^{4}\right) \tag{7.31}$$

The convective heat transfer coefficient on the outer surface of the brick that is exposed to the ambient is defined as $h_{out} = 20$ W/m$^2$ K. On the other hand, the emissivity of the exterior surface of the brick is assumed to be $\varepsilon = 0.8$. On the inner surface, the only heat transfer mechanism is the convection. The indoor air

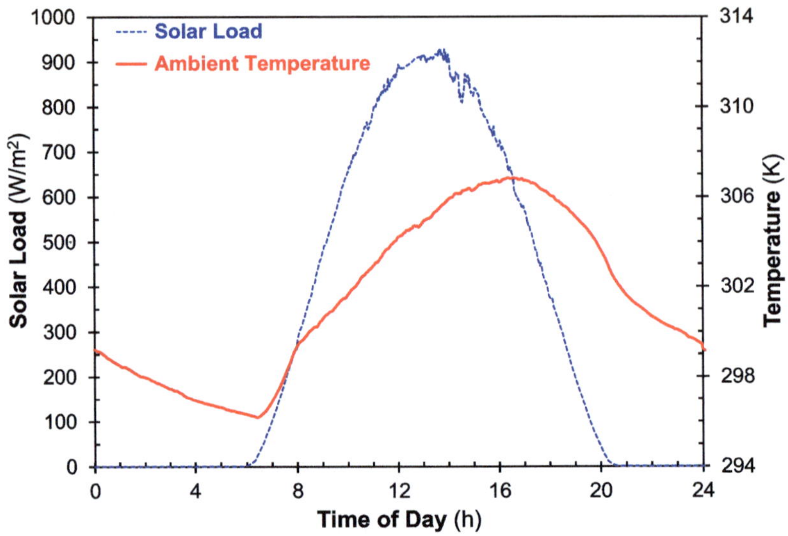


**Fig. 7.40** Weather data of Izmir City in July 2016

**Table 7.7** Thermophysical properties of brick and PCM

| Property | PCM (n-octadecane) | Brick |
|---|---|---|
| Melting temperature $(T_m)$ | 300 K | – |
| Thermal conductivity $(k)$ | Solid, 0.358 W/mK; liquid, 0.148 W/mK | 0.7 W/mK |
| Specific heat $(c_p)$ | Solid, 1934 J/kgK; liquid, 2196 J/kgK | 840 J/kgK |
| Density $(\rho)$ | Solid, 865 kg/m$^3$, liquid, 780 kg/m$^3$ | 1600 kg/m$^3$ |
| Latent heat of fusion $(L)$ | 243.5 kJ/kg | – |

Adapted from Haghshenaskashani and Pasdarshahri (2009)

temperature and the convective heat transfer coefficients are assumed to be kept constant throughout the simulations as $T_\infty = 296.5$ K and $h_{in} = 10$ W/m$^2$ K. Each side of the brick is assumed to be cyclic, and symmetric boundary condition is applied to these surfaces. The following reductions are made by following the numerical work of Haghshenaskashani and Pasdarshahri (2009):

- Thermal properties of the brick are constant.
- Thermal properties of PCM do not vary in each phase.
- Natural convection inside the liquid phase of PCM is not considered.

Under these assumptions, the energy equation is reduced to the following form:

$$\frac{\partial}{\partial t}(\rho c T) = \frac{\partial}{\partial x}\left[\frac{\partial}{\partial x}(kT)\right] + \frac{\partial}{\partial y}\left[\frac{\partial}{\partial y}(kT)\right] \tag{7.32}$$

The thermophysical properties of the brick and PCM (n-octadecane) are given in Table 7.7. Notice that the latent heat of fusion of PCM is not considered in 2D heat diffusion equation. Instead, effective heat capacity approach is used to include the solid-liquid phase change enthalpy. In this approach, the volumetric heat capacity ($=\rho c$) is defined as piecewise functions as given below:

$$(\rho c)_{PCM} = \begin{cases} (\rho c)_{solid} & T < T_m & \text{Solid} \\ \frac{1}{2}\left[(\rho c)_{solid} + (\rho c)_{liquid}\right] + \frac{\rho_{solid} + \rho_{liquid}}{2}\left(\frac{L}{\Delta T_m}\right) & T_m \le T \le T_m + \Delta T_m & \text{Mushy} \\ (\rho c)_{liquid} & T > T_m + \Delta T_m & \text{Liquid} \end{cases} \tag{7.33}$$

Notice that, in addition to the solid and liquid phases of the PCM, the mushy phase is also defined. Mushy range stands for an artificial temperature range in which the latent heat of the material is defined to improve the numerical stability of the solution method. The temperature range of the mushy phase significantly affects the accuracy and the convergence of the mathematical model. That is, a preliminary survey should be conducted to achieve a reasonable mushy region which is adequate for the phase change problem. In this simple 2D heat conduction problem, the mushy region temperature range is selected as $\Delta T_m = 1$ °C. The domain is divided into 10,000 uniform control volumes. Time-step size is defined as 10 s. For each time step, the residual of the energy equation is dropped below 1E-9 to achieve convergence. It is assumed that initially there is a uniform temperature profile inside the brick which is identical to the ambient temperature. Such an assumption is not realistic and significantly affects the variation of the temperature throughout the day. That is, analyses are conducted for five consecutive days to be sure whether a cyclic variation is constructed. In Fig. 7.41, the timewise variations of the brick temperatures are compared for four configurations. Notice that the variations significantly differ for the first 2 days due to the uniform temperature assumption at the initial time. Beyond the third day, the variations become cyclic. The fifth-day results

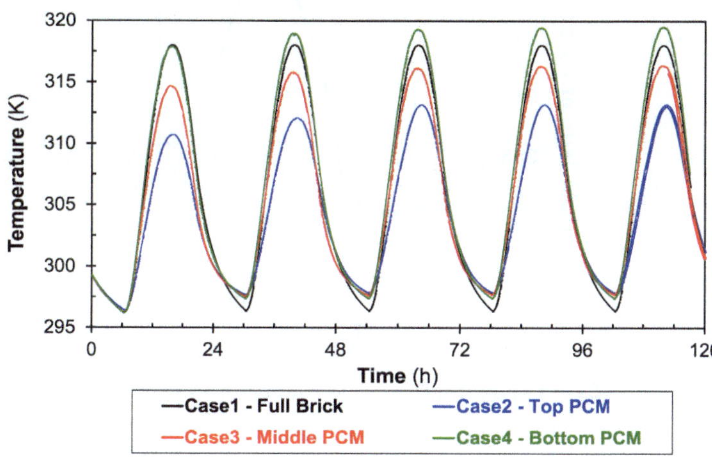


**Fig. 7.41** Variation of brick temperatures for different configurations

are considered to examine the influence of PCM location on the transient heat transfer inside the brick. The highest and lowest temperatures are observed for the top- and bottom-PCM configurations, respectively. For the top-PCM design, the maximum brick temperature reduces almost 5 K in comparison with the full-brick design.

Figure 7.42 compares the timewise temperature variations of the indoor surface temperature, indoor heat flux, and the mean PCM temperature. Utilization of PCM inside the brick significantly reduces the indoor surface temperature and the heat flux through the indoor space. Figure 7.42a, b reveals that for the case in which the PCM

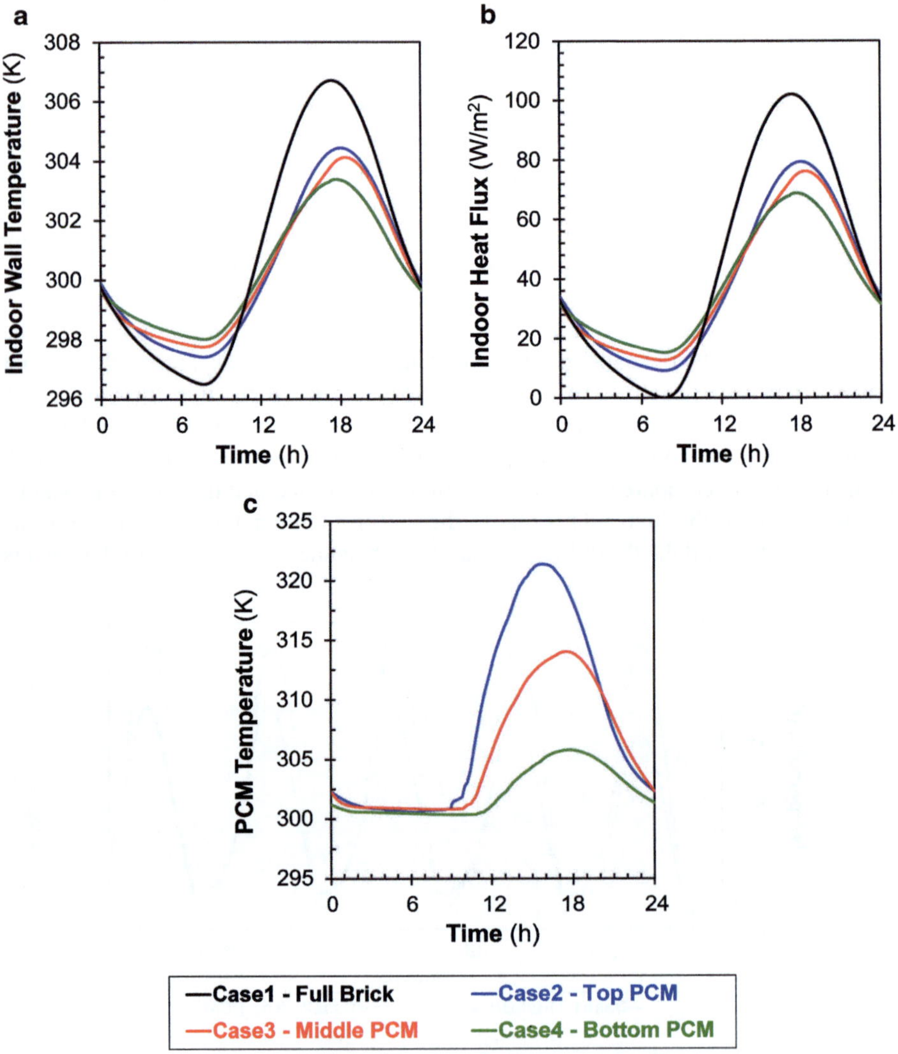


**Fig. 7.42** Influence of location of PCM on the heat transfer inside the brick

is placed close to the indoor ambient, the lowest surface temperature and heat gain are achieved. The reduction in surface temperature is nearly 4 K in comparison with the reference case. Moreover, the bottom-PCM case induces the maximum heat from nearly 40%. Top- and middle-PCM configurations have similar temperature and heat flux variations. The reductions in the surface temperature and heat gain are nearly 2 K and 20%, respectively. In Fig. 7.42c, the average PCM temperatures are compared for each configuration. The highest temperature is observed in the top-PCM design. The heat gain from the incident solar radiation increases the PCM temperature for the top-PCM design. The lowest PCM temperatures are evaluated for the bottom-PCM configuration. The stored energy within the PCM releases through the indoor and outdoor beyond sunset. As clearly seen in Fig. 7.41, for each case the minimum temperature fall below 300 K, which is the melting temperature of PCM.

Local temperature distributions are also examined to discuss the influence of PCM on the heat transfer within the brick. Figure 7.43 compares the temperature distributions at 10 am, 2 pm, and 6 pm. To achieve better performance from a latent heat TES inside building elements, the PCM should undergo complete melting and solidification processes throughout the day. In the current conditions, the PCM may not transform into the solid phase completely. Environmental conditions and indoor comfort ranges should be considered while selecting the melting temperature, the mass, and location of the PCM.

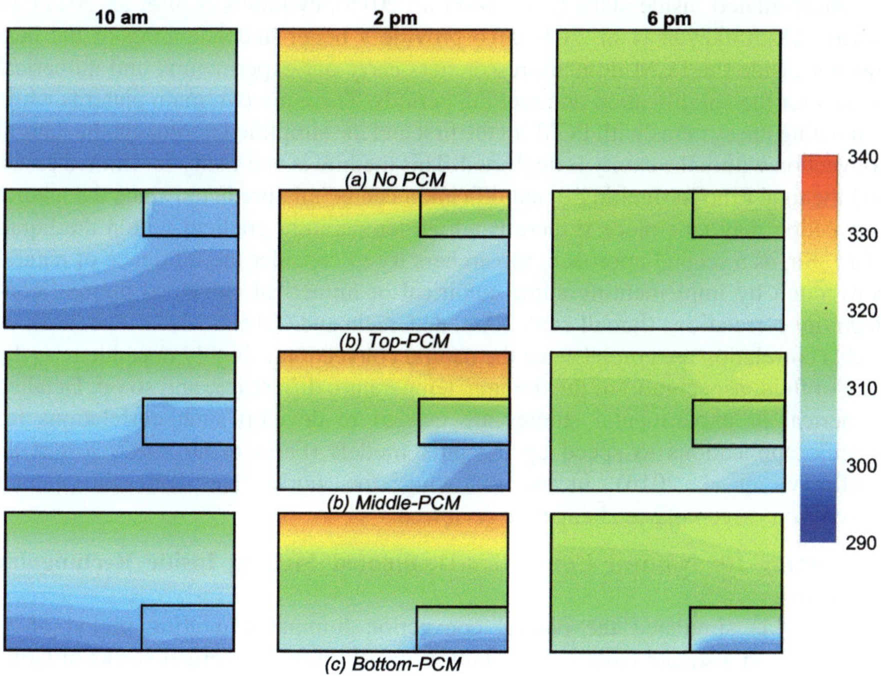


**Fig. 7.43** Isotherms for different brick configurations

## 7.5 Natural Convection-Driven Phase Change

Natural convection can be observed around a body or inside an enclosure if there is a finite temperature difference between the wall and the fluid. The temperature gradients within the fluid domain form density variations since in most of the fluids, both liquids and gases, the density is highly dependent on the temperature changes. The density gradients cause buoyancy forces, and the fluid starts to move without requiring any additional external interferences. Natural convection is an important topic since it can provide heat removal from the systems that have limited spacings, such as electronic cooling, without consuming power and without using any additional equipment. Numerous works deal with the experimental or numerical investigation of natural convection within a confined spacing in Cartesian, cylindrical, or spherical coordinates.

Thermal energy storage (TES) is a favorite subject in this era due to the well-known advantages that can be achieved by implementing the TES units into conventional heating or cooling applications. Latent heat storage makes it possible to store a significant amount of energy within a small volume in comparison with the sensible heat storage mode. There are several critical aspects in designing an appropriate latent heat thermal energy storage (LHTES) unit to meet the demand by the user as the type of PCM and the heat exchanger (HEX). The design of the PCM-HEXs is a popular subject, and there are various designs in which the PCMs are encapsulated inside slabs (Navarro et al. 2015), cylinders (Liu et al. 2017), or spheres (Karthikeyan et al. 2014). To provide a better understanding of the heat transfer inside the PCM domain, researchers carry out experimental and numerical studies for the solidification and melting periods. There are two main patterns while simulating the systems with PCM; in the first and the simplified approach, the natural convection within the cavity is neglected. This method is valid only for limited cases, and the total time for melting or solidification can be mis-predicted since the natural convection may take place if there is a high-temperature gradient within the liquid PCM. So, as a second approach, researchers try to consider the influence of natural convection by implementing some empirical or numerical enhanced thermal conductivity correlations (Ismail et al. 2003; de Gracia and Cabeza 2017). The limits of such a correlation are crucial since the natural convection is highly dependent on the type of fluid, aspect ratio of the domain, temperature difference, and so on. Detailed numerical or experimental studies are needed to develop such correlations for specific applications to speed up the bulk models (Farid et al. 1989; Xia et al. 2010; Wu et al. 2016). In the following, the works that deal with natural convection-driven phase change are reviewed.

***Case Study 10:* Natural Convection-Dominated Melting Inside Rectangular Enclosure**

Ezan (2011) simulated the natural convection-dominated melting period of *n*-octadecane in a square cavity. This study is based on the numerical works of Gong and Mujumdar (1998). They simulated the melting problem using the streamline upwind/Petrov-Galerkin finite element method in combination with a fixed grid primitive variable method for various Rayleigh numbers.

The current problem investigates the heat transfer characteristics of natural convection-driven inward melting inside a square enclosure. Pure *n*-octadecane is filled in a square enclosure with an initial dimensionless temperature of $\theta_i = -0.0256$. While the bottom surface is kept at a constant temperature, the remaining three surfaces of the enclosure are assumed to be well insulated. Suddenly, the temperature of the bottom surface ($y = 0$) of the container increases to a dimensionless temperature of $\theta_{bottom} = 1$, which is above the melting temperature of the PCM. Therefore, in time, there will be an inward melting through *y*-direction. A schematic representation of the problem is given in Fig. 7.44, and the dimensionless thermophysical and geometric parameters of the current problem are defined in Table 7.8. In this problem, the dimensionless temperature, Rayleigh number, Fourier number, and Stefan number are defined as follows:

$$\theta = \frac{T - T_m}{T_{bottom} - T_m} \tag{7.34}$$

$$Ra = \frac{\rho^2 c_1 g \beta L^3 (T_{bottom} - T_m)}{\mu k_l} \tag{7.35}$$

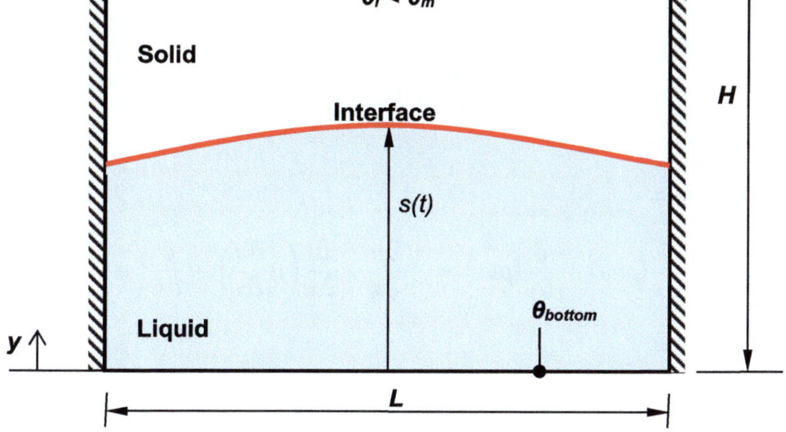


**Fig. 7.44** Illustration of the mathematical model for inward melting inside the cavity. (Ezan 2011)

**Table 7.8** Dimensionless properties for inward melting inside the cavity

| Parameter | Definition | Value |
|---|---|---|
| $R$ | Aspect ratio $L/H$ | 1.0 |
| $Ra$ | Rayleigh number | $2.844 \times 10^4$–$2.844 \times 10^5$ |
| $Pr$ | Prandtl number | 46.1 |
| $Ste$ | Stefan number | 0.138 |
| $c_s/c_l$ | The ratio of solid/liquid specific heat | 0.964 |
| $k_s/k_l$ | The ratio of solid/liquid thermal conductivity | 2.419 |
| $\theta_i$ | Dimensionless initial temperature | −0.0256 |

$$Fo = \frac{t\alpha_l}{L^2} \tag{7.36}$$

$$Ste = \frac{c_l(T_{bottom} - T_m)}{h_{sf}} \tag{7.37}$$

Numerical analyses are performed using the commercial CFD code ANSYS-FLUENT. In numerical analyses, the following main assumptions are adopted:

1. *n-octadecane* is treated as the Newtonian and incompressible fluid.
2. Natural convection of *n-octadecane* is laminar without viscous dissipation and radiation effects.
3. Thermophysical properties of PCM differ for solid and liquid phases, and properties are temperature independent for the same phases. Only density and dynamic viscosity of liquid water are defined as temperature dependent.
4. No-slip conditions ($u = 0$, $v = 0$) are valid for all boundaries.
5. Except for the constant temperature boundaries, all other surfaces are adiabatic if there is no contrary designation.

Under these assumptions, the time-dependent governing equations for the two-dimensional Cartesian coordinate system can be expressed as

*For mass:*

$$\frac{\partial}{\partial t}(\rho) + \frac{\partial}{\partial x}(\rho u) + \frac{\partial}{\partial y}(\rho v) = 0 \tag{7.38}$$

*For x-momentum:*

$$\frac{\partial}{\partial t}(\rho u) + \frac{\partial}{\partial x}(\rho uu) + \frac{\partial}{\partial y}(\rho uv) = -\frac{\partial p}{\partial x} + \frac{\partial}{\partial x}\left(\mu \frac{\partial u}{\partial x}\right) + \frac{\partial}{\partial y}\left(\mu \frac{\partial u}{\partial y}\right) \tag{7.39}$$

*For y-momentum:*

$$\frac{\partial}{\partial t}(\rho v) + \frac{\partial}{\partial x}(\rho uv) + \frac{\partial}{\partial y}(\rho vv) = -\frac{\partial p}{\partial y} + \frac{\partial}{\partial x}\left(\mu \frac{\partial v}{\partial x}\right) + \frac{\partial}{\partial y}\left(\mu \frac{\partial v}{\partial y}\right) + \rho g(T - T_m) \tag{7.40}$$

*For energy:*

$$\frac{\partial}{\partial t}(\rho cT) + \frac{\partial}{\partial x}(\rho ucT) + \frac{\partial}{\partial y}(\rho vcT) = \frac{\partial}{\partial x}\left(k\frac{\partial T}{\partial x}\right) + \frac{\partial}{\partial y}\left(k\frac{\partial T}{\partial y}\right) \tag{7.41}$$

In the numerical analysis, the nonuniform mesh structure is applied near the bottom boundary. The computational domain is divided into 6400 computational elements. After making sensitivity analyses, the optimum time-step size is found as $\Delta t = 0.1$ s. Iterations are continued until the convergence criteria of $\varepsilon = 10^{-4}$ is satisfied for all transport equations, and it needs at least 800 iterations for each time step.

Temperature and streamline predictions inside the cavity are evaluated for two different values of Rayleigh number, $2.844 \times 10^4$ and $2.844 \times 10^5$. In Fig. 7.45, the predicted isotherms and streamlines are given together for $Ra = 2.844 \times 10^4$ at three dimensionless $Fo$ values: 1.08, 1.296, and 1.62. Numerical results indicate that buoyancy forces become clear after $Fo = 1.08$. In the earlier periods of melting, the solid-liquid interface moves as conduction dominated. The reason is that the actual Rayleigh number is relatively small, due to the small height of melted PCM. In time, with increasing height of melting, the Rayleigh number increases. Two separated circulation cells are observed after $Fo = 1.08$, and in time, these circulation cells grow in the upper direction. Flow directions of the left and the right circulation cells are counterclockwise and clockwise, respectively. In progressing time, the shape of the solid-liquid interface becomes parabolic owing to the formation of two separated circulation cells. Relatively hot fluid flows from the bottom to the upper side of the cavity and causes a plume at the center of the cavity. On the other hand, for $Ra = 2.844 \times 10^5$, the results are represented in Fig. 7.46 at four dimensionless $Fo$ values: 0.302, 0.454, 0.605, and 0.756. Due to the relatively higher Rayleigh number, natural convection develops much earlier times in comparison with the $Ra = 2.844 \times 10^4$. Four independent circulation cells are observed after $Fo = 0.302$, and in time, these circulation cells grow in the upper direction. Two symmetrical cups form on the solid-liquid interface due to the formation of four separated circulation cells. This cup-like interface formation is observed because of the flow direction of the melted PCM. From left to right, flow cell directions vary as clockwise and counterclockwise. For the first circulation cell on the left-hand side, hot fluid flows up in the clockwise direction and melts interface as plume near the wall. Unlike the first circulation cell, in the second one, hot fluid flows up in the counterclockwise direction, and plume occurs at the center of the cavity. This effect can also be seen from the isotherms.

***Case Study 11:* Natural Convection-Dominated Melting Inside Spherical Capsule**

In this case, natural convection-driven phase change within a spherical capsule is numerically investigated by using the commercial CFD solver ANSYS-FLUENT. Yavuz (2017) compared the predictions with the work of Tan et al. (2009). The influence of wall temperature on the transient phase change process is monitored by

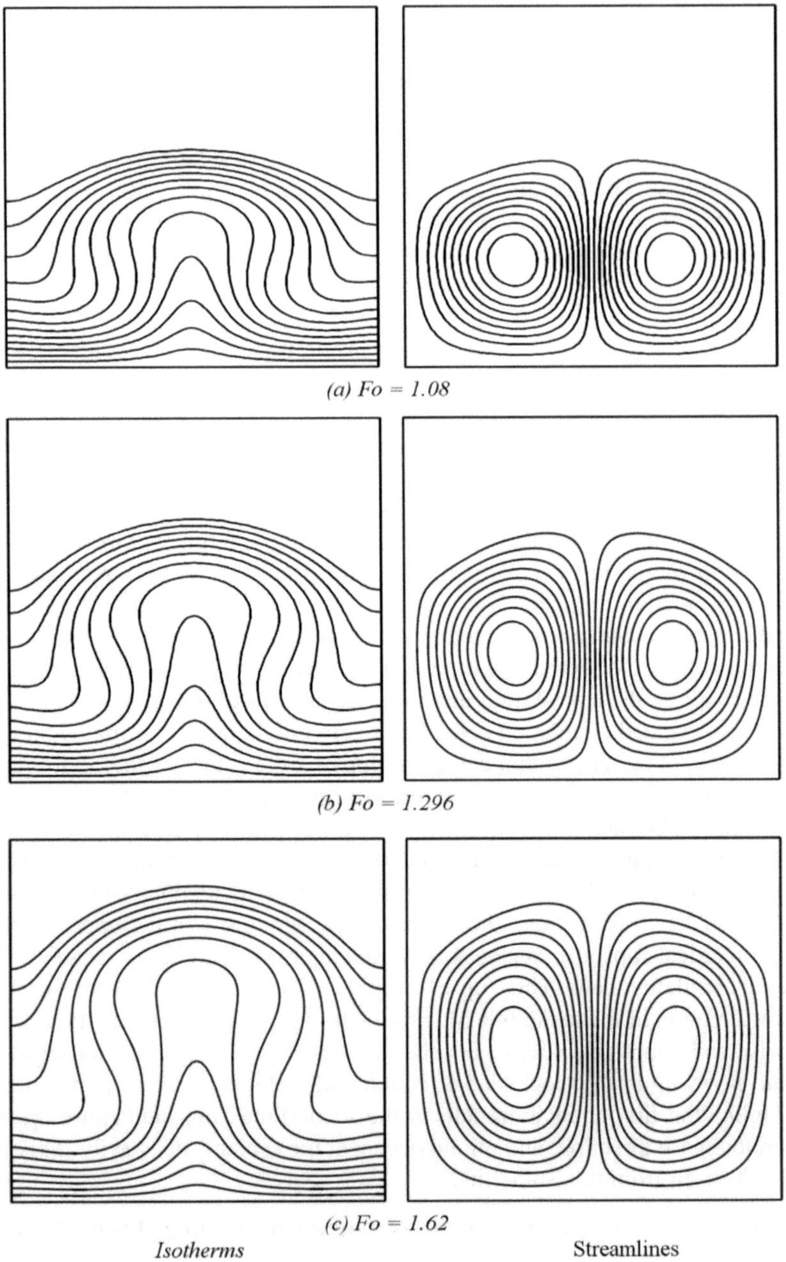


**Fig. 7.45** Isotherms and streamlines for $Ra = 2.844 \times 10^4$ (Counterclockwise circulation cell is on the left-hand side, and clockwise circulation cell is on the right-hand side). (Ezan 2011)

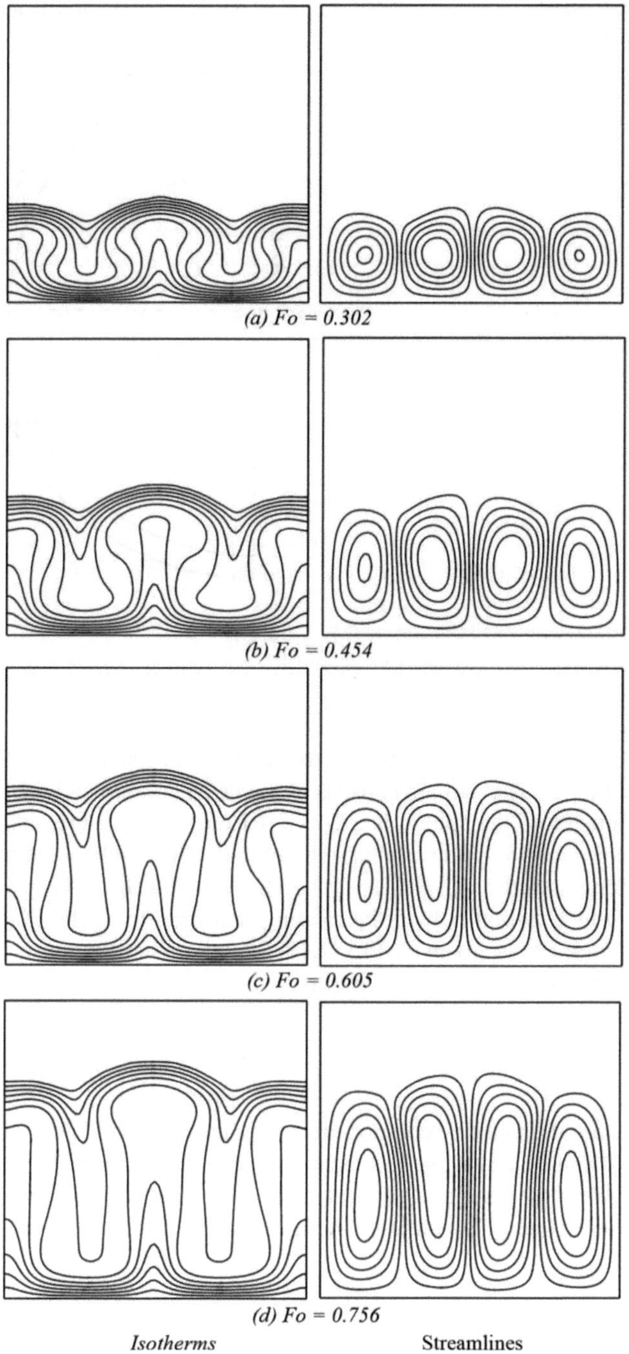


**Fig. 7.46** Isotherms and streamlines for $Ra = 2.844 \times 10^5$ (Ezan 2011)

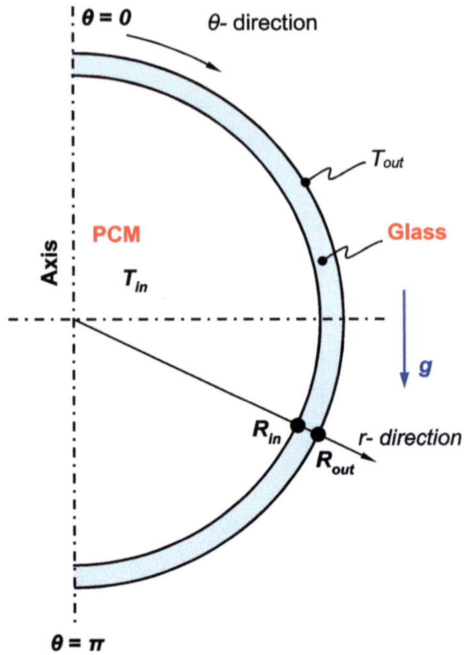


**Fig. 7.47** Mathematical model. (Yavuz 2017)

comparing the local temperature variations and contour plots. The natural convection-driven phase change inside the spherical capsule is reduced to a 2D mathematical model. The domain consists of the PCM domain and the glass domain. The inner diameter of the sphere is $2R_{in} = 101.66$ mm, and the wall thickness is $R_{out} - R_{in} = 1.5$ mm. It is assumed that initially the PCM is subcooled by 1 °C below its melting temperature, $T_{in} = T_m - 1$. The outer wall is kept at a constant temperature, which is higher than the melting temperature of the PCM. The schematic of the problem is given in Fig. 7.47. Initially, the domain is at a uniform temperature with the following conditions:

*Initial conditions:*

$$T(r,\theta) = 27^{\circ}\mathrm{C} \qquad u(r,\theta) = 0 \qquad v(r,\theta) = 0$$

where $u$ and $v$ are the velocity components along the radial and polar directions. On the axis, $\theta = 0$ and $\theta = \pi$, as well as the radial velocity component, the gradients of polar velocity and the temperature along the polar direction are zero. The mathematical representations of boundary conditions on the axis are defined as

$$\text{Axis} \rightarrow \quad v = 0, \frac{\partial T}{\partial \theta} = 0, \frac{\partial u}{\partial \theta} = 0 \quad \text{at} \quad \theta = 0 \text{ and } \theta = \pi$$

**Table 7.9** Thermophysical properties of n-octadecane

| Melting temperature (°C) | Specific heat (J/kgK) | Density (kg/m$^3$) | Thermal conductivity (W/mK) | Kinematic viscosity (m$^2$/s) | Latent heat (J/kg) | Thermal expansion coff. (1/K) |
|---|---|---|---|---|---|---|
| 27.5 | 2330 | 772 | 0.1505 | 5E-6 | 243.5 | 0.00091 |

Adapted from Tan et al. (2009) and Tan (2008)

On the walls and the solid domains of the sphere, the no-slip boundary condition is defined as follows:

$$\text{Walls} \rightarrow \quad u = v = 0 \quad \text{at} \quad r = R_{out} \&\text{Solid domains}$$

The outer wall temperature is maintained at $T_{out} = 40$ °C and $T_{out} = 45$ °C in the current analyses. By following the work of Tan et al. (2009), n-octadecane is used as PCM. The thermophysical properties of the PCM are given in Table 7.9.

The following reductions are considered to simplify the problem:

- Geometry is two-dimensional and axis-symmetric.
- PCM is incompressible, and the type of flow is laminar.
- Materials are isotropic, and except for the density, the thermophysical properties are independent of the temperature variation.
- The influence of radiation is neglected.
- During the inward melting, the PCM is not floating; it is constrained.

The governing equations for the spherical coordinates can be reduced to the following form:

*For continuity:*

$$\frac{1}{r^2}\frac{\partial}{\partial r}\left(r^2 u\right) + \frac{1}{r\sin\theta}\frac{\partial}{\partial\theta}(\sin\theta v) = 0 \tag{7.42}$$

*For r-momentum*:

$$\begin{aligned}\frac{\partial}{\partial t}(\rho u) + \frac{\partial}{\partial r}(\rho uu) + \frac{1}{r}\frac{\partial}{\partial\theta}(\rho vu) - \frac{\rho v^2}{r} = &-\frac{\partial p}{\partial r} + \frac{1}{r^2}\frac{\partial}{\partial r}\left(\mu r^2\frac{\partial u}{\partial r}\right) \\ &+ \frac{1}{r^2\sin\theta}\frac{\partial}{\partial\theta}\left(\mu\sin\theta\frac{\partial u}{\partial\theta}\right) \\ &- \rho g\beta(T - T_m)\sin\theta - \frac{C(1-\lambda)^2}{\lambda^3}u\end{aligned} \tag{7.43}$$

*For θ-momentum*:

$$\frac{\partial}{\partial t}(\rho v)+\frac{\partial}{\partial r}(\rho u v)+\frac{1}{r}\frac{\partial}{\partial \theta}(\rho v v)+\frac{\rho u v}{r}=-\frac{1}{r}\frac{\partial p}{\partial \theta}+\frac{1}{r^2}\frac{\partial}{\partial r}\left(\mu r^2\frac{\partial v}{\partial r}\right)+\frac{1}{r^2\sin\theta}\frac{\partial}{\partial \theta}\left(\mu\sin\theta\frac{\partial v}{\partial \theta}\right)+\frac{2\mu}{r^2}\frac{\partial u}{\partial \theta}-\rho g\beta(T-T_m)\cos\theta-\frac{C(1-\lambda)^2}{\lambda^3}v \tag{7.44}$$

*For energy*:

$$\frac{\partial}{\partial t}(H)+\frac{\partial}{\partial r}(uH)+\frac{1}{r}\frac{\partial}{\partial \theta}(vH)=\frac{k}{\rho c}\frac{1}{r^2}\frac{\partial}{\partial r}\left(kr^2\frac{\partial H}{\partial r}\right)+\frac{k}{\rho c}\frac{1}{r^2\sin\theta}\frac{\partial}{\partial \theta}\left(\mathrm{k}\sin\theta\frac{\partial H}{\partial \theta}\right)-\frac{\partial}{\partial t}(\Delta H)-\frac{1}{r^2}\frac{\partial}{\partial r}\left(r^2u\Delta H\right)-\frac{1}{r\sin\theta}\frac{\partial}{\partial \theta}(v\sin\theta\Delta H) \tag{7.45}$$

The enthalpy-porosity method is implemented to simulate the phase change problem. The additional terms in the momentum equations, last terms, come from the Darcy law. The details of the enthalpy-porosity method can be found elsewhere (Tan et al. 2009; Khodadadi and Zhang 2001). ANSYS-FLUENT software is used to resolve the governing equations. The domain is divided into small control volumes to capture the variation of the interface front and convection current more precisely. The domain is divided into 31,216 triangular and quadrilateral cells. Since the natural convection-driven phase change is a highly nonlinear problem, the time-step size is defined as 0.1 s. The SIMPLE algorithm is used to resolve the discretized equations iteratively. For each time step, the iterations proceed until the residuals drop below 1E-5.

The validity of the current solution method is checked by comparing the current predictions against the work of Tan et al. (2009). In the reference work of Tan et al. (2009), several thermocouples along the axis monitor the transient temperature variations. The positions of the thermocouples are given in Table 7.10.

The timewise variations of the temperature at points A, B, D, and G are given in Fig. 7.48. Here, the square markers correspond to the experimental work of Tan (2008); the solid black curves are the numerical predictions of Tan et al. (2009). Solid red curves, on the other hand, indicate the current predictions. At point A,

**Table 7.10** Thermocouple positions

| Distance | A | H | G | B | F | E | C | D |
|---|---|---|---|---|---|---|---|---|
| Below (−) or above (+) centerline (mm) | (−) | (+) | (+) | (−) | (+) | 0 | (−) | (−) |
| | 44 | 37.5 | 25 | 37.5 | 12.5 | | 25 | 12.5 |

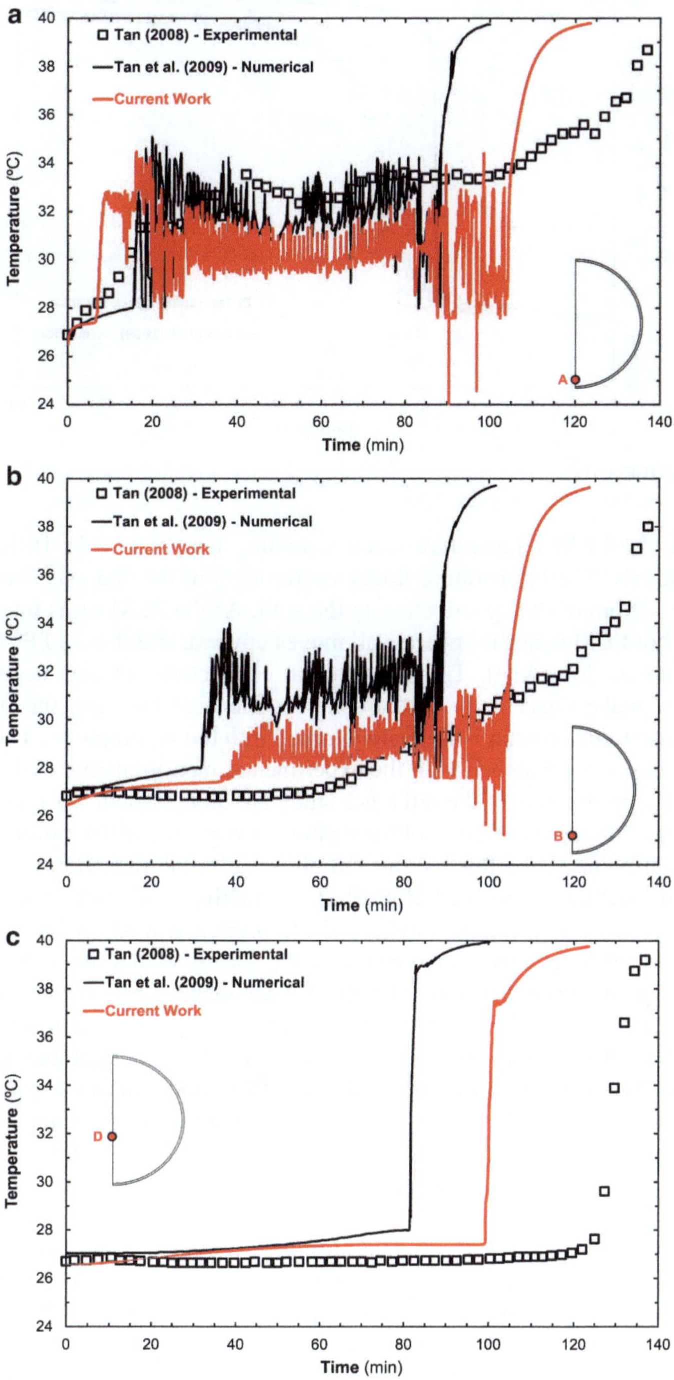


**Fig. 7.48** Timewise variations of the PCM temperature at several positions along the axis. (Yavuz 2017)

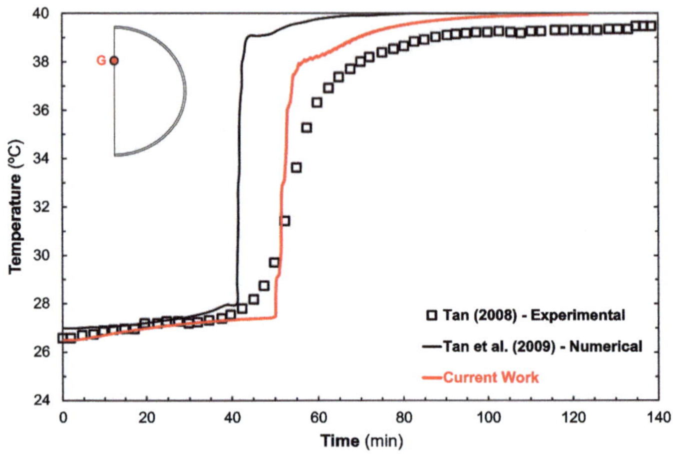


**Fig. 7.48** (continued)

Fig. 7.48a, the PCM temperature reaches melting temperature in 10 min for the current analysis. The temperature fluctuates throughout the analysis since position A is at the bottom of the sphere close to the wall. As the PCM turns into the liquid phase, the hot PCM near the sphere wall moves upward, and the cold PCM near the interface moves downward. Transient natural convection currents inside the thin liquid layer make these parts of the sphere quite unstable, and the fluctuations disappear when the domain is ultimately turned into the liquid phase. The initiation of the melting is not matched with the experimental measurements of the reference work, but the predictions of the reference study are quite close to current results. At point B, Fig. 7.48b, the initiation of the melting occurs around 40 min in the current model. As in point A, the fluctuations are observed in the temperature values until the complete melting is achieved. Regarding the melting time of PCM at point B, the difference between the current model and the reference work is less than 5 min. Point D, Fig. 7.48c, is close to the center of the sphere so that the melting time for this position is quite late. The time of melting is not matching for three cases, and the current predictions stay between the numerical and experimental results from the literature. It is interesting to note that there are no significant temperature fluctuations at this point. This is because the melting front of PCM does not move symmetrically; instead it is eccentric. At the top of the sphere, there is stagnant conduction-dominated melting because the density gradients cause temperature stratification. Besides, at the bottom side of the sphere, the density gradients cause natural convection as described above. Consequently, one can say that while the top side of the sphere is stable, the bottom side is unstable. At point G, Fig. 7.48d, the melting time is observed at around 50 min for the current simulation. The results are in harmony with the experimental work of Tan (2008). Like at point D, the transient temperature variations are stable at point G.

Further comparison is given in Fig. 7.49 regarding the timewise variations of the liquid fraction of PCM. Squares indicate the experimental measurement of Tan

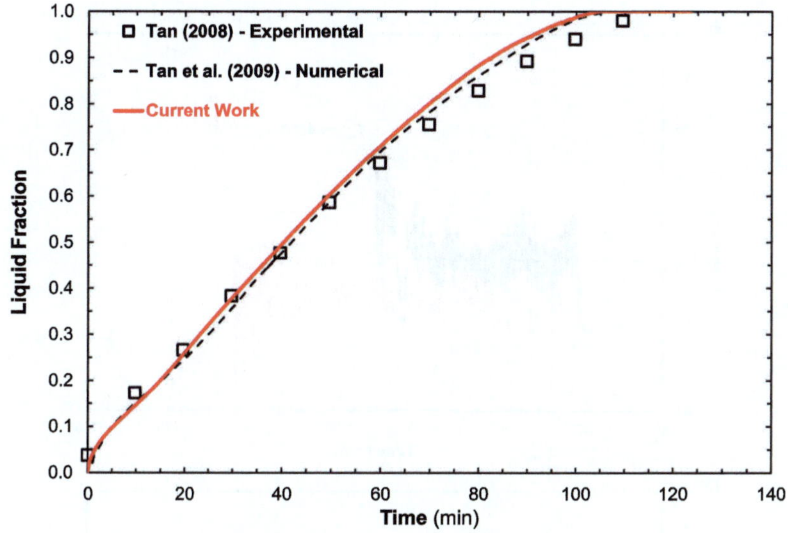


**Fig. 7.49** Timewise variations of the liquid fraction. (Yavuz 2017)

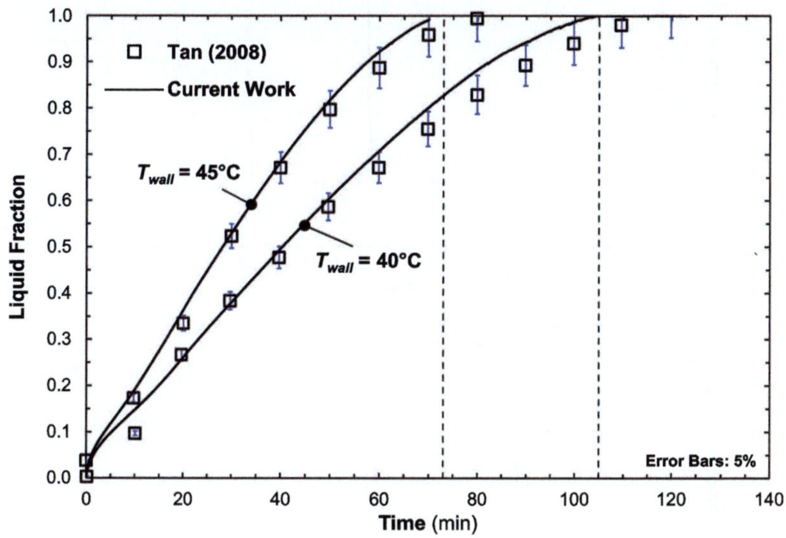


**Fig. 7.50** Influence of wall temperature on the liquid fraction. (Yavuz 2017)

(2008), and the dashed line is the numerical data of Tan et al. (2009). Solid red line, on the other hand, is the liquid fraction variation that is obtained from the current analysis. It is clear that the current predictions are close to the reference results.

To see the influence wall temperature on the inward melting process inside the spherical capsule, timewise variations of the liquid fraction and the temperature at the selected points are compared in Figs. 7.50 and 7.51, respectively. As seen from

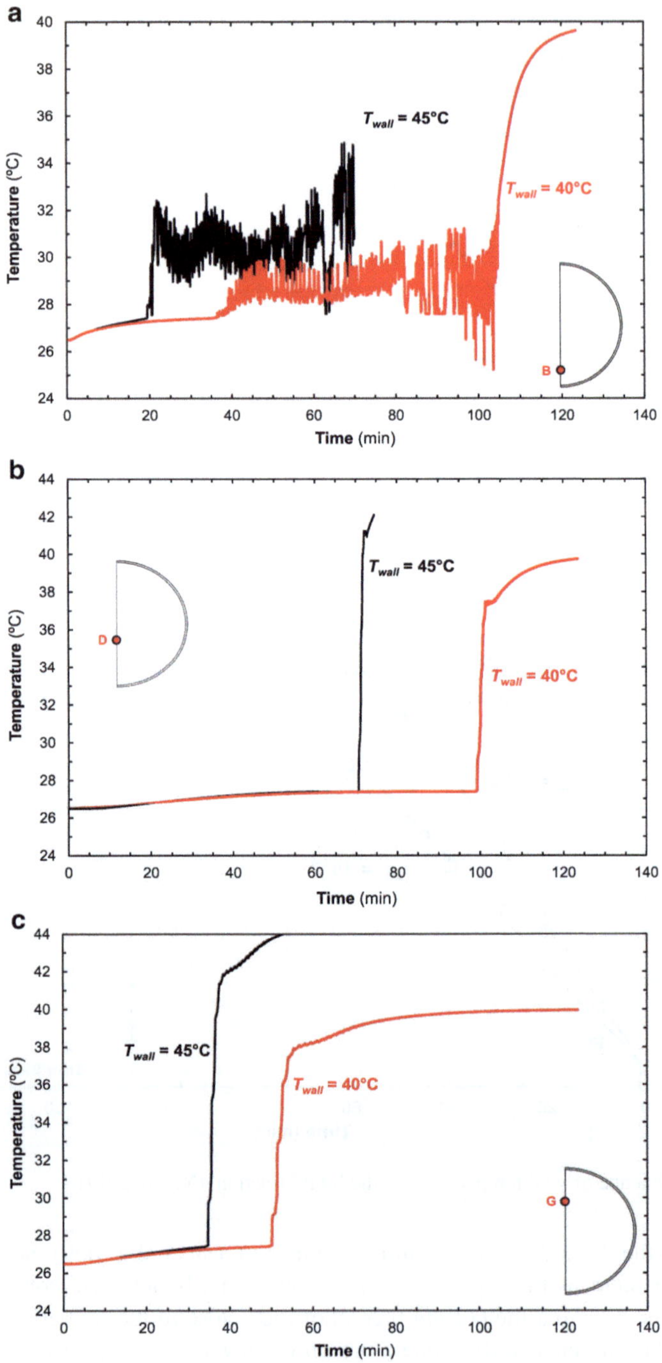


**Fig. 7.51** Influence of wall temperature on the temperature variations. (Yavuz 2017)

Fig. 7.50, increasing the wall temperature from 40 °C to 45 °C enhances the heat transfer within the PCM, and complete melting is achieved more quickly for higher wall temperature. Time for complete melting is obtained at 72.7 min and 105 min, for $T_{wall} = 45$ °C and $T_{wall} = 40$ °C, respectively. Here, the squares correspond to the experimental measurements of Tan (2008), and the predicted liquid fraction variations are in harmony with the reference work with a discrepancy of less than 5%.

In Fig. 7.51, the variations of the PCM temperatures are compared for $T_{wall} = 40$ °C and $T_{wall} = 45$ °C at selected thermocouple points. It is clear that, for each point, the increment on the wall temperature speeds up the melting process. At point B, Fig. 7.51a, the initiation of the melting process is reduced from 40 min to 20 min by increasing the wall temperature by 5 °C. Similarly, for point D, the influence of wall temperature on the temperature variation is very significant. However, at point G, the increment of wall temperature does not shift the curve the same order as in the previous nodes. This may be due to the difference between heat transfer mechanisms on the top and bottom regions of the sphere. At the top side of the sphere, there is a conduction-dominated phase change, and the increment of the wall temperature shifts the curves by only 10 min. Besides, at the bottom of the sphere, there is a natural convection-driven phase change and the heat transfer significantly influenced by the wall temperature.

In Fig. 7.52, the isotherms (*left*) and streamlines (*right*) are given for two different wall temperatures at $t = 10$ min and $t = 30$ min. At the early stages of the melting process, $t = 10$ min, at the top side of the sphere, the temperature of the PCM is warmer than the bottom side. At the bottom region, the hot PCM moves upward, and the cold liquid PCM drops down, and this fluid motion creates several circulation cells. Increasing the wall temperature enhances the rate of heat transfer, and at the top region, the liquid PCM penetrates deeper than at the bottom. At $T_{wall} = 40$ °C, there are five circulation cells at the bottom of the sphere, and 5 °C increment in the wall temperature strengthens the circulations and units. There are four circulation cells at $T_{wall} = 45$ °C at the same flow time. At $t = 30$ min, there is only a single circulation cell at the bottom of each wall temperature. For $T_{wall} = 45$ °C, the liquid PCM regions at the bottom and the top side of the sphere are more significant than the case in which the wall is kept at 40 °C.

## 7.6 Aquifers with TES

Aquifers are freshwater sources that contain a significant amount of water with substantial thermal energy storage capacity. Aquifer TES (ATES) allows storing a significant amount of thermal energy for long durations. An ATES is composed of two discrete well groups. The central concept of ATES is straightforward. In summertime, the water with low temperature from the cold well is pumped through the HVAC unit to remove heat from the building. The water with increased temperature is then reinjected back into the second well in which the high-

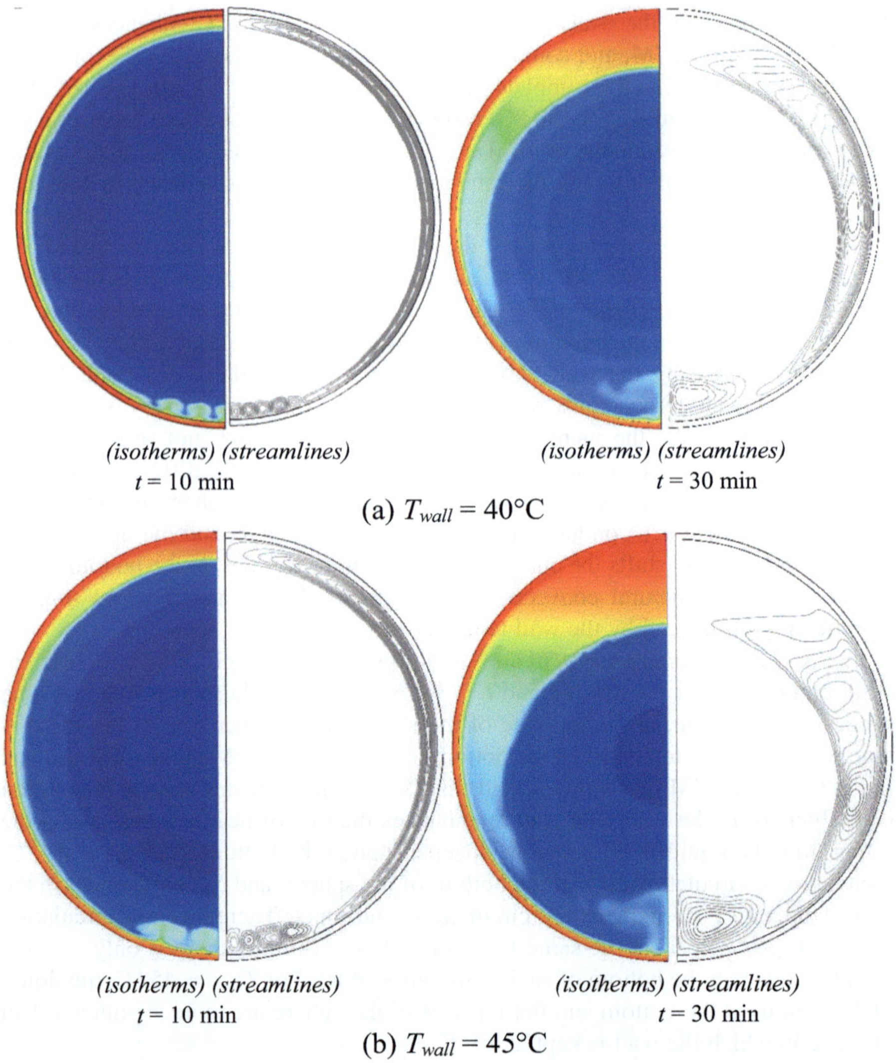


**Fig. 7.52** Isotherms and streamlines at selected instances. (Yavuz 2017)

temperature water is stored (i.e., hot well). In contrast, in winter, the water with a high temperature from the hot well is circulated through the air conditioning unit to heat up the building. ATES systems are integrated into the commercial buildings or discrete heating/cooling facilities which are responsible for the vast amount of heating/cooling demand. In the following, the highlights of the four selected cases on the ATES applications from different countries are introduced.

***Case Study 12:* Numerical Simulation of Auburn University Field ATES Experiments (USA)**

Tsang et al. (1981) introduced the results of the numerical simulations for the second field experiments of the ATES unit that was built at Auburn University. The first set of experiments were completed in 1976, and the results were published by Molz et al. (1978). A numerical model was developed by Papadopulos and Larson (1978) by using the finite difference method. The results of the second set of experiments were published by Molz et al. (1981). It is to be noted that in comparison to the first set of experiments, the net quantity of the hot water that was injected into the ATES unit was significantly improved. Tsang et al. (1981) developed a three-dimensional numerical model to simulate the second set of Auburn University field experiments. It is reported that in the first 6 months, injection and storage processes were conducted cyclically. Approximately 55,000 $m^3$ of water was heated to a mean temperature of 55.2 °C. The supply temperature of the ambient water was 20 °C. At the end of 79.2 days of injection, the warm water was pumped out from the well with an average mass flow rate of 15.65 kg/s until the temperature of the warm water dropped to 32.8 °C. The efficiency of the process was obtained as 66%. The durations of the injection, storage, and the recovery processes were 1900 h, 1213 h, and 987 h, respectively. In the second 6 months, a similar approach was followed. For the second cycle of the ATES unit, the efficiency of the system was obtained as 76%. The injection, storage, and recovery periods of the second cycle were 1521 s, 1502 s, and 1328 s, respectively.

Tsang et al. (1981) compared the experimental measurements for injection mass flow rate and temperature with the predicted average numerical results. The timewise and spanwise variations of the temperature values are evaluated with the model and compared with the experimental measurements for the first and second cycle for the ATES unit. In Fig. 7.53, the temperature contours are given together with the measured temperature values that are obtained at the end of the injection process for the first cycle ($t = 1900$ s). Here, the solid lines correspond to the wells. It is noted that due to the influence of buoyancy effects, the thermal disturbances are observed at around $r = 45$ m.

Figure 7.54 compares the timewise variation of the predicted production temperature with the measured one. The maximum discrepancy between the measured and the predicted temperatures is less than 1 °C. The recovery efficiency is defined as the ratio of the net energy recovered to the net energy injected into the wells for the cycle. The predicted and measured efficiencies for the cycle are 68% and 66%, respectively. The comparative results suggest that the numerical model that was developed by Tsang et al. (1981) has a good consistency with the experimental observations.

Table 7.11 summarizes the energetic outputs of the first and second cycles of the ATES system. It is interesting to note that the efficiency of the system significantly improved as the cycle of the ATES unit increases. Tsang et al. (1981) stated that the efficiency of the system would further improve for subsequent cycles of the ATES system.

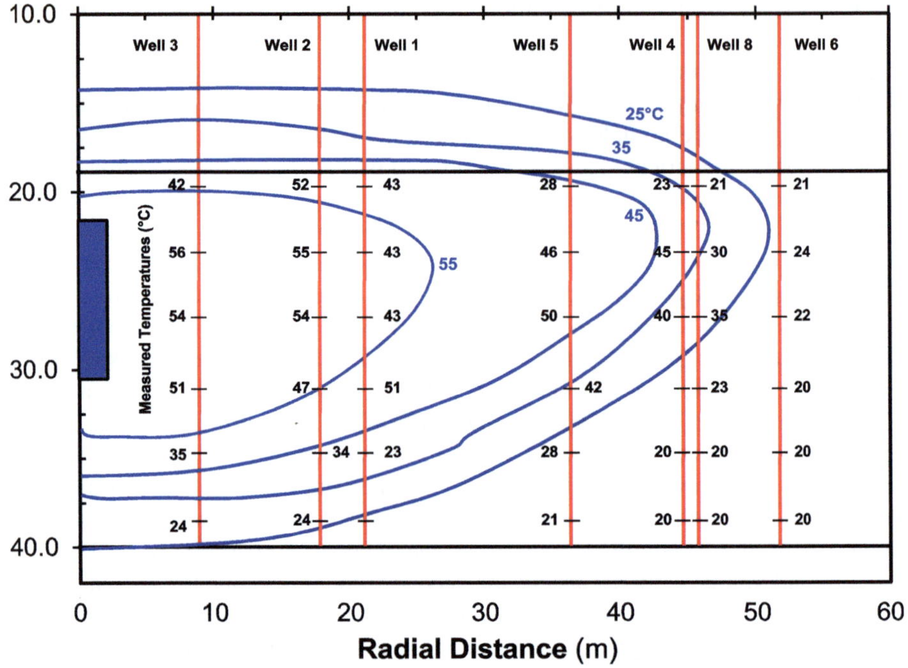


**Fig. 7.53** Temperature distribution at the end of the injection process of the first cycle. (Adapted from Tsang et al. 1981)

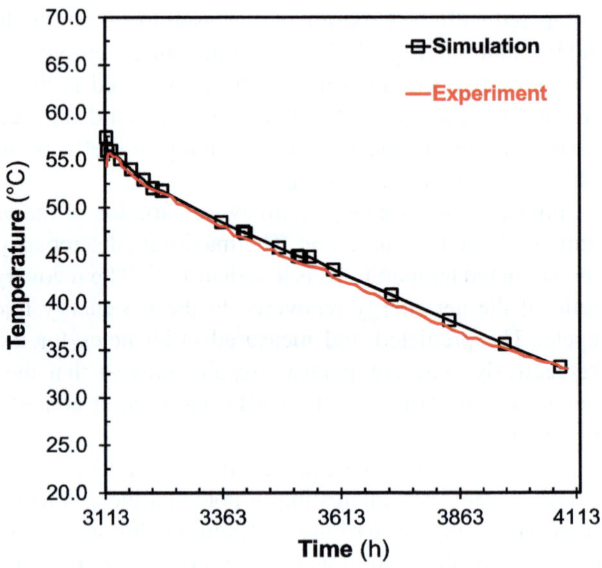


**Fig. 7.54** Timewise variation of the predicted and measured production temperatures from ATES system for the first cycle. (Data from Tsang et al. 1981)

**Table 7.11** Energetic outputs of the ATES unit

| Cycle | Injected energy (J) | Produced energy (J) | Efficiency (%) |
|---|---|---|---|
| First cycle | $0.721 \times 10^{13}$ | $0.486 \times 10^{13}$ | 68 |
| Second cycle | $0.765 \times 10^{13}$ | $0.591 \times 10^{13}$ | 78 |

Adapted from Tsang et al. (1981)

***Case Study 13:*** **Analysis of a Real Case of Multiple ATES Systems (Netherlands)**
Bakr et al. (2013) developed a model to simulate coupled flow and heat transfer processes in porous media and implementation of the model to ATES systems. *Conductive*, *convective*, and *dispersive* heat fluxes through a porous medium are defined as follows:

$$H_C = -\lambda_b \nabla T \quad (\text{in W/m}^2) \tag{7.46}$$

$$H_a = q\rho_w c_w T \quad (\text{in W/m}^2) \tag{7.47}$$

$$H_d = -\rho_w c_w \alpha q \nabla T \quad (\text{in W/m}^2) \tag{7.48}$$

For a transient system, Eqs. (7.46), (7.47), and (7.48) are combined with the source/sink mixing term to yield the conservation of energy in the following form:

$$(\rho c)_b \frac{\partial T}{\partial t} = \nabla[(-\rho_w c_w \alpha q + \lambda_b)\nabla T] - \rho_w c_w \nabla q T - \rho_w c_w q_s T_s \quad (\text{in W/m}^3) \tag{7.49}$$

The balance equation is then reduced by using the porosity ($\theta$) and the specific discharge ($q$) to obtain

$$R_T \frac{\partial(\theta T)}{\partial t} = \nabla \cdot (D_T \cdot \nabla T) - \nabla(vT) - \frac{q_s}{\theta} T_s \quad (\text{in K/m}) \tag{7.50}$$

For the mass transport within the porous medium, the governing equation is evaluated which has a similar structure to Eq. (7.50). Further details about the governing equations can be found in the reference work by Bakr et al. (2013). A numerical code is developed by Bakr et al. (2013) to simulate the thermal behavior of the ATES system. Annual performance of the ATES systems is assessed by evaluating the thermal efficiency of the interference. The efficiency of the ATES system is defined regarding the recovered and the injected rate of thermal energies. Bakr et al. (2013) considered the ATES systems that are installed in the city of the Hague, the Netherlands. It is reported that there is a total of 19 ATES systems in use in this city within an area of about 3.8 km$^2$. A total of 76 wells are under operation. Table 7.12 summarizes the characteristics of the aquifer system that is considered. In the model, the total number of stress periods (injection and recovery periods) is 21 of half-a-year length each. It is assumed that the steady-state condition is valid for the fluid flow

**Table 7.12** Characteristics of the ATES system

| Parameter | Value | Unit |
|---|---|---|
| Effective porosity ($\theta$) | 0.35 | – |
| Specific heat capacity of water ($C_w$) | 4183 | J/(kg°C) |
| Density of water ($\rho_w$) | 1000 | kg/m$^3$ |
| Bulk thermal conductivity ($\lambda_{bulk}$) | 2.55 | W/(m°C) |
| Molecular diffusion coefficient ($D_m$) | 0.125 | m$^2$/d |
| Longitudinal dispersivity ($\alpha_L$) | 0 | m |
| Transverse dispersivity ($\alpha_T$) | 0 | m |
| Thermal retardation factor ($R_T$) | 2 | – |

Adapted from Bakr et al. (2013)

after a short time of starting the pump. In Table 7.13, on the other hand, mean pumping and injection rates of each well during the cold and warm periods are listed.

In Figs. 7.55 and 7.56, illustrative results are represented for interference among temperature distributions for the wells at the end of stress period 20 (winter) and 21 (summer), respectively. It is noted that there are merging temperature contours for neighboring wells either below or above the ambient groundwater temperature. Long-term operation time is also evaluated with a mathematical model to introduce the efficiency of the system. It is interesting to note that the efficiencies of the ATES systems are increased over time of operation. As an instance, the efficiency of a system is improved from 68% to 87% in a 10-year working period. This observation is consistent with the expectation of Tsang et al. (1981). The influence of interference on the efficiencies of the ATES systems is also revealed. It is denoted that the interference affects the efficiency both positively and negatively. The interference may improve or reduce the efficiency as much as 20%. It is suggested that a high-performance ATES system could be achieved by optimizing the distribution of the wells, especially the proximity of wells to each other. The working parameter of the ATES system, namely, the pumping or rejection rates, also significantly influences the performance.

***Case Study 14:* Heating and Cooling of a Hospital with Solar Energy Integrated ATES (Turkey)**

Paksoy et al. (2000) considered an ATES unit for heating and cooling of the Cukurova University, Balcali Hospital in Adana, Turkey. Two conceptual designs were considered in the study. In the first one, two heat exchangers (HEX1 and HEX2) with two wells which were separated from each other by a suitable distance (see Fig. 7.57a). In the winter mode, the low-temperature cold was stored in the cold well to maintain a comfortable temperature by rejecting the heat from the hospital during summertime. In the summer mode, on the other hand, high-temperature heat was stored in the warm well for preheating the air during winter time. The ATES system was placed nearby the Seyhan Lake to provide low-temperature cold energy from the lake water. In the second system, solar collectors were used to increase further the temperature of the groundwater that was stored in the warm well (see Fig. 7.57b). It is reported that the temperature of the warm well reached as much as 40 °C before initiating discharging heat from the warm well.

**Table 7.13** Pumping and injection rates of the wells during cold and warm periods

| System Id | Well Id | Type | $Q$ (m$^3$/d) |
|---|---|---|---|
| S01 | W01 | Cold | 730.6 |
| S01 | W02 | Warm | −730.6 |
| S02 | W01 | Cold | 547.9 |
| S02 | W02 | Cold | 547.9 |
| S02 | W03 | Cold | 547.9 |
| S02 | W04 | Warm | −547.9 |
| S02 | W05 | Warm | −547.9 |
| S02 | W06 | Warm | −547.9 |
| S03 | W01 | Cold | 566.2 |
| S03 | W02 | Cold | 566.2 |
| S03 | W03 | Cold | 566.2 |
| S03 | W04 | Cold | 566.2 |
| S03 | W05 | Cold | 566.2 |
| S03 | W06 | Warm | −566.2 |
| S03 | W07 | Warm | −566.2 |
| S03 | W08 | Warm | −566.2 |
| S03 | W09 | Warm | −566.2 |
| S03 | W10 | Warm | −566.2 |
| S04 | W01 | Cold | 547.9 |
| S04 | W02 | Cold | 547.9 |
| S04 | W03 | Warm | −547.9 |
| S04 | W04 | Warm | −547.9 |
| S06 | W01 | Cold | 200.9 |
| S06 | W02 | Warm | −200.9 |
| S07 | W01 | Cold | 526.0 |
| S07 | W02 | Warm | −526.0 |
| SO8 | W01 | Cold | 13.7 |
| SO8 | W02 | Warm | −13.7 |
| S09 | W01 | Cold | 33.2 |
| S09 | W02 | Warm | −33.2 |
| S10 | W01 | Cold | 730.6 |
| S10 | W02 | Cold | 730.6 |
| S10 | W03 | Cold | 730.6 |
| S10 | W04 | Warm | −730.6 |
| S10 | W05 | Warm | −730.6 |
| S10 | W06 | Warm | −730.6 |
| S11 | W01 | Cold | 712.3 |
| S11 | W02 | Cold | 712.3 |
| S11 | W03 | Warm | −712.3 |
| S11 | W04 | Warm | −712.3 |
| S12 | W01 | Cold | 749.5 |
| S12 | W02 | Cold | 749.5 |

(continued)

**Table 7.13** (continued)

| System Id | Well Id | Type | $Q$ ($m^3$/d) |
|---|---|---|---|
| S12 | W03 | Cold | 749.5 |
| S12 | W04 | Warm | −749.5 |
| S12 | W05 | Warm | −749.5 |
| S12 | W06 | Warm | −749.5 |
| S13 | W01 | Cold | 305.9 |
| S13 | W02 | Warm | −305.9 |
| S14 | W01 | Cold | 1424.7 |
| S14 | W02 | Warm | −1424.7 |
| S15 | W01 | Cold | 1369.9 |
| S15 | W02 | Warm | −1369.9 |
| S16 | W01 | Cold | 599.1 |
| S16 | W02 | Cold | 599.1 |
| S16 | W03 | Cold | 599.1 |
| S16 | W04 | Warm | −599.1 |
| S16 | W05 | Warm | −599.1 |
| S16 | W06 | Warm | −599.1 |
| S17 | W01 | Cold | 730.6 |
| S17 | W02 | Cold | 730.6 |
| S17 | W03 | Cold | 730.6 |
| S17 | W04 | Warm | −547.9 |
| S17 | W05 | Warm | −547.9 |
| S17 | W06 | Warm | −547.9 |
| S17 | W07 | Warm | −547.9 |
| S18 | W01 | Cold | 949.8 |
| S18 | W02 | Warm | −949.8 |
| S19 | W01 | Cold | 803.7 |
| S19 | W02 | Warm | −803.7 |
| S20 | W01 | Cold | 353.1 |
| S20 | W02 | Cold | 353.1 |
| S20 | W03 | Cold | 353.1 |
| S20 | W04 | Warm | −264.8 |
| S20 | W05 | Warm | −264.8 |
| S20 | W06 | Warm | −264.8 |
| S20 | W07 | Warm | −264.8 |

Adapted from Bakr et al. (2013)

A commercial software (CONFLOW) is used to evaluate the thermal behavior of the ATES unit. The software allows to investigate the effects of the number of wells, distances between wells, and also the thermal conditions of the surrounding during the charging (pumping) and discharging (reinjection) processes. It is reported that to obtain an optimum thermal front, the distance between straight well groups should be in the order of 300–350 m. On the other hand, the distance between the wells which are in the same group should be in the range of 60–80 m. It is denoted that to

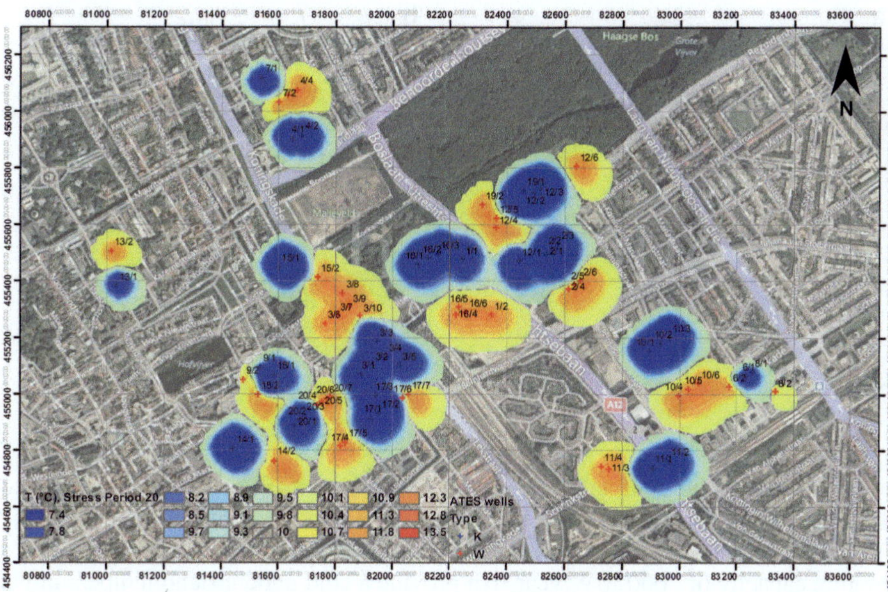


**Fig. 7.55** Temperature contours at the end of stress period 20 (*winter*). (Bakr et al. 2013)

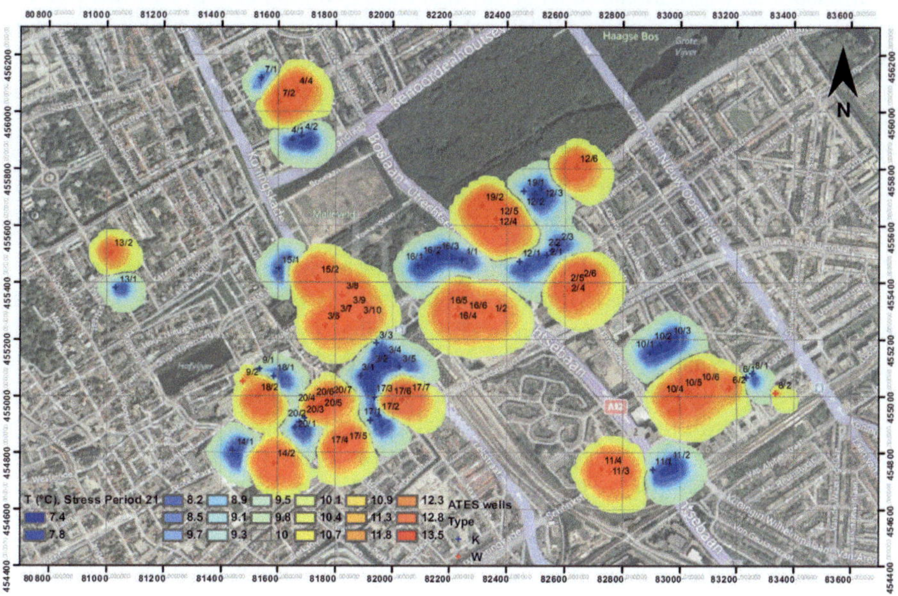


**Fig. 7.56** Temperature contours at the end of stress period 21 (*summer*). (Bakr et al. 2013)

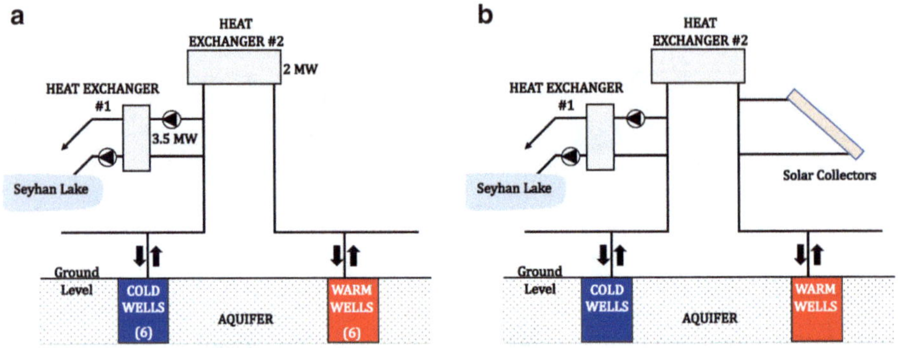


**Fig. 7.57** Conceptual ATES system. (Reproduced from Paksoy et al. 2000)

provide heating and cooling at +10 °C, a field with 350 m × 350 m is enough to achieve storage of 14,000 MW/year. Economic and environmental benefits of the ATES unit are also discussed in detail by Paksoy et al. (2000). An ATES with a cooling capacity of 7000 MW/year will provide a cold underground temperature at 9 °C at the end of the charging process. Due to the extended storage duration, there will be heat gain, and the temperature of the cold well will improve to 10 °C. In summertime, the ATES unit will provide 6500 MW of cooling load for 3000 h. It corresponds to significant energy savings as the cooling capacity of the conventional air-conditioning unit will be reduced. Besides, using the high-temperature water that is stored in the warm well during the winter time will also reduce the energy consumption that is responsible for preheating the air. It is reported that the approximate oil savings are in the order of 1000 $m^3$/year with the implementation of the ATES during wintertime. Consequently, the usage of ATES reduces the $CO_2$ (reduced by 2100 tons/year), $SO_x$ (reduced by 7 tons/year), and $NO_x$ (reduced by 8 tons/year) emissions and ozone depletion as the capacities are significantly limited.

***Case Study 15:*** **Heating and Cooling of a Building with ATES (Canada)**

AlZahrani and Dincer (2016) developed a thermodynamic model to perform the energy and exergy analyses of an ATES unit that is integrated in a building in Oshawa, ON, Canada. As a first step, they reviewed the variation of the outdoor temperature throughout the year to decide the heating and cooling demand of the building. As shown in Table 7.14, the ATES should be worked in cooling mode for 4 months. On the other hand, ATES should be worked in heating demand for 6 months. For 2 months, as there is no heating or cooling demand, ATES is in the storage mode.

The experimental data of Sykes et al. (1982) is used to evaluate the discharging temperature profile. A linear-fitted equation is proposed to define the variation of the discharging temperature. Based on the climatic conditions of Oshawa, the parameters that are provided in Table 7.15 are used in the analyses.

**Table 7.14** Operating modes of the ATES throughout the year

| Operating mode | January | February | March | April | May | June | July | August | September | October | November | December |
|---|---|---|---|---|---|---|---|---|---|---|---|---|
| *Charging* | Cooling | Cooling | Cooling | Cooling | Storing | Heating | Heating | Heating | Heating | Storing | Cooling | Cooling |
| *Discharging* | Heating | Heating | Heating | Heating | Storing | Cooling | Cooling | Cooling | Cooling | Storing | Heating | Heating |

Adapted from AlZahrani and Dincer (2016)

**Table 7.15** Operating modes of the ATES throughout the year

| Parameter | Heating mode | Cooling mode |
|---|---|---|
| Charging temperature | 358 K | 275 K |
| Storing time | 30 days | 30 days |
| Temperature drop/rise during storage process | 5, 10, 15, and 20 K | 1, 2, 3, and 4 K |
| Discharging temperature | Linear profile | Linear profile |
| Charging mass flow rate | 1 kg/s | 1 kg/s |
| Discharging mass flow rate | 1 kg/s | 1 kg/s |
| Average ambient temperature | 298 K | 285 K |
| Charging time | 150 days | 150 days |
| Discharging time | 150 days | 150 days |

Adapted from AlZahrani and Dincer (2016)

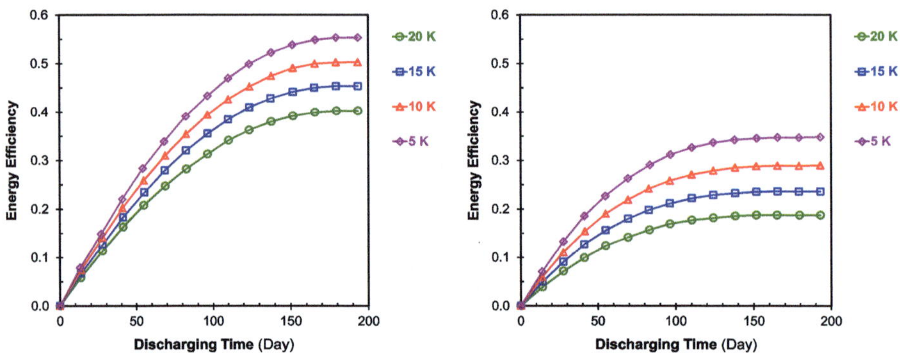


**Fig. 7.58** Evolution of energy and exergy efficiencies for the heating mode. (Data from AlZahrani and Dincer 2016)

As a result, variations of the exergy quantities, energy, and exergy efficiencies are provided by varying the temperature drop and rise during the discharging periods for heating and cooling modes of an ATES system. Figure 7.58 provides the variations of the energy and exergy efficiencies during the discharging period of heating mode. Here, the temperature drop varies from 5 K to 20 K to assess the influence of the heat loss during the storage period on the system performance. The efficiency of the system increases as the discharging time increases and reaches the maximum value at the end of the discharging process. For the case in which the heat loss is minimum, $T_{loss} = 5$ K, the efficiency is approximately 55%. Efficiency drops to 40% as the temperature reduction is increased to 20 K. The energy-based analysis does not give information about the usefulness of the discharged energy. Exergy analyses consider the quality of the energy by taking into account the temperature difference between the system and its surroundings. The maximum exergy efficiency is about 35%, and it reduces with increasing heat loss during the storage periods. The exergy efficiency drops below 20% in the worst scenario.

## 7.7 Greenhouse with TES

Thermal energy storage is widely used in greenhouses to maintain indoor temperature within a predefined temperature range. During the daytime, the excessive amount of energy is stored in the water tanks, barrels, or ground tubes to meet the heating load of the greenhouse at nighttime. Such a system can include solar collectors to supply thermal energy to the storage medium, but a typical approach is placing the storage tanks or barrels around the greenhouse facing the solar radiation and collecting the thermal energy without using any dedicated solar collector. There are various experimental works on investigating the performance of sensible heat and latent heat TES systems with water, rock, or PCM for greenhouse heating. Santamouris et al. (1994) prepared a comprehensive review which includes 95 of the completed studies worldwide. In Tables 7.16, 7.17, and 7.18, the passive solar greenhouse heating applications with water storage, rock storage, and latent heat thermal energy storage are compared in many aspects. The performances of the systems are evaluated by considering the increments in the average indoor temperature against the ambient temperature that is achieved by using the storage unit. In the following, four selected cases on the greenhouse heating systems with TES applications are briefly introduced.

***Case Study 16:*** **Heat Storage with Water Mass in a Greenhouse (India)**
Gupta and Tiwari (2002) developed a computer model to simulate the transient thermal behavior of a greenhouse with a sensible heat thermal energy storage unit. A storage tank, with 0.55 m in diameter and 0.9 m in height, was placed inside the greenhouse to store the solar energy during the sunshine hours. It is reported that the tank was filled with water mass and the outer surface of the tank was blackened to increase the fraction of the absorbed solar energy. During the sunshine hours, a portion of the solar radiation was transmitted inside the greenhouse. The solar radiation that passed through the greenhouse was absorbed by the storage tank, floor, and the other components within the greenhouse. Conductive, convective, and radiative modes of heat transfer take place between the components of the greenhouse, and the room air increases. After sunset, the energy that is stored in the water tank and the ground releases and prevents a sudden drop in room temperature. In Fig. 7.59, the modes of heat transfer are schematically illustrated.

In the study, a modified IARI mode greenhouse, with length, width, and height of 5.03 m × 4.13 m × 2.42 m, is considered. A numerical model is developed, and the validity of the model is proven by comparing the predictions with the experimental measurement. In the model, energy balance equations are written for each component of the greenhouse as

*Water tank:*

$$\alpha_T \tau \sum_{i=1}^{8} I_{ci}(t) A_{ci} F_T(t) = (mc)_W \frac{dT_W}{dt} + h_W (T_W - T_r) A_T \tag{7.51}$$

**Table 7.16** Thermal energy storage in the greenhouse with water mass

| Location | Ground area ($m^2$) | Cover material | Cultivation | Storage medium | Installed | Results | References |
|---|---|---|---|---|---|---|---|
| Adana, TU | 120 | Glass | Tomatoes | Ground tubes | 1986 | 1–1.5 °C higher | Baytorum (1987) |
| Ayia Napa, CY | – | P.E. | Vegetables | Ground tubes | 1985 | 2–4 °C higher | Fotiades (1988) |
| Acheleia, CY | – | P.E. | Vegetables | Ground tubes | 1985 | 2–4 °C higher | Fourcy (1981) |
| Almeria, SP | 72 | P.E. | Strawberries | Ground tubes (1.5 $m^3$) | 1988 | 2–4 °C higher | Montero et al. (1987) |
| Arizona, USA | 22 | Filon | Plants | Water tanks (2.25 $m^3$) | 1976 | 16–22 °C > $T_{atr}$ | Mac Kinnon (1981) |
| Arlington, USA | 40 | Glass | Plants | Water tanks (18.2 L) | 1977 | 4–5 °C higher | Straub et al. (1978) |
| Atalia, TU | 350 | Glass | Vegetables | Ground tubes | 1986 | 4 °C higher | Sallanbas et al. (1987) |
| Atalia, TU | 350 | Glass | Vegetables | Ground tubes | 1987 | 10 °C higher | Sallanbas et al. (1987) |
| Athens, GR | 150 | P.E. | Tomatoes | Ground tubes (5 $m^3$) | 1987 | 2–4 °C higher | Kyritsis and Mavroyiannopoulos (1987) |
| Avignon, FR | 231 | P.E. | Plants | Ground tubes (5.4 $m^3$) | 1978 | 2–4 °C higher | Baille (1987) |
| Beit Dagan, IL | 1000 | P.E. | Roses | Water tanks (400 $m^3$) | 1979 | 11 °C higher | Levav and Zamir (1987) |
| Copenhagen, DN | 12 | Glass | Tomatoes | Water barrels (3.2 $m^3$) | 1986 | 4 °C higher | Sørensen (1989) |
| Deryneia, CY | – | P.E. | Vegetables | Ground tubes | 1985 | 2–4 °C higher | Fotiades (1988) |
| Dordogne, FR | 190 | Glass | Plants | Water barrels (4 $m^3$) | 1979 | 10 °C higher | Mercier (1982) |
| Fanar, RL | 287 | P.E. | Melons | Ground tubes (25.6 $m^3$) | 1985 | 2 °C higher | Farah (1987) |
| Flagstaff, USA | 167 | Filon | Vegetables | Water barrels (0.22 $m^3$) | 1979 | 13–22 °C > $T_{atr}$ | Flerking (1981) |
| Grenoble, FR | 72 | Double glass | Plants | Water tank (1.7 $m^3$) | 1979 | – | Fourcy (1982) |
| Grenoble, FR | 150 | Double P.E. | Plants | Water tank | 1976 | 70% cover | Fourcy (1981) |
| Hannover, D | 218 | P.E. | Plants | Ground tubes (4.4 $m^3$) | 1985 | 3 °C higher | Von Zabeltitz and Rosocha (1987) |

| | | | | | | | |
|---|---|---|---|---|---|---|---|
| Hannover, D | 230 | Glass | Vegetables | Water tanks (21 $m^3$) | 1976 | 18.3% cover | Damrath and Von Zabeltitz (1987) |
| Helsinki, SF | 30 | Double PVC | Tomatoes | Water tanks (6 $m^3$) | 1980 | 2 °C higher | Yiannoulis (1990) |
| Hérault, FR | 104 | P.E. | Plants | Water barrels | 1988 | 2–10 °C higher | Campiotti et al. (1988) |
| Israel | 260 | P.E. | Melons | Ground tubes (12 $m^3$) | 1985 | 2.5 °C higher | Esquira et al. (1987) |
| Israel | – | Polycarbonate | Plants | Water tanks | 1985 | 50% cover | Zeroni (1990) |
| Lepa, FL | – | P.E. | Vegetables | Water barrels | 1982 | 5 °C higher | Liskola (1987) |
| Lisbon, P | 300 | P.E. | Melons | Ground tubes (15 $m^3$) | 1985 | 2–4 °C higher | Pacheco et al. (1987) |
| Nashville, USA | 30 | P.E. | Plants | Water tanks (1 $m^3$) | 1976 | 2–3 °C higher | Nash and Williamson (1978) |
| Nicosia, CY | 300 | P.E. | Tomatoes | Ground tubes | 1985 | 2–4 °C higher | Fotiades (1987) |
| Prague, CZ | 135 | P.E. | Tomatoes | Ground tubes | 1985 | – | Jelinkova (1987) |
| Prague, CZ | 135 | P.E. | Cucumber | Ground tubes | 1985 | – | Jelinkova (1987) |
| Quebec, CDN | 95 | P.E. | Tomatoes | Ground tubes | 1980 | 34 °C > min $T_{atr}$ | Woodston (1982) |
| Rome, I | 300 | Polycarbonate | Cucumber | Water tank (5.8 $m^3$) | 1983 | 2–10 °C higher | Campiotti et al. (1988) |
| Rome, I | 100 | Polycarbonate | Vegetables | Water barrels | 1988 | – | Picciurro (1988) |
| Salonika, GR | 500 | P.E. | Vegetables | Water tanks (60 $m^3$) | 1983 | 5–6 °C higher | Grafiadellis (1987) |
| Salonika, GR | 500 | P.E. | Flowers | Ground tubes (100 $m^3$) | 1984 | 2–4 °C higher | Grafiadellis (1985) |
| Sotira, CY | – | P.E. | Vegetables | Ground tubes | 1985 | 2–4 °C higher | Fotiades (1988) |
| Tarn, FR | 60 | Glass | Flowers | Water barrels (14.2 $m^3$) | 1980 | 70% cover | Mercier (1982) |
| Tennessee, USA | 235 | Fiberglass | Flowers | Water barrels (31 $m^3$) | 1979 | 75% cover | Gowan and Black (1980) |
| Tunisia | – | P.E. | Melons | Ground tubes | 1986 | – | Mougou and Verlodt (1987) |

Adapted from Santamouris et al. (1994)

**Table 7.17** Thermal energy storage in the greenhouse with a rock bed

| Location | Ground area ($m^2$) | Cover material | Cultivation | Storage medium | Installed | Results | Ref. |
|---|---|---|---|---|---|---|---|
| Aranjuez, SF | 500 | | Flowers | Pebble bed | 1989 | | Vonarburg and Gallacher (1982) |
| Barald, ALG | 240 | P.E. | | Rock bed (50 mm gravel) | 1988 | 4–6 °C higher | Bouhdgar and Boulbing (1990) |
| Bonn, D | 161 | Double glass | Flowers | Rock bed | 1979 | 20% cover | Eggers and Vickermann (1983) |
| Budapest, HUN | 100 | P.E. | | Bricks | 1984 | | Kavin and Kurtan (1987) |
| Destelbergen, B | 1000 | | | Pebble bed | 1989 | | Deforche (1990) |
| Hannover, D | 1700 | Polycarbonate | Pot plants | Rock bed | 1981 | 30% cover | Bredenbeck (1987) |
| Montreal, CAN | 2850 | P.E. | Tomatoes | Rock bed (40 mm gravel) | 1982 | 40% cover | Bricault et al. (1982) |
| Murcia, Spain | 50 | P.E. | Melons | Rock bed | 1984 | | Garcia (1987) |
| Nicosia, CY | 300 | R.E. | | Rock bed | 1982 | 76% cover | Fotiades (1987) |
| Novoli, USSR | 500 | Fiberglass | Plants | Solar geometry | 1986 | 3–4 °C higher | Saidov and Akhtamov (1987) |
| Oregon, USA | 19 | Filon | | Rock bed | 1980 | 10–20 °C mean $T$ | Mazria and Baker (1981) |
| Prague, CZ | 432 | Glass | | Rock bed | 1984 | | Jelinkova (1987) |
| Tascend, USSR | 40 | Double P.E. | Plants | Rock bed (2 cm gravel) | 1978 | 13 °C higher | Arezov and Niyazov (1980) |
| Toulouse, FR | 30 | Glass | Vegetables | Rock bed | 1980 | | Bonrehi (1982) |

Adapted from Santamouris et al. (1994)

**Table 7.18** Thermal energy storage in the greenhouse with a rock bed

| Location | Ground area ($m^2$) | Cover material | Cultivation | Storage medium | Storage mass (kg) | Installed | Results | Ref. |
|---|---|---|---|---|---|---|---|---|
| Antibes, FR | 445 | Glass | Plants | $NaOH + Cr_2N$ | | 1978 | Gain 5000 L oil | Paris (1981) |
| Avignon, FR | 176 | Polycarbonate | Tomatoes | $CaCl_26H_2O$ | 3000 | 1985 | 30% cover | Boulard and Baille (1987) |
| Canberra, AUS | 20 | Glass | Plants | $CaCl_26H_2O$ | 100 | 1983 | 1 °C > Tair | Brandstetter (1987) |
| Israel | 200 | Glass | Roses | $CaCl_26H_2O$ | 3000 | 1979 | | Groves (1984) |
| Israel | 66 | Glass | Vegetables | $CaCl_26H_2O + CaBr_26H_2O$ | | 1983 | 21% cover | Groves (1984) |
| Nice, FR | 5000 | Glass | Roses | $CaCl_26H_2O$ | 135,000 | 1979 | 75% cover | Jaffrin and Cadier (1982) |
| Nice, FR | 500 | Glass | Roses | $CaCl_26H_2O$ | | 1986 | 51% cover | Jaffrin et al. (1982) |
| N. Carolina, USA | 100 | Fiberglass | Plants | $CaCl_26H_2O$ | | 1982 | 2 °C higher | Huang et al. (1986) |
| Patras, GR | 4 | P.E. | Plants | $CaCl_26H_2O$ + acetic acid | 32 | 1983 | 2–3 °C higher | Yoshioka (1989) |
| Rosignano, IT | 200 | Glass | Flowers | $CaCl_26H_2O$ | 2800 | 1983 | 22% cover | Balducci (1985) |
| Japan | 352 | Glass | Tomatoes | $0.45Na_2\ SO_410H_2O/0.45Na_2\ CO_310H_2O/0.1NaCl$ | 2500 | 1983 | 8 °C higher | Machida et al. (1985) |

Adapted from Santamouris et al. (1994)

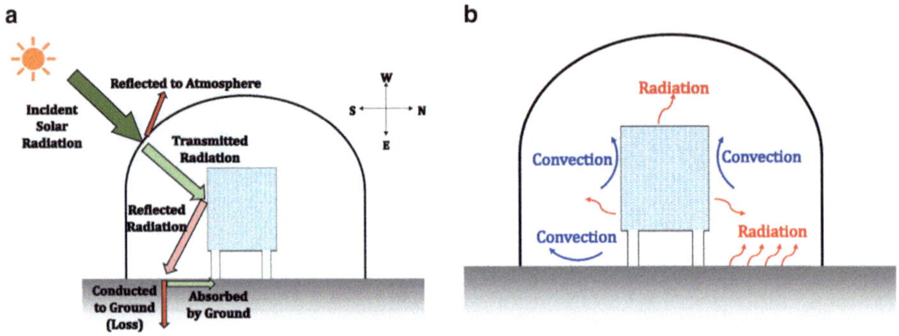


**Fig. 7.59** Sketch of the greenhouse for (**a**) sunshine hours and (**b**) off sunshine hours. (Reproduced from Gupta and Tiwari 2002)

*Floor:*

$$\alpha_g[1 - F_T(t)]\tau \sum_{i=1}^{8} I_{ci}(t)A_{ci} = U_g\left(T_g\big|_{x=0} - T_a\right)A_g + h_g\left(T_g\big|_{x=0} - T_r\right)A_g \quad (7.52)$$

*Room air:*

$$h_W(T_W - T_r)A_T + h_g\left(T_g\big|_{x=0} - T_r\right)A_g$$
$$= U_c(T_r - T_a)A_c + U_d(T_r - T_a)A_d + 0.33NV(T_r - T_a) \quad (7.53)$$

Equations (7.51), (7.52), and (7.53) are reorganized to obtain explicit expressions for the water temperature, room temperature, and ground temperature. A dimensionless thermal load leveling (TLL) index is defined to evaluate the optimum design condition for the greenhouse:

$$Thermal\ Load\ Leveling(TLL) = \frac{(T_{r,\max} - T_{r,\min})}{(T_{r,\max} + T_{r,\min})} \quad (7.54)$$

It is stated that the TLL value is a significant parameter to assess the thermal condition of a greenhouse. During summer, the TLL should be maximum, and during winter TLL should be minimum to achieve the best condition for crop growth. In Fig. 7.60, the effect of the mass of storage water on the TLL is shown for 2 typical days that are selected in the winter and summer seasons. Increasing the mass of water reduces the TLL for both winter and summer. The amount of mass could be optimized by considering the summer and winter conditions.

***Case Study 17:* Packed-Bed Heat Storage Unit for a Greenhouse (Turkey)**
Öztürk and Başçetinçelik (2003) carried out energy and exergy analyses for a greenhouse with packed-bed heat storage unit. The tunnel greenhouse had a floor area of 120 m$^2$, and an external heat collection unit was developed to collect and

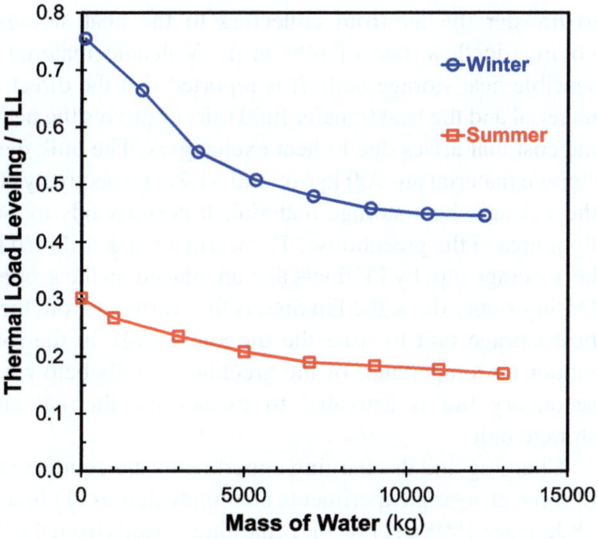


**Fig. 7.60** Influence of the mass of water on the TLL. (Data from Gupta and Tiwari 2002)

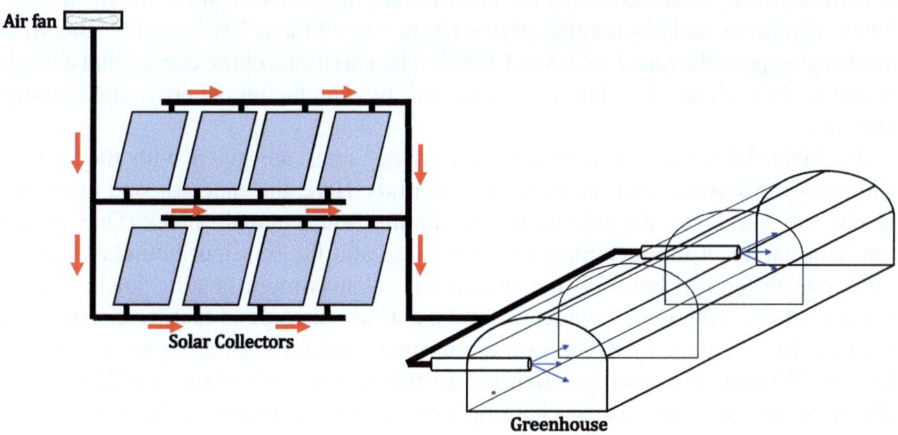


**Fig. 7.61** The tunnel shape greenhouse with the solar collectors. (Reproduced from Öztürk and Başçetinçelik 2003)

store the solar energy. The dimensions of the greenhouse were 20 m in length, 6 m in width, and 3 m in height. The greenhouse had a semicylindrical shape, and it was aligned in the north-south direction. The south-facing solar air collectors had 55° tilt angle with a total surface area of 27 $m^2$. The schematic of the heat collection system with the tunnel greenhouse is provided in Fig. 7.61.

The heat storage unit is placed underneath the greenhouse. The dimensions of the storage tank are 6 m × 2 m × 0.6 m. The tank is insulated with 0.2 mm of PE film and 5 cm of glass wool to prevent heat loss through the soil. A centrifugal fan is used to transfer the air from collectors to the heat storage unit. The fan provides a volumetric flow rate of 600 $m^3/h$. Volcanic material is used in the packed-bed sensible heat storage tank. It is reported that the direct contact between the storage material and the heat transfer fluid (air) improves the heat transfer rate and minimizes the cost that arises due to heat exchangers. The bulk density and the porosity of the storage material are 900 $kg/m^3$ and 41.22, respectively. The tank includes 6480 kg of the volcanic heat storage material. It corresponds to 54 kg of storage material per floor area of the greenhouse. The warm air is distributed into the greenhouse from the heat storage unit by PE ducts that are placed on the ground surface of the greenhouse. During sunny days, the fan directs the warm air from the solar collectors through the heat storage unit to store the thermal energy in the form of sensible heat. As the indoor air temperature of the greenhouse falls below a preset temperature value, a secondary fan is activated to extract the thermal energy that is stored in the storage unit.

Charging and discharging experiments are conducted in different periods. A total of three charging experiments are conducted as (1) first charging period (from 13 to 18 January 1998), (2) second charging period (from 4 to 9 March 1998), and (3) third charging period (from 1 to 7 April 1998). Similarly, a total of three discharging experiments are conducted as (1) the first discharging period (from 13 to 18 January 1998), (2) the second discharging period (from 4 to 9 March 1998), and (3) the third discharging period (from 1 to 7 April 1998). The variations of the energy and exergy contents throughout the day are evaluated by conducting energy and exergy analyses.

In Table 7.19, the stored energy and exergy rates are given with the energy and exergy efficiencies for three charging periods. Here, the energetic and exergetic results are given for the minimum, maximum, and average values. Due to the transient nature of the incoming solar radiation and the transient boundary conditions, the difference between the minimum and maximum heating load is more than two times. Although the three charging periods correspond to different months, January, March, and April, the average heating load for each charging run varies in a small band. The energy efficiency of the system varies between 22.6% and 45.3% for the first and second charging experiments. However, in the third experiment, the range of the minimum and maximum values are extended through 4.1% and 52.9%. Regarding the energy efficiency, the average values stand around 40%. Considering the second law of thermodynamics, the efficiency of the system is

**Table 7.19** Energy-based and exergy-based results for the *charging periods*

| Charging period | Heat (W) | | | Energy efficiency (%) | | | Exergy (W) | | | Exergy efficiency (%) | | |
|---|---|---|---|---|---|---|---|---|---|---|---|---|
| | Min. | Max. | Av. | Min. | Max. | Av. | Min. | Max. | Av. | Min. | Max. | Av. |
| #1 | 519 | 1470 | 1150 | 22.6 | 45.3 | 38.8 | 1.2 | 23.1 | 14.3 | 0.07 | 1.3 | 0.8 |
| #2 | 734 | 1470 | 1160 | 28.6 | 45.3 | 40.0 | 14.6 | 46.6 | 33.7 | 0.80 | 2.6 | 1.9 |
| #3 | 78.4 | 2020 | 1400 | 4.1 | 52.9 | 40.4 | 6.0 | 89.9 | 61.0 | 0.34 | 4.9 | 3.4 |

Adapted from Öztürk and Başçetinçelik (2003)

less than 4%. Öztürk and Başçetinçelik (2003) stated that the average exergy efficiency of the system during the charging period is 2%. It is concluded that the current sensible heat thermal energy storage is inefficient regarding the exergetic aspect.

In Table 7.20, the minimum, maximum, and average values of the heating demand and the rate of heat load that is supplied by the heat storage unit are given. The ratio of the supplied heat from the storage unit to the heating demand varies in the range of 10–43%. As an average, the fraction of the heat that is supplied by the storage unit is 18.9%.

***Case Study 18:* LHTES with Solar Collectors in a Greenhouse (Turkey)**

Benli and Durmuş (2009) integrated a latent heat thermal energy storage system (LHTESS) with phase change material (PCM) into a greenhouse system. The system mainly consists of ten solar collectors, a PCM tank, greenhouse, and airflow control system to maintain the circulation of the warm air through the system components. The experimental system was built in Elazig Turkey, and the system performance was evaluated during the winter season of 2005.

$CaCl_26H_2O$ is used as PCM within the LHTES tank. The thermophysical properties and the cost of the PCM are given in Table 7.21. The LHTES tank has a capacity of 300 kg, and it is reported that 6 kg of $KNO_3$ is dispersed into the PCM to

**Table 7.20** Energy-based results during the discharging period

| | Heating demand of the greenhouse | | | Heat supplied by the storage unit (W) | | | Heat supplied by the storage unit (%) | | |
|---|---|---|---|---|---|---|---|---|---|
| Discharging period | Min. | Max. | Av. | Min. | Max. | Av. | Min. | Max. | Av. |
| #1 | 7530 | 8600 | 8250 | 716 | 873 | 810 | 9.8 | 10.1 | 9.8 |
| #2 | 5940 | 6040 | 5950 | 591 | 794 | 690 | 9.9 | 13.2 | 11.6 |
| #3 | 788 | 1170 | 828 | 189 | 383 | 304 | 28.9 | 43.0 | 35.4 |

Adapted from Öztürk and Başçetinçelik (2003)

**Table 7.21** Properties of $CaCl_26H_2O$

| Property | Value |
|---|---|
| Melting point | 29 °C |
| Density (*solid*) | 180 kg/m$^3$ |
| Density (*liquid*) | 1560 kg/m$^3$ |
| Specific heat (*solid*) | 1460 J/kgK |
| Specific heat (*liquid*) | 2130 J/kgK |
| Thermal conductivity (*solid*) | 1.088 W/mK |
| Thermal conductivity (*liquid*) | 0.539 W/mK |
| Latent heat of fusion | 187.49 kJ/kg |
| Number of thermal cycling | 3000–5000 |
| Price ($/kg in 2006) | 2 |
| Toxic effect | No |

Adapted from Benli (2006)

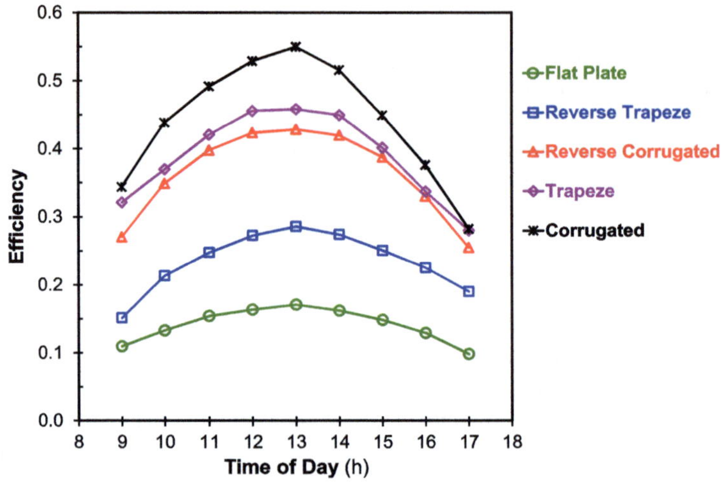


**Fig. 7.62** Greenhouse heating with the solar collectors and LHTES unit. (Data from Benli and Durmuş 2009)

restrain the subcooling effect and achieve successful crystallization during the solidification of the PCM. The work of Benli and Durmuş (2009) mainly focused on the evaluation of the influence of the type of collector on the system performance. To do so, five different types of solar collectors are placed in serial in the experimental systems. The types of the solar collectors are *flat*, *corrugated*, *reverse corrugated*, *trapeze*, and *reverse trapeze*. In each solar collector, the pressure drops, and the temperature differences are measured to evaluate the efficiency of each type of collector that are connected serially. Figure 7.62 shows the variation of collector efficiency for a typical day. It is certain that the type of absorber geometry has a significant influence on the system performance. The maximum efficiency is obtained to be 55% in the corrugated absorber design. The worst design is found to be the flat plate configuration. The maximum efficiency is observed to be 17% in the case of the flat plate collector. The corrugated design improves the heat transfer surface area between the absorber plate and the heat transfer fluid (*air*). Moreover, the corrugated surface disturbs the boundary layer development and causes wakes. Disturbing the flow field along the flow direction improves the mean convective heat transfer coefficient on the absorber plate. It is reported that the average Nusselt number improves almost two times when the corrugated design is used instead of the flat plate one.

Even though disturbing the flow field improves the heat transfer, the pressure drop across the collector will also increase as the surface roughness of the absorber is modified. Benli and Durmuş (2009) reported that the pressure loss across the collector improves nearly 2.5 times as the flow Reynolds number is increased from 3000 to 5000. Moreover, regarding the pressure drop, the worst case is the collector with trapeze absorber surface. In comparison with the flat plate collector, the pressure drop increases nearly five times in the trapeze design. The corrugated

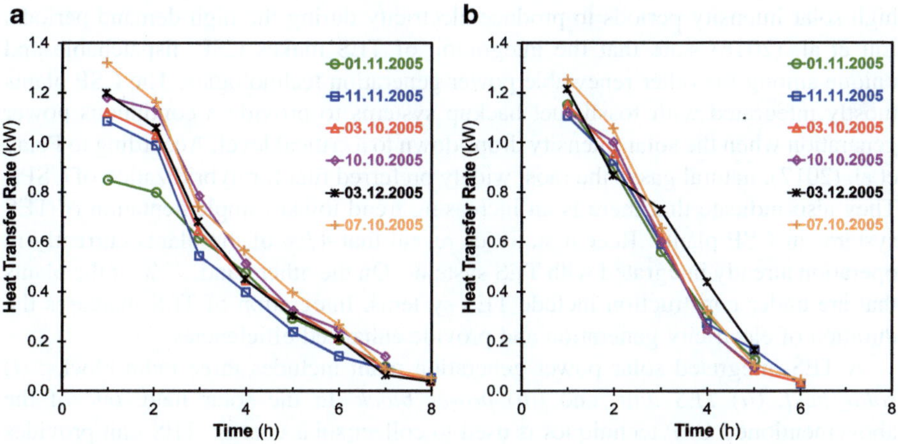


**Fig. 7.63** Variations of heat transfer rates during charging and discharging. (Data from Benli and Durmuş 2009)

design is the second worst design regarding the pressure loss. One should perform an optimization study to determine the optimum working and design parameters of the combined heat collection and storage unit.

Benli and Durmuş (2009) also revealed the variations of the rate of heat transfer (in kW) throughout the day during the charging and discharging periods. Figure 7.63 shows the timewise variations of the heat transfer rates for six selected days in three different months. Due to the different outdoor conditions (i.e., the irradiation, ambient temperature, and wind speed), the maximum rate of heat transfer values vary in the range of 0.82–1.32 kW for the charging process. It takes 7 h to reach complete charging (i.e., to reach zero heat transfer rate) condition. On the other hand, the discharging period is 5 h, and the rate of heat transfer during the discharging period has a similar tendency as in the charging period.

## 7.8 High-Temperature TES for Solar Thermal Energy

Solar energy that reaches the surface of the Earth has excellent potential and is sufficient to meet the energy requirement of humanity. The technological developments in solar energy seek to convert a significant portion of solar energy into useful power or thermal energy. Among various solar energy conversion technologies, the concentrated solar power (CSP) receives particular attention (Pelay et al. 2017; Liu et al. 2016; Mao 2016). There are different technologies to convert solar energy into useful thermal energy such as parabolic trough solar collectors, solar towers, and solar dishes. Some significant advantages of the CSP technologies are listed as (*i*) providing high efficiency, (*ii*) low operating cost, and (*iii*) easy integration of TES techniques. In a TES-integrated CSP plant, solar energy is simply stored during the high solar intensity periods to produce electricity during the high-demand periods. Liu et al. (2016) state that the integration of TES makes CSP dispatchable and unique among the other renewable power generation technologies. The CSP plants mostly integrated with fossil fuel backup systems to provide a continuous power generation when the solar intensity drops down to a critical level. According to Pelay et al. (2017), natural gas is the most widely preferred fuel for hybridization of CSPs. They also indicate that there is an increasing trend toward implementation of TES systems in CSP plants. Recent statistics reveal that 47% of the plants currently in operation already integrated with TES systems. On the other hand, 72% of the plants that are under construction include TES systems. Integration of TES increases the duration of electricity generation and provide enhanced efficiencies.

A TES-integrated solar power generation plant includes three main blocks: (*i*) *solar field*, (*ii*) *TES unit*, and (*iii*) *power block*. In the solar field, one of the abovementioned CSP techniques is used to collect solar energy. TES unit provides a continuous energy supply to the power block. The power block is responsible for producing the mechanical power from the thermal energy. Organic Rankine cycle (ORC) is a widely used approach to produce electricity in CSP plants. The Rankine cycle with steam is useful for high-temperature power production with steam. Alternatively, in the ORC, organic working fluids are used to produce electricity from a low- or medium-temperature heat source. ORC allows to recover low-grade waste heat and improve the overall efficiency of a plant. The overall efficiency of the plant could be improved further by integrating heating and cooling facilities to the power generation plants. Such a "combined cooling, heating, and power" generation systems are mostly called CCHP. Either for heating, cooling, or power generation purposes, one of the most critical components of a CCHP is the TES since it allows flexibility in the meeting of the heating, cooling, or electricity demands. In the following case studies, the implementation of TES units in solar thermal energy systems is reported.

***Case Study 19:* CCHP System with TES Unit**

Al-Sulaiman et al. (2012) proposed a novel TES-integrated trigeneration system and developed a thermodynamic model to investigate different working modes on the overall performance of the plant. The schematic of the plant is illustrated in Fig. 7.64. The system includes four blocks: (*i*) parabolic solar trough collectors (PTSCs), (*ii*) hot and cold TES tanks, (*iii*) organic Rankine cycle, and (*iv*) single-effect absorption chiller. PTSC has a parabolic shape and focuses the incoming solar radiation with reflective mirrors to a receiver. The receiver is a long pipeline which is covered by a shield to reduce heat losses through the ambient via convective or radiative heat transfer. In this way, the efficiency of the PTSC improves significantly. Commercial heat transfer oil (Therminol-66) circulates in the PTSC loop. The heat that is extracted from the PTSC is then transferred through either storage block or the power block. Al-Sulaiman et al. (2012) considered three working modes: (*i*) low-solar radiation mode, (*ii*) high-solar radiation mode, and (*iii*) storage mode. Each mode of working is discussed below. TES block involves hot and cold storage tanks. In the storage mode, the heat transfer fluid (Therminol-66) circulates through

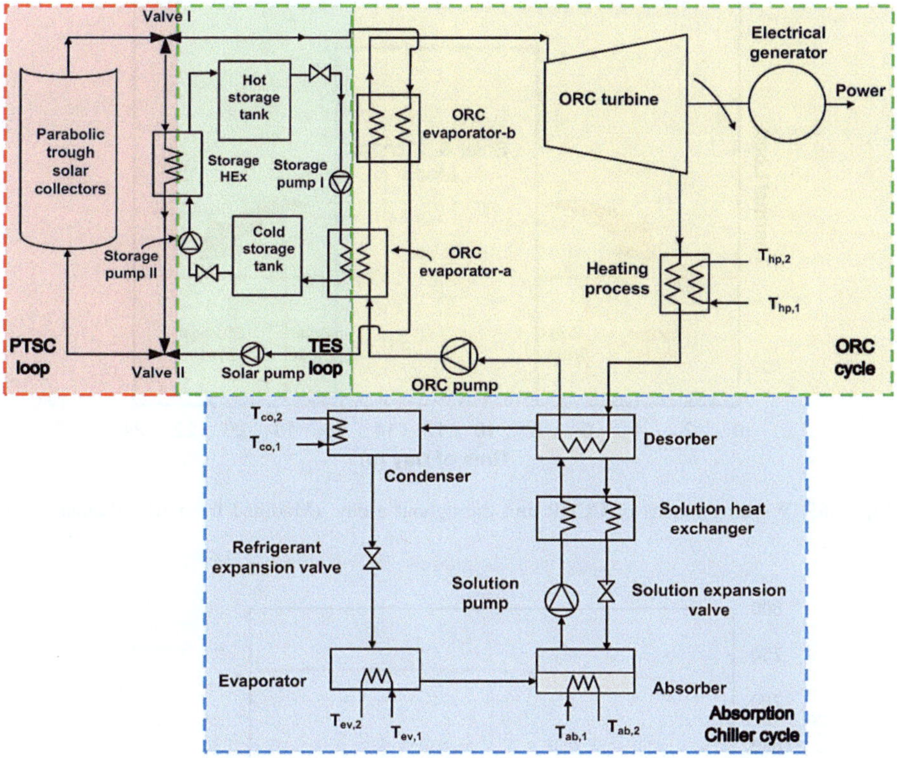


**Fig. 7.64** Schematic overview of the combined heating, cooling, and power generation system. (Modified from Al-Sulaiman et al. 2012)

the storage HEX and transfers the high-temperature thermal energy through the hot storage tank. The stored energy in hot tank is then used to supply the ORC cycle. The HTF cools down across the evaporator of the ORC, and the low-temperature thermal energy is stored in the cold storage tank. ORC consists of four components, a turbine, heat exchangers, pump, and electric generator. n-octane is used as working fluid in the ORC cycle.

Four modes of the CCHP unit are illustrated in Fig. 7.65. In the first mode, from 6 am to 8 am, and from 16 to 18, solar collectors work without thermal energy storage. In the solar mode, the solar collectors directly meet the requested thermal energy by the ORC. From 8 am to 16, a portion of the thermal energy that is collected in the solar collectors is transferred through the storage tanks for further usage. In the case study of Al-Sulaiman et al. (2012), according to the preliminary performance analyses, it is decided to store 70% of the collected thermal energy in the tanks. In the third mode, from 18 to 6 am, the storage unit works to meet the requested thermal energy in the CCHP plant.

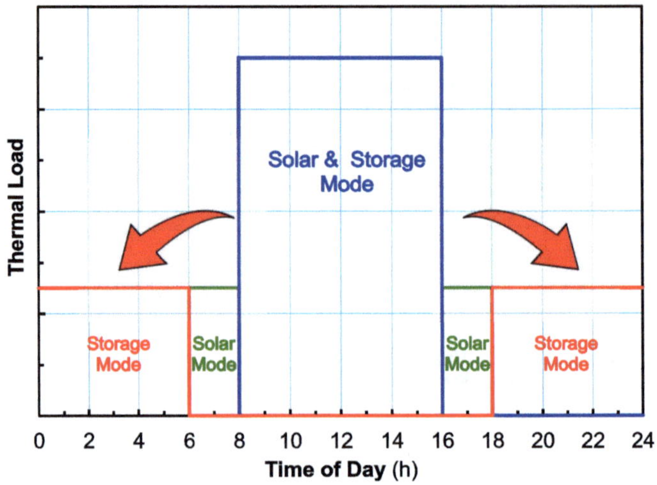


**Fig. 7.65** Working modes of CCHP unit throughout a day. (Modified from Al-Sulaiman et al. 2012)

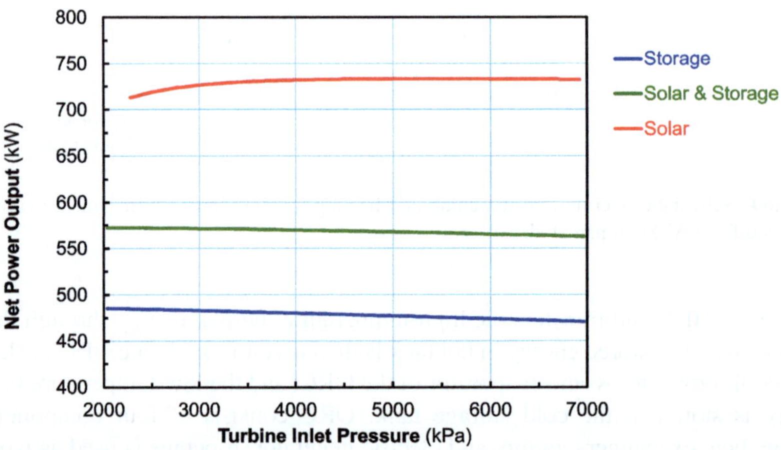


**Fig. 7.66** Influence of turbine inlet pressure on the net electrical power output (Data from Al-Sulaiman et al. 2012)

For each subsystem, i.e., PTSC, TES unit, and ORC, governing equations are defined in Engineering Equation Solver (EES) software, and parametric analyses are conducted to assess the influences of several working scenarios on the energetic outputs such as first law efficiency, power output, and heating/cooling ratios. Figures 7.66, 7.67, and 7.68 illustrate the results of the parametric analyses. In Fig. 7.66, the effect of turbine inlet pressure on the net power output is achieved from the CCHP under three working modes. The maximum power output is achieved in the solar mode which is around 710 kW. In this mode of working, the

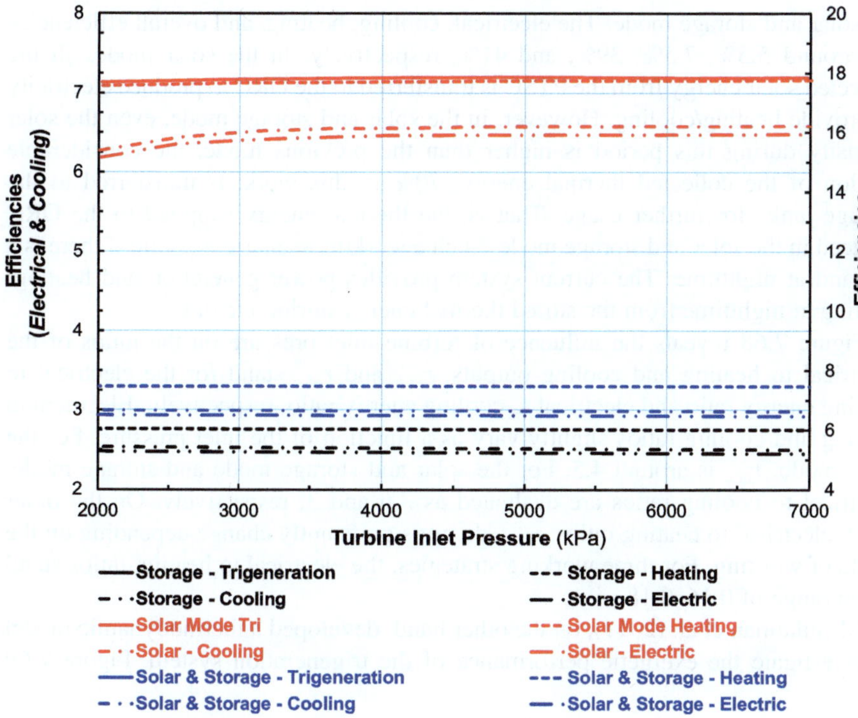


**Fig. 7.67** Influence of turbine inlet pressure on first law efficiencies. (Data from Al-Sulaiman et al. 2012)

power generation increases as the inlet pressure of the turbine rises. For the solar and storage mode, the power output reduces almost 20%. With increasing inlet pressure of turbine, the power output slightly reduces for solar and storage mode and storage mode. In the storage mode, the power output is around 480 kW, and regarding the solar mode, the reduction is nearly 50%. Figure 7.67 reveals a comprehensive comparison for the CCHP. For each mode of working, i.e., solar, solar and storage, and storage, the first law efficiencies of the combined system and each subsystem are represented. It is interesting to note that the inlet pressure of the turbine does not cause significant variation in the current design and working conditions. For the selected range of inlet pressure, the variation of efficiency is far below 1%. This is an important point since working at high pressures is costly and requires additional expenses for the plant. That is, working at low pressure may provide significant benefits for a CCHP without any significant reduction in outputs. Highest efficiencies are achieved in the solar mode. The electrical, cooling, heating, and overall efficiencies are obtained as 13%, 16%, 88%, and 91%. On the other hand, for the solar and storage mode, the electrical, cooling, heating, and overall efficiencies are evaluated as 6.5%, 8%, 44%, and 46%, respectively. In comparison with the solar mode, the efficiencies drop by half. For the storage mode, the efficiencies are close to the solar and storage mode. The electrical, cooling, heating, and overall efficiencies are around 5.5%, 7.5%, 39%, and 41%, respectively. In the solar mode, all the collected solar energy from the PTSC is transferred to the ORC to produce electricity or provide heating/cooling. However, in the solar and storage mode, even the solar intensity during this period is higher than the previous mode, the considerable portion of the collected thermal energy, 70% in this work, is transferred to the storage tanks for further usage. That is, the thermal energy supplied to the ORC reduced in the solar and storage mode. Such a working scenario is useful if there is a demand at nighttime. The current system provides power generation and heating/cooling at nighttime from the stored thermal energy during the day.

Figure 7.68 reveals the influence of turbine inlet pressure on the ratios of the electrical to heating and cooling outputs. $r_{el,h}$ and $r_{el,c}$ stand for the electrical to heating energy ratio and electrical to cooling energy ratio, respectively. Electrical to heating and cooling ratios slightly vary as a function of the inlet pressure. For the solar mode, $r_{el,c}$ is around 4.5. For the solar and storage mode and storage mode, electrical to cooling ratios are evaluated as 3.6 and 3, respectively. On the other hand, electrical to heating ratios, $r_{el,h}$, do not significantly change depending on the mode of working. For three working strategies, the electrical to heating ratios stand in the range of 0.16–0.18.

Al-Sulaiman et al. (2011), on the other hand, developed a thermodynamic model to investigate the exergetic performance of the trigeneration system. Figure 7.69

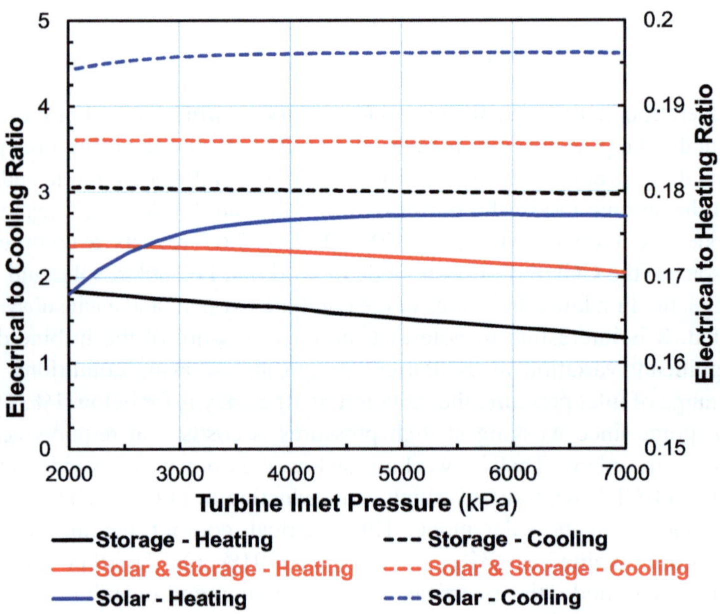


**Fig. 7.68** Influence of turbine inlet pressure on electrical to cooling and heating ratios. (Data from Al-Sulaiman et al. 2012)

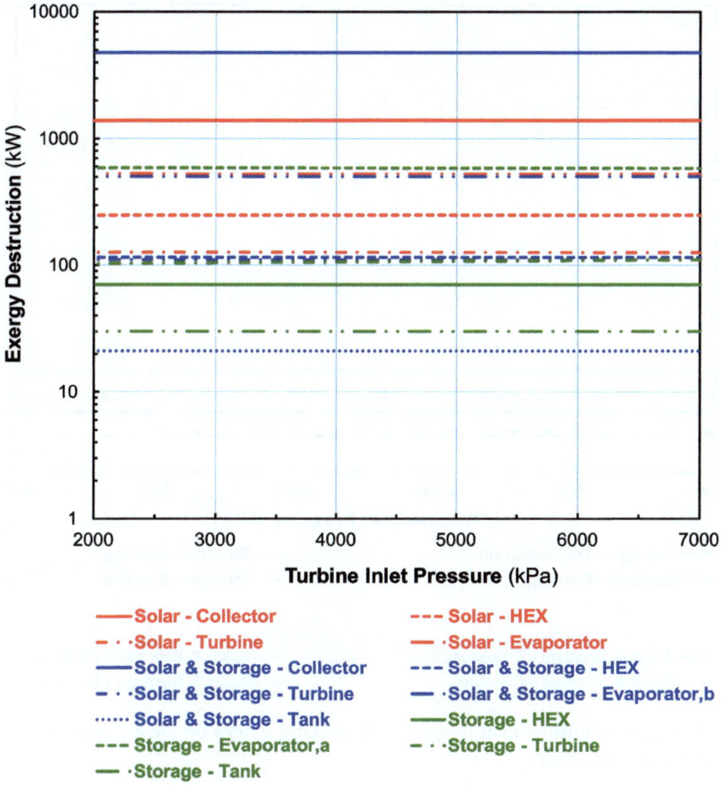


**Fig. 7.69** Influence of turbine inlet pressure and working mode on the rate of exergy destruction. (Data from Al-Sulaiman et al. 2011)

represents the variations of the rate of exergy destruction as a function of turbine inlet pressure for the solar mode, solar and storage mode, and storage mode. For each mode of working, the inlet turbine pressure does not cause a significant change in the exergy destruction. The variation is below 30 W. However, the mode of CCHP has a significant role in the exergy destruction. In the figure, the destructed exergy rates are given for the solar collectors ($\dot{E}x_{d,collector}$), heating process heat exchanger ($\dot{E}x_{hp,hex}$), turbine ($\dot{E}x_{turbine}$), heat storage tank ($\dot{E}x_{d,storage}$), and evaporators ($\dot{E}x_{evap,a}$ or $\dot{E}x_{evap,b}$). In the case of the solar mode, the rate of exergy destructions for the solar collector, evaporator-b, heating process HEX, and turbine are 1400 kW, 520 kW, 260 kW, and 130 kW, respectively. The exergy destruction inside the solar collector is more than two times higher than the one in evaporator-b. In the solar and storage mode, the exergy destruction rate significantly increases as shown in Fig. 7.70. The corresponding exergy destruction rates for the solar collectors, evaporator-b, heating process HEX, turbine, and storage tank are 4800 kW, 500 kW, 115 kW, 110 kW, and 21 kW, respectively. In comparison to the solar mode, the exergy destruction rate of solar collector is increased almost four times.

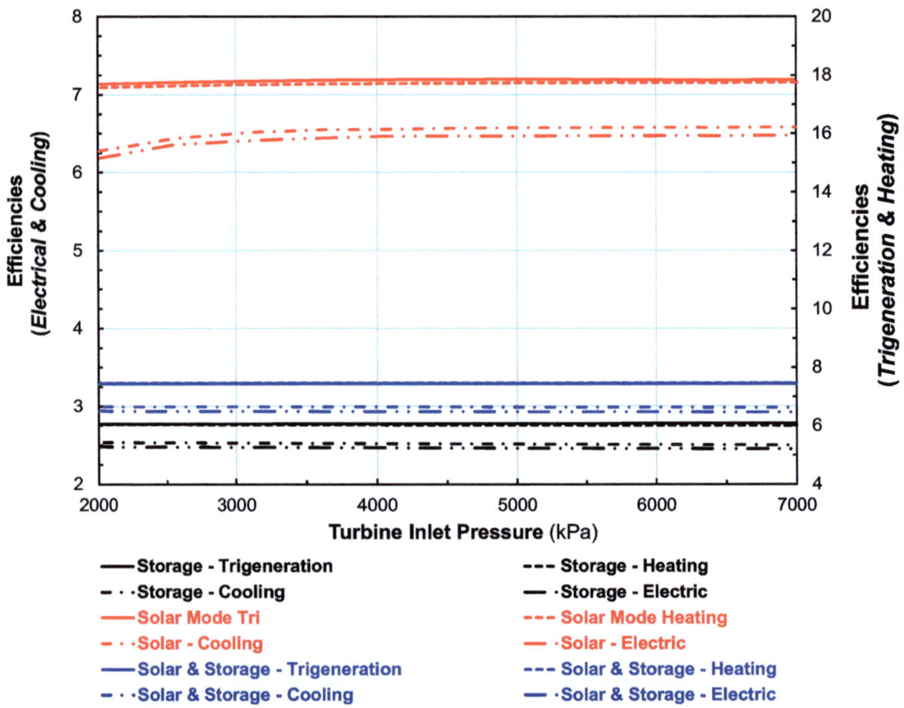


**Fig. 7.70** Influence of turbine inlet pressure and working modes on the exergy efficiencies. (Data from Al-Sulaiman et al. 2011)

This increment is due to the high incoming solar irradiation for the solar and solar storage mode of working. In the case of the storage mode, the exergy destruction rates of evaporator-a, heating process HEX, turbine, and storage tank are approximately 585 kW, 70 kW, 100 kW, and 20 kW, respectively. It is noted that the solar collectors and ORC evaporators are responsible for the significant portion of exergy destruction; that is, care should be taken during the design or selection periods of these components to minimize exergy destruction and improve the overall exergy efficiency of the system.

Figure 7.70 illustrates the exergetic efficiencies of each component for three modes of working. The highest exergy efficiencies are evaluated for the solar mode, as the collected solar energy is directly used to produce electricity and meet the heating/cooling demand. The electrical efficiencies are evaluated as 6%, 3%, and 2.5% for the solar mode, solar and storage mode, and storage mode, respectively. The solar mode has the highest electrical efficiency. On the other hand, the overall efficiencies of the trigeneration units are around 18%, 7.5%, and 6.5% for the three operation modes. One can deduce that solar mode has the highest exergy efficiencies compared with the other two alternative working modes. The inlet pressure of the turbine does not have a significant role in the energetic or exergetic system performance indicators. Working at low pressures may reduce the investment and operational costs of the CCHP.

***Case Study 20:* Solar-Powered Multigeneration System with TES and Hydrogen Production**

In the previous case study, results of a TES-integrated combined heating, cooling, and power generation plant are discussed. Recently Almahdi et al. (2016) proposed a multigeneration system which includes three ORCs, low-temperature and high-temperature TES units, hydrogen production, biomass dryer, heat pump, and chillers. The schematic of the proposed multigeneration system is given in Fig. 7.71. Such a system produces five outputs simultaneously: electrical power, space heating and cooling, hydrogen, and drying of biomass. During the daytime, the solar energy that is collected in the PTSC is transferred through the ORC1 and ORC2 to produce electricity. Isobutene is used as heat transfer fluid in the organic Rankine cycles. The output of the evaporator of the ORC1 is transferred to the thermal energy storage tank. The TES tank stores the waste heat of the ORC1 during the daytime. Thermal energy stored within the TES unit is then rejected by ORC3, during the

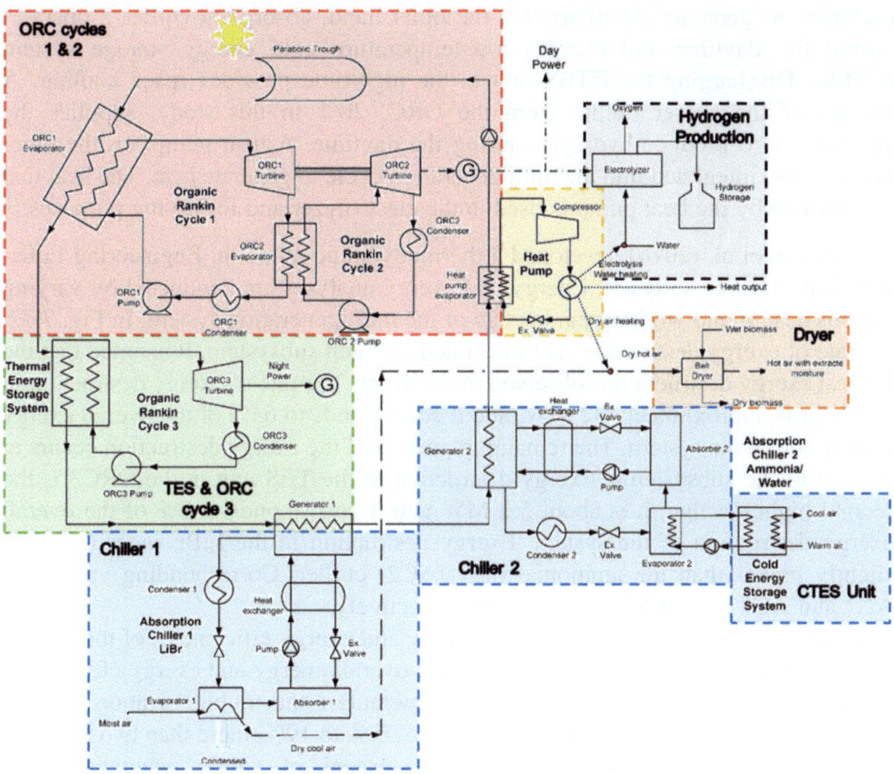


**Fig. 7.71** Schematic overview of the multigeneration system with high-temperature and low-temperature TES units. (Adapted from Almahdi et al. 2016)

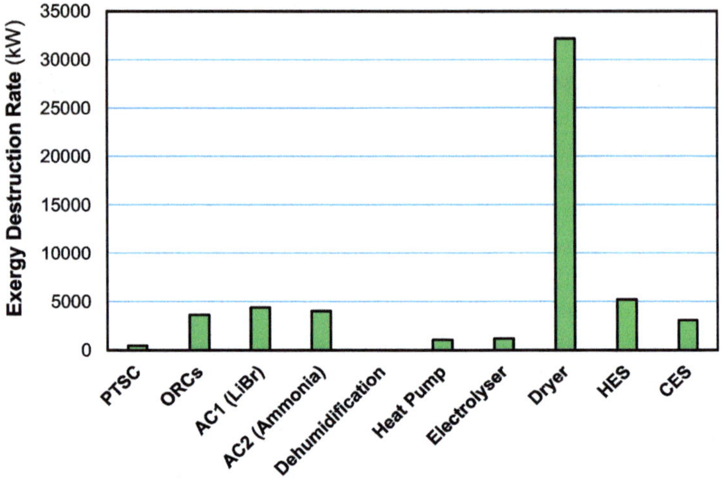


**Fig. 7.72** The rate of exergy destruction in each subsystem. (Data from Almahdi et al. 2016)

nighttime to generate electricity. On the other hand, absorption chiller 2 operates during the daytime and charges low-temperature cold energy storage system (CTES). Discharging the CTES unit at the nighttime provides space cooling. A portion of the power output from the ORC, 20% in this study, supplies the electrolyzer to produce hydrogen during the daytime. A heat pump, on the other hand, is also integrated into the multigeneration cycle to generate heat. The heat that is produced by the heat pump is used in the electrolyzer and for drying purposes.

Almahdi et al. (2016) developed a thermodynamic model in Engineering Equation Solver (EES) software. Energy and exergy analyses are conducted by varying various design and working parameters of the multigeneration system. In Fig. 7.72, the rate of exergy destructions are illustrated for each subsystem. It is noted that the highest exergy destruction is observed in the dryer. The rate of exergy destruction in the dryer is approximately 32 MW, and it corresponds to 64% of the overall exergy destruction of the system. The remaining portion of the exergy destruction occurs in the rest of the subsystems. Exergy destruction in the TES unit in the ORC3 is the second highest value; it is about 5.2 MW and it corresponds to 8% of the overall exergy destruction of the system. Exergy destruction in the LiBr water (AC1) is slightly higher than the ammonia-water (AC2) chiller. Corresponding values for AC1 and AC2 are 4.4 MW and 4 MW, respectively.

Figures 7.73 and 7.74 compare the energy and exergy efficiencies of the system working at different modes. In Fig. 7.73, the overall energy and exergy efficiencies of the single-generation, cogeneration, trigeneration, and multigeneration systems are shown. Energy efficiency improves from 8.78% to 19%, more than two times, as the system works multigeneration mode instead of single generation. The corresponding energy efficiencies, on the other hand, for the tri- and multigeneration systems are 20.2% and 20.7%, respectively. Besides, the exergy efficiency of the

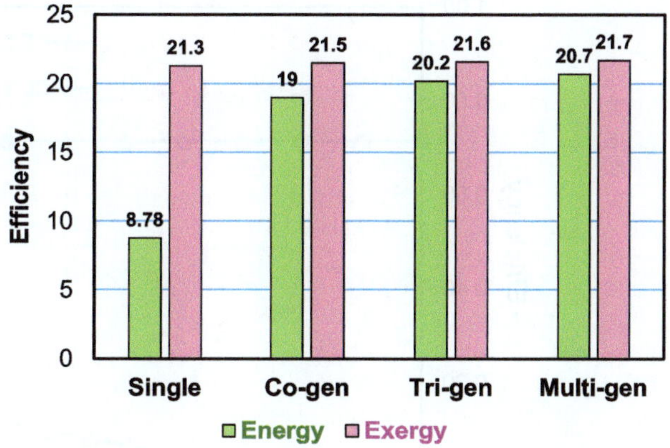


**Fig. 7.73** Overall energy and exergy efficiencies of the systems. (Data from Almahdi et al. 2016)

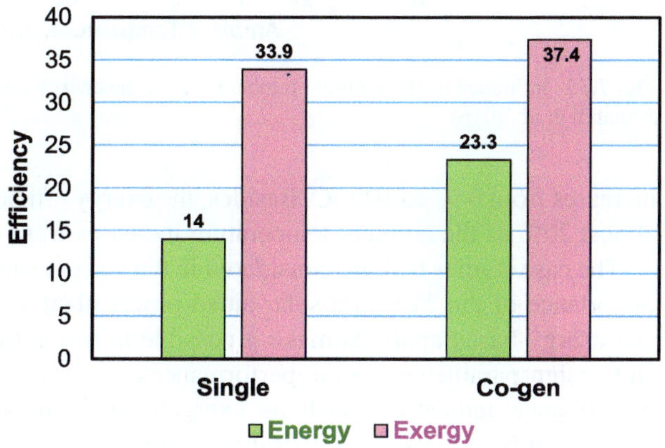


**Fig. 7.74** Energy and exergy efficiencies for nighttime operation. (Data from Almahdi et al. 2016)

system does not vary significantly depending on the working mode of the system. The exergetic efficiency of the system varies in the range of 21.3% and 21.7%. Figure 7.74 compares the energy and exergy efficiencies of the systems for nighttime operation. The energy efficiency of the single-generation system is evaluated as 14%. The energy efficiency of the cogeneration system is 23.3%. On the other hand, regarding the exergy-based analyses, the single- and cogeneration system efficiencies are evaluated as 33.9% and 37.4%, respectively. Results reveal that considering the energetic or exergetic point of the system, the multigeneration systems provide better performance than the single generation system.

Figure 7.75 illustrates the influence of the ambient temperature (or the dead state), on the energy and exergy efficiencies of the HES (high-temperature heat energy storage system) and CES (low-temperature cold storage system) units. No significant change is observed regarding the energy efficiency by varying the ambient temperature. However, the exergy efficiency hardly depends on the ambient temperature. The exergy efficiency of the HES reduces from 40% to 0% as the temperature

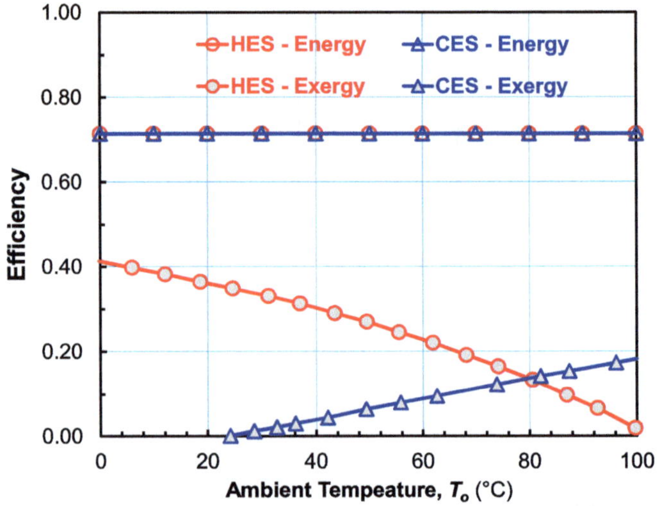


**Fig. 7.75** Influence of the ambient temperature on the energy and exergy efficiencies. (Data from Almahdi et al. 2016)

increases from 0 °C to 100 °C. Besides, the exergy efficiency of the CES unit reaches almost 20% as the ambient temperature increases from 0 °C to 100 °C.

The case studies that are considered in the current subsection show the significant importance of the TES in a solar-aided power plant with multigeneration. Energy- and exergy-based analyses make it possible to reveal the influences of the working and design parameters on the performances of each component and on the overall performance indicators, such as exergetic and energetic efficiencies. Once the component-based results are evaluated, further improvements could be achieved by implementing alternative subsystems.

## 7.9 Passive Thermal Control of Battery Cells

Electric or hybrid vehicles offer several unique advantages over gas engines with low emission and reduced noise. Driving with low emission and noise in urban regions makes electric vehicles (EVs) and hybrid vehicles (HVs) good alternatives. Li-ion batteries have very high energy density, 200 Wh/kg, and are compact and light, and so they are potential candidates in EVs. Several topics are currently under research for EVs, and one of the most critical problems that these vehicles are facing is the safety of the battery packs. The temperature of the battery packs should remain below a critical limit to prevent possible damages to the cell. According to a recent review of Malik et al. (2016), significant heat generation occurs inside the cell depending on the drawn power from the battery. The heat generated inside the cell should be extracted from the cell. Otherwise, the temperature of the battery increases and causes thermal runaway or an explosion. That is, the thermal management of battery packs, especially for the case in which the battery works at high ambient temperature with high power rates. In such a case, the thermal management of the battery pack becomes crucial to prevent thermal runaway.

Active and passive thermal management strategies are used to keep the temperature of the battery pack below a critical temperature. In the active strategies, a fan (air or gas cooling) or pump (liquid cooling) is used with a channel or piping system to maintain the circulation of the heat transfer fluid across the battery packs. The performance of the active thermal management hardly depends on the geometrical limitations. If the working space allows the forced convection of the fluid across the battery cells, a significant amount of thermal energy could be extracted in this approach. However, in the case of EVs, the battery packs are mostly placed in narrow spacings. Additionally, running a fan or pump to cool down the pack may reduce the efficiency of the battery. Implementation of PCM, on the other hand, is a more direct and straightforward way to extract the excessive heat that is generated within the pack. In such a case, PCM-filled cavities or packs cover the battery cells to store the thermal energy via melting. The geometry of the PCM enclosure and thermophysical properties of the PCM should be selected considering the dynamic working parameters of the battery pack. In the following, a case study of the application of PCM on a battery cell is numerically investigated.

***Case Study 21:* Electrical Vehicle Battery Pack with PCM**

Javani et al. (2014) developed a numerical model to simulate the transient heat transfer within the battery cell with PCM. Recently Celik et al. (2017) considered the same problem and investigated the influence of various working conditions on the timewise variation of the cell temperature. The current case study represents the solution method and significant outcomes of the works of Javani et al. (2014) and Celik et al. (2017). Heat generation rate of a battery cell depends on several factors such as internal resistance and charging/discharging rate (C-rate) of the battery. In Table 7.22, the heat generation rates of Li-ion cells are represented with the operating conditions.

The geometry of the Li-ion cell is illustrated in Fig. 7.76, and the corresponding geometrical dimensions are listed in Table 7.23. Initially, the battery cell is in thermal equilibrium with the surrounding air. The ambient air is at $T_\infty = 294.15$ K. The convective heat transfer coefficients on each surface of the cell are assumed to be constant throughout the process. The heat transfer coefficient is defined as $h = 7$ W/m$^2$K. n-octadecane, with a melting temperature of 302.15 K, is used as

**Table 7.22** Volumetric heat generation rates under different operating conditions

| Volumetric heat generation rate (W/m$^3$) | Li-ion cell operating conditions |
|---|---|
| 6855 | Standard US06 |
| 22,800 | Max. 135 Amps (150 kW), 2.6 W/cell |
| 63,970 | 2C, 4.45 W/cell |
| 200,000 | Full power, uphill condition |

Adapted from Javani et al. (2014)

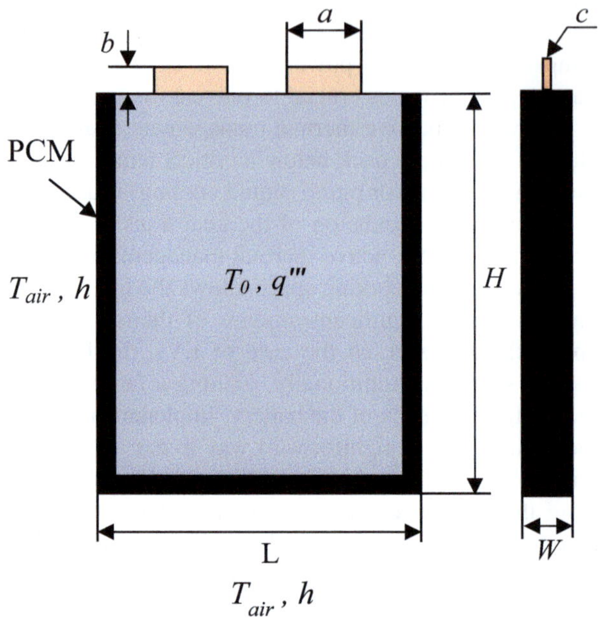


**Fig. 7.76** PCM-covered battery cell geometry and boundary conditions

**Table 7.23** The dimensions of the battery cell and terminals

| | | | |
|---|---|---|---|
| Battery cell | L | H | W |
| | 146 mm | 194 mm | 5.4 mm |
| Terminals | a | b | c |
| | 35 mm | 15 mm | 0.6 mm |

Adapted from Celik et al. (2017)

PCM. The boundary conditions and geometrical dimensions of Javani et al. (2014) and Celik et al. (2017) are identical. In the work of Javani et al. (2014) the effect of natural convection is neglected within the PCM domain, so this work is called the *conduction-dominated case* in the rest of this section. Celik et al. (2017), on the other hand, resolved the flow field equations for the PCM domain to reveal the influence of natural convection, so this work is called the *natural convection-driven case*. Both studies use the control volume approach to discretize the differential equations into algebraic form. Conduction-dominated case is resolved in the commercial CFD solver ANSYS-FLUENT; besides the numerical simulation of the natural convection-driven case is conducted in open-source CFD package OpenFOAM. For the convection-dominated case continuity, momentum and energy equations are resolved. A set of preliminary analyses are conducted to reveal the accuracy of the numerical model. Figure 7.77 compares the results of convective and conductive models with the predictions of Moraga et al. (2016). Here, results are represented by two different designs. The temperature variations are evaluated for reference cell and PCM-integrated cell unit. Notice that for both approaches, either conduction or convection based, the evolution of cell temperatures stands close to the variations

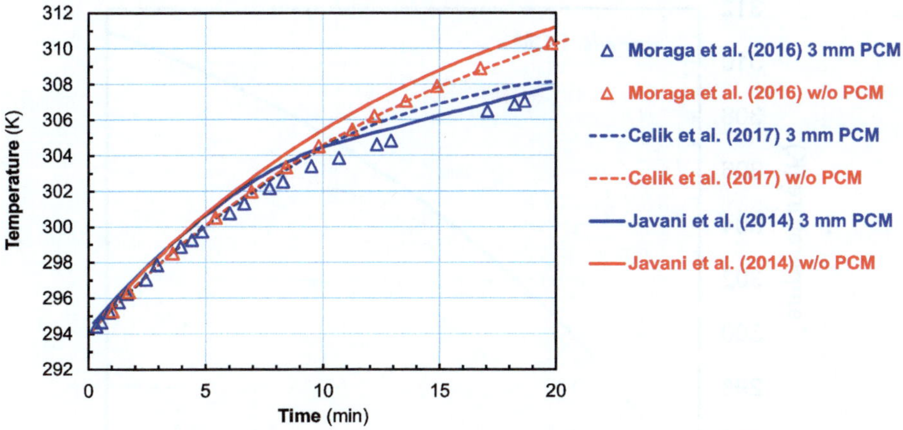


**Fig. 7.77** Comparison of the timewise variations of the cell temperature. (Adapted from Celik et al. 2017)

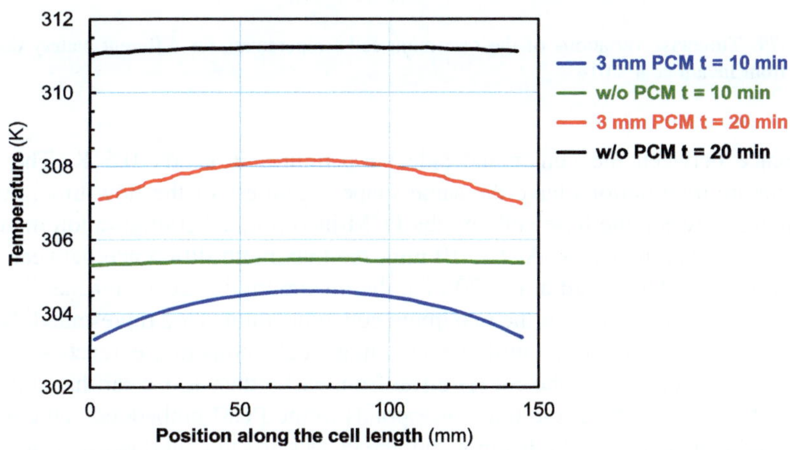


**Fig. 7.78** Spanwise temperature distributions inside the cell at different flow times. (Data from Javani et al. 2014)

of the reference work. One should also note that the implementation of 3 cm of PCM around the cell slightly restrains the temperature rise. At the end of 1200 s, the temperature of the cell with PCM is 2 K below than the cell *w/o* PCM.

In the conduction-dominated analyses (Javani et al. 2014), timewise and spanwise temperature variations are evaluated to introduce the influence of PCM thickness. The local temperature distributions along the cell length are given in Fig. 7.78 for two different cell configurations, *w/o* PCM and 3 mm PCM, at two different instants. For the plane cell, *w/o* PCM, the temperature along the cell length is almost uniform. However, for the 3 mm PCM-covered cell configuration, the

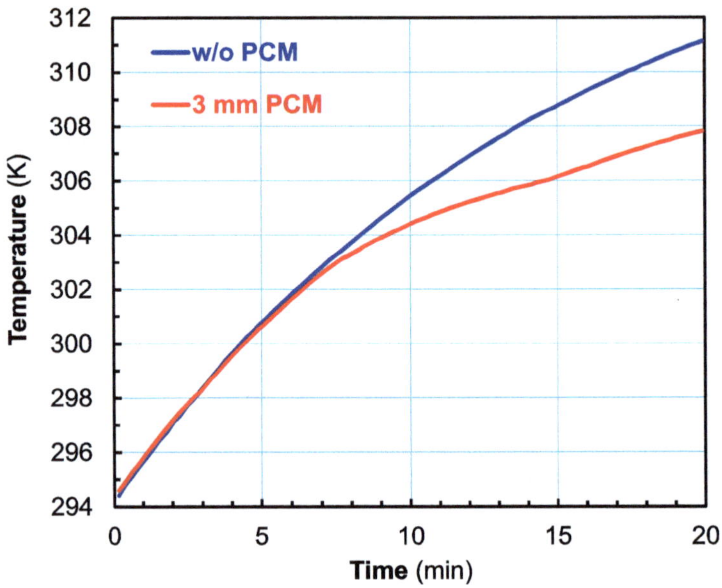


**Fig. 7.79** Timewise variations of the maximum cell temperatures for different battery designs. (Data from Javani et al. 2014)

difference between the center and side temperatures is nearly 0.5 K. The local temperature distributions have the same shape regardless of the flow time, but the difference between the base cell and the PCM-incorporated configuration increases as the flow time increases. At $t = 10$ min, the maximum difference between each design is about 0.9 K, and at $t = 20$ min, the maximum difference reaches 3 K.

In Fig. 7.79, variations of the maximum cell temperatures are represented. Notice that the variations overlap until the maximum cell temperature reaches 303 K. Beyond this temperature, the curves differ from each other. It is clear that at around 5 min, the slope of the curve that corresponds to the PCM-embedded cell configuration suddenly changes. At this moment, the PCM starts to melt down and stores the excess heat that is dissipated from the battery. The temperature of the cell with 3 mm of PCM reaches 308 K at the end of 20 min. For the plane cell, *w/o* PCM, there is no significant variation in the slope of the curve throughout the process, and the maximum temperature of the cell reaches 311 K. For the current case, implementation of 3 mm of PCM provides almost 3 K temperature reduction at the end of 20 min.

One may wonder what will happen if the thickness of the PCM is further increased. The thickness of the PCM cavity varies from 3 mm to 12 mm with 3 mm intervals to evaluate the influence of the PCM thickness on the local and temporal temperature variations. Figure 7.80 compares the temperature variations along the cell length for four different PCM thicknesses with the plane cell without PCM. The temperature variations are close to each other for each PCM thickness. A

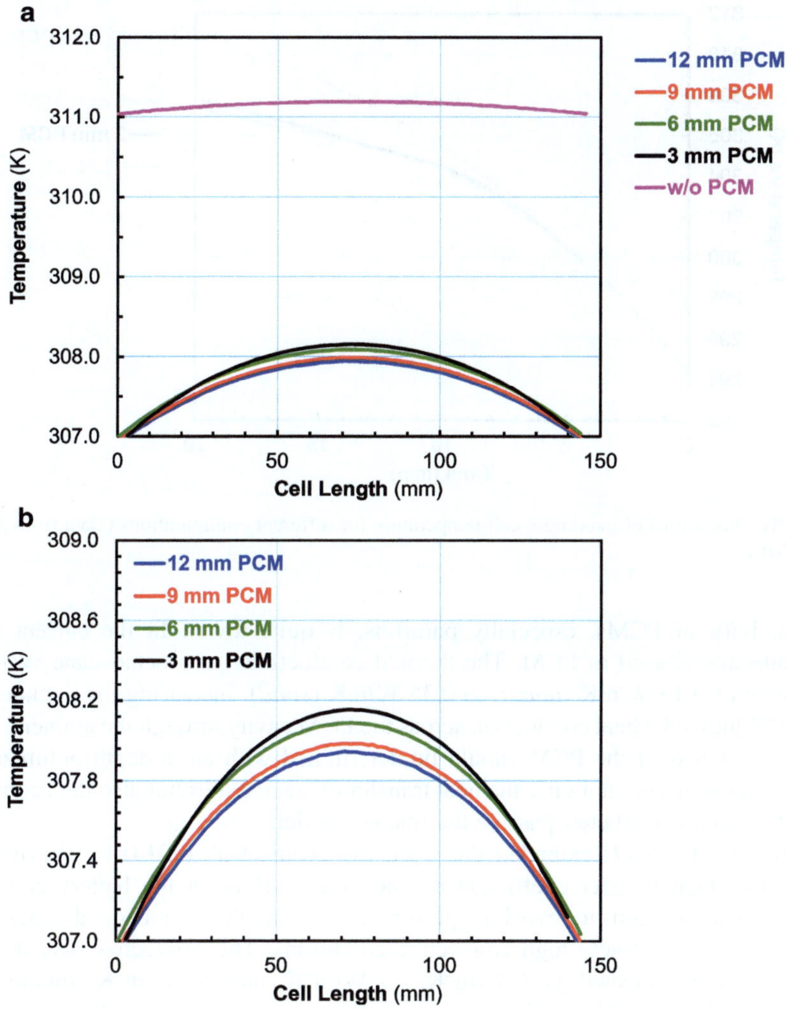


**Fig. 7.80** Temperature variations along the cell length for different configurations. (Data from Javani et al. 2014)

close look at the spanwise temperature variations for different PCM thicknesses is given in Fig. 7.80b. It is clear that increasing the thickness of the PCM from 3 mm to 12 mm, by four times, drops the maximum temperature of the cell less than 0.5 K.

In Fig. 7.81, the evolution of the maximum cell temperatures is represented for the same configurations. No significant difference is observed for the configurations that involve PCM. The results are essential to introduce that the amount of PCM, or the thermal mass, is not the only parameter to achieve better thermal management for the systems. One should keep in mind that the dynamic response of the PCM has a significant role in the transient heat transfer. It is a well-known fact that the thermal

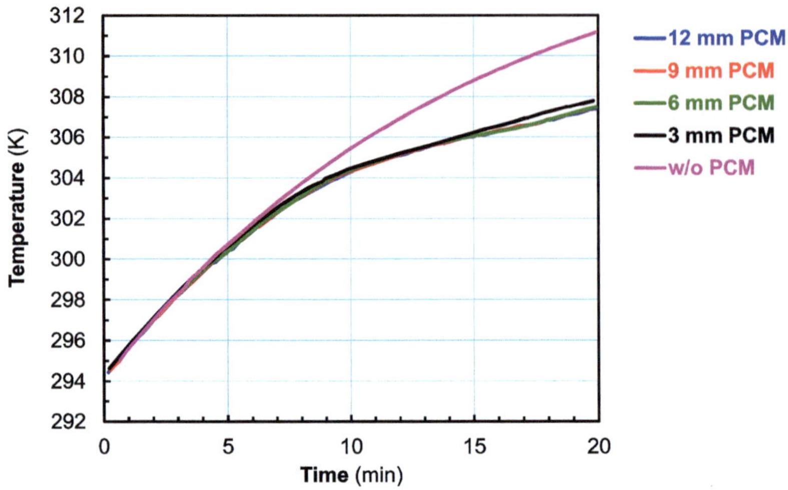


**Fig. 7.81** Variations of maximum cell temperature for different configurations (Data from Javani et al. 2014)

conductivity of PCMs, especially paraffins, is quite lower. In the current case, n-octadecane is used as PCM. The thermal conductivity of n-octadecane varies in the range of 0.15 W/mK (*liquid*) to 0.36 W/mK (*solid*). Increasing the thickness of the PCM limits the heat conduction across the PCM cavity through the ambient. That is, the thickness of the PCM should be determined with an in-depth optimization work by considering not only the heat transfer characteristics but also the economic aspects to achieve a better passive thermal controller.

Celik et al. (2017) extended the work of Javani et al. (2014) by varying the convective heat transfer coefficient on the outer surface of the battery cell. The numerical survey aims to reveal the performance of the PCM-embedded battery cell temperature at relatively high convective conditions. The convective heat transfer coefficients are selected as 7 W/m$^2$K, 15 W/m$^2$K, and 40 W/m$^2$K. Figure 7.82 compares the evolution of maximum cell temperatures for plane cell and PCM-embedded cell under three different convective boundary conditions. The results indicate that the incorporation of PCM around the battery cell is an effective thermal management strategy when heat transfer between the cell and surrounding ambient is limited. Increasing the convective heat transfer reduces the temperature difference between the plane cell and PCM-embedded cell. Figure 7.83 illustrates variations of the temperature difference between cells with PCM and *w/o* PCM configurations. At the highest convective heat transfer coefficient, $h = 40$ W/m$^2$K, there is no remarkable difference between each configuration. As the convective heat transfer coefficient reduces, the effect of PCM becomes relevant. At $h = 15$ W/m$^2$K and $h = 7$ W/m$^2$K, the maximum temperature differences are evaluated as 0.2 K and 2 K, respectively. Table 7.24 compares the maximum cell temperatures at the end of 20 min for three battery configurations and different surrounding conditions. The

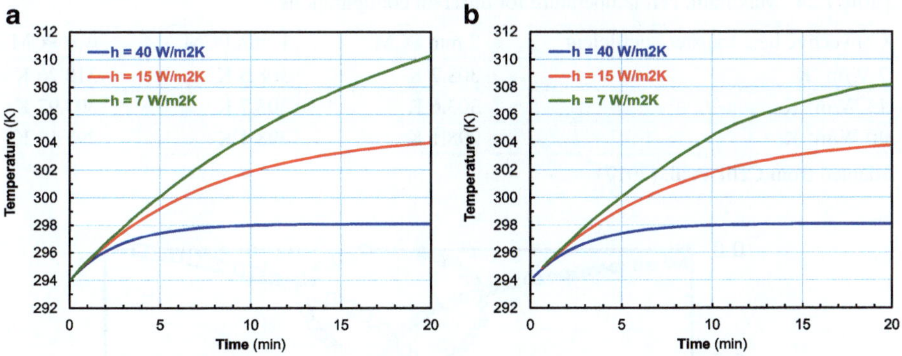


**Fig. 7.82** Variations of maximum cell temperature under different ambient conditions. (**a**) w/o PCM, (**b**) 3 mm PCM (Data from Celik et al. 2017)

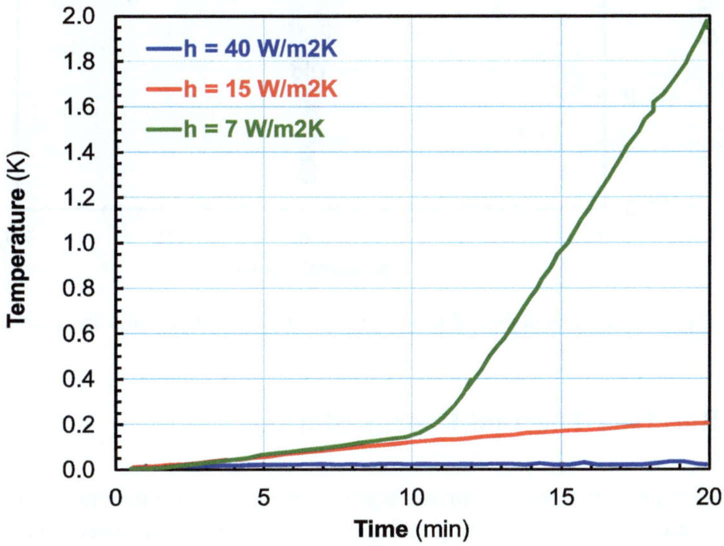


**Fig. 7.83** Variations of the temperature difference between the reference cell and PCM-embedded cell. (Data from Celik et al. 2017)

cell temperature reduces as the heat transfer coefficient on the outer wall of the cell increases. It is obvious that at higher convective heat transfer coefficients, the implementation of PCM around the battery cell becomes insignificant. That is, to evaluate the performance of the PCM-embedded battery cell, long-term analyses, or experiments, should be conducted under various dynamic boundary conditions and at different charging/discharging rates. Notice that the type of the PCM and the design of the PCM enclosure also significantly affect the transient heat transfer of the battery cell.

**Table 7.24** Maximum cell temperature for different configurations

| Convective heat transfer coefficient | 3 mm PCM | 1 mm PCM | w/o PCM |
|---|---|---|---|
| 7 W/m$^2$ K | 308.2 K | 308.5 K | 310.26 K |
| 15 W/m$^2$ K | 303.6 K | 303.7 K | 303.97 K |
| 40 W/m$^2$ K | 298.1 K | 298.1 K | 298.13 K |

Adapted from Celik et al. (2017)

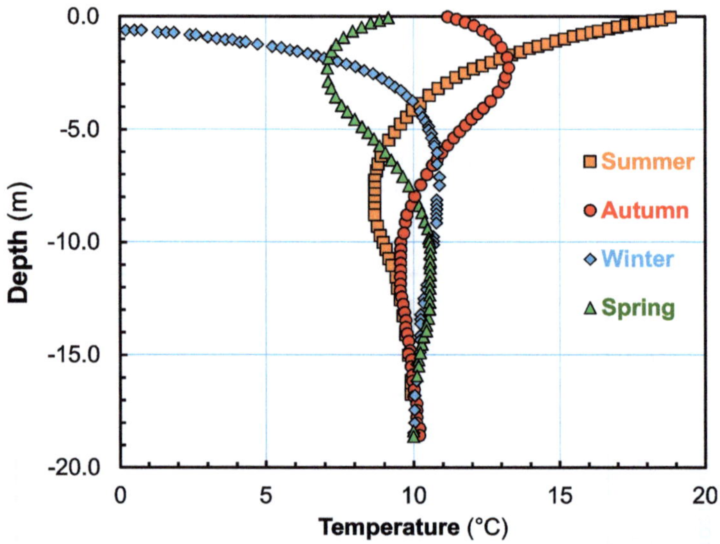


**Fig. 7.84** Approximate annual ground temperatures in the UK. (Data from Cibsejournal 2017)

## 7.10 Borehole Thermal Energy Storage

The temperature of the ground surface significantly changes throughout the year due to the variations in the incoming solar radiation, the wind speed, and ambient temperature. However, such variations are observed to a certain depth, and beyond a critical depth, there are no significant fluctuations. Ground, with a substantial thermal mass, has almost uniform temperature throughout the year beyond 10–15 m. Figure 7.84 represents the variations of ground temperature as a function of depth throughout the year. The seasonal temperature fluctuations are observed from surface to a depth of 10 m, and beyond this depth, the temperature of ground remains almost constant. Thermal stability of the ground makes it an attractive heat source or sink in HVAC systems. Ground source heat pumps (GSHPs) are widely in use mainly in Europe and North America as they provide specific benefits by improving the COP of the heating/cooling units and energy savings.

In a GSHP, borehole heat exchangers (BHE) are used to absorb (or release) thermal energy from the ground for the air-conditioning applications of residential or commercial buildings. In Scandinavia, groundwater is often used to fill the space between the borehole wall and collector wall, while otherwise, it is more common to backfill with some grouting material. There are many advantages of using water such as being cheap for installations and more easy access to the collector if it is required. Grouting, on the other hand, is required in many countries by national legislation to prevent groundwater contamination or is used to stabilize the borehole wall (Kizilkan and Dincer 2012). The thermal performance of BHEs is site specific, as the thermal properties of the ground, the size, and configurations of BHEs and backfill materials of boreholes have significant influences on the heat transfer characteristics of BHEs. The temperature of the ground mainly depends on the thermal conductivity, geothermal gradient, water content, and water flow rate through the borehole. Borehole thermal energy storage (BTES) systems, on the other hand, use the ground as a heat source or sink for space conditioning in residential and commercial buildings. BTES includes boreholes that are drilled into the ground. Vertical U-tubes are inserted into the boreholes and usually backfilled with grout or water to ensure proper thermal contact with the ground. In a vertical U-tube, a pump maintains water circulation. The circulating fluid mostly includes antifreeze to avoid flow blockage during cold winter periods. Borehole heat exchangers mostly consist of pipes with 10–15 cm diameter and drilled into the ground with a depth of 20–300 m. A borehole system includes many individual boreholes.

In the following case study, the thermodynamic model of the underground borehole thermal energy storage system at the University of Ontario Institute of Technology is represented.

***Case Study 22:* Thermodynamic Analyses of a Borehole Thermal Energy Storage System**

Recently Kizilkan and Dincer (2012, 2015) developed thermodynamic models to evaluate the performance of a closed BTES unit at the University of Ontario Institute of Technology (UOIT) in Oshawa, Canada, for cooling and heating purposes, respectively. UOIT campus consists of several new buildings and renewable sources that are integrated into the heating/cooling system to reduce greenhouse gas emissions. According to the preliminary drilling test programs, the feasibility of the underground TES system is assessed. From 55 m to 200 m below the ground level, almost an impermeable limestone formation was found. A total of 370 boreholes, with 200 m depth each, were used to meet the energy demand of the buildings. The schematic of the BTES unit at the UOIT is shown in Fig. 7.85. According to reference works (Kizilkan and Dincer 2012, 2015), the heating and cooling loads of the campus buildings are 6800 kW and 7000 kW, respectively.

The components of the BTES systems are illustrated in Figs. 7.86 and 7.87 for the cooling and heating modes, respectively. Here, A1–A10 represent ten campus buildings of the UOIT. In summertime, chillers operate to reject the thermal energy from the buildings and pump through the BTES unit. A glycol solution circulates inside the polyethylene tubing through an underground network. The temperature of the solution at the inlet and outlet sections of the ground piping system is 29.4 °C and

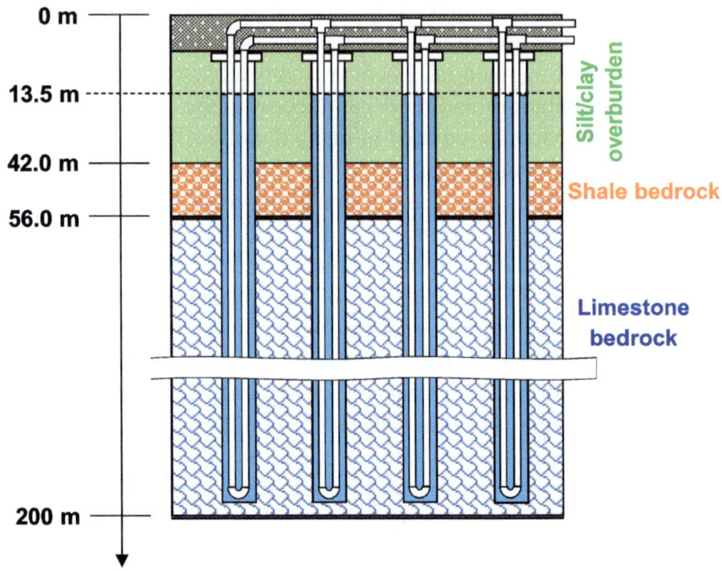


**Fig. 7.85** Schematic of the BTES system at UOIT. (Modified from Kizilkan and Dincer 2012)

35 °C, respectively. There is a secondary glycol loop between the system and buildings to transfer heat for cooling of the buildings. The glycol solution temperature at the inlet and outlet sections of the fan coils is 5.5 °C and 14.4 °C, respectively. In summertime, the system rejects heat from the buildings and transmits it to the ground. In winter time, the system is used to heat the buildings. The inlet and outlet temperatures of the solution are 5.6 °C and 9.3 °C, respectively. The evaporator water (15% glycol solution) goes into the borehole field, and heat energy is absorbed from the borehole water by the evaporator and transferred to the refrigerant. Heat pumps transfer the thermal energy to the secondary fluid to supply the heat through the buildings. The secondary circulation loop, 30% glycol solution, is used to provide the heat transfer between the heating system and the buildings. In the heating mode, the inlet and exit temperatures of the solution to/from the fan coils in buildings are 52 °C and 41.3 °C, respectively. Natural gas boilers are integrated into the BTES system to meet the heating load when additional heating demand is requested from the buildings.

In Fig. 7.88, the influence of ambient temperature (or the dead state) on the exergy destruction and exergy efficiency of the BTES system is represented in cooling mode. The exergy destruction rate and the exergy efficiency vary linearly as a function of the ambient temperature. The rate of exergy destruction increases as the reference ambient temperature increases. From 20 °C to 30 °C, the rate of exergy destruction rises almost 20%. Besides, the exergy efficiency of the BTES unit reduces from 65% to 62.5% by increasing the ambient temperature from 20 °C to 30 °C.

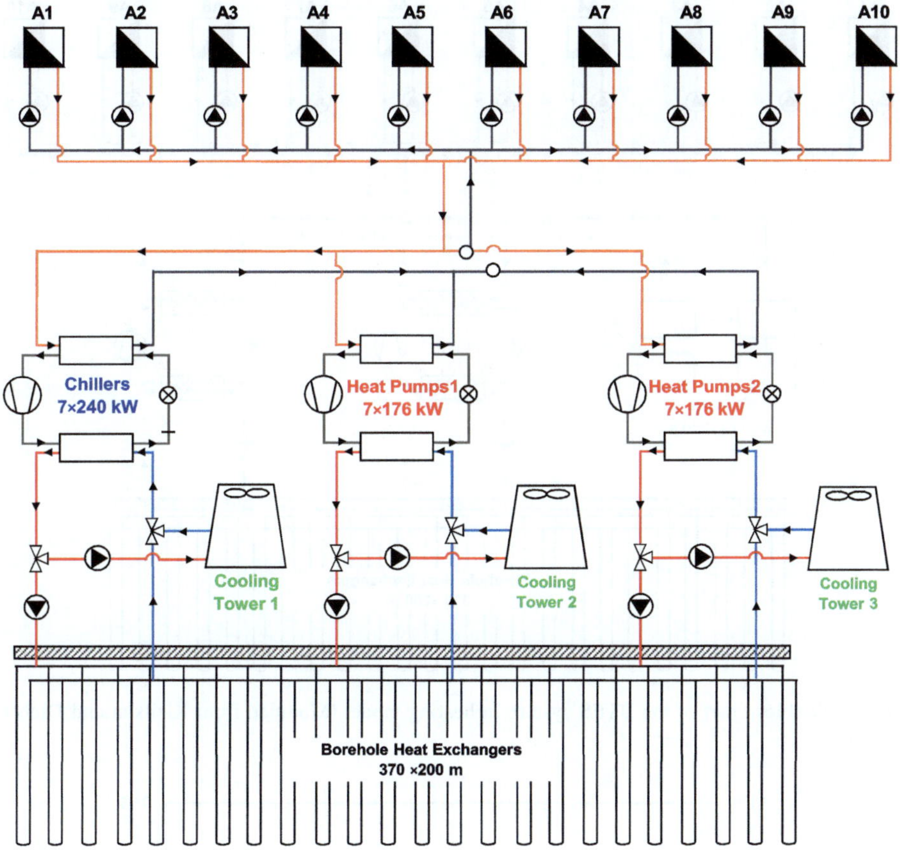


**Fig. 7.86** Flowchart of the BTES system in cooling mode. (Modified from Kizilkan and Dincer 2012)

Figure 7.89 illustrates the effect of the inlet temperature of glycol solution on the exergetic outputs for the cooling mode of a BTES unit. Kizilkan and Dincer (2012) stated that the inlet temperature of the glycol solution to the condenser is one of the most representative parameters for a BTES unit. The inlet temperature of the glycol solution to the condenser of the refrigeration unit is higher than the ground temperature during the summer season. So, the inlet temperature of the solution varies between 28 °C and 34 °C. Figure 7.89 depicts that increasing the inlet temperature of the glycol solution from 28 °C to 32 °C slightly affects the exergetic outputs. Beyond 32 °C, however, varying the inlet temperature of the glycol solution leads to considerable changes in the exergy efficiency and the rate of exergy destruction. The rate of exergy destruction increases almost 40% as the inlet temperature of the glycol solution varies from 32 °C to 34 °C.

Figure 7.90 shows the influence of glycol concentration on the exergetic efficiency and the rate of exergy destruction for the cooling mode of the BTES unit. The

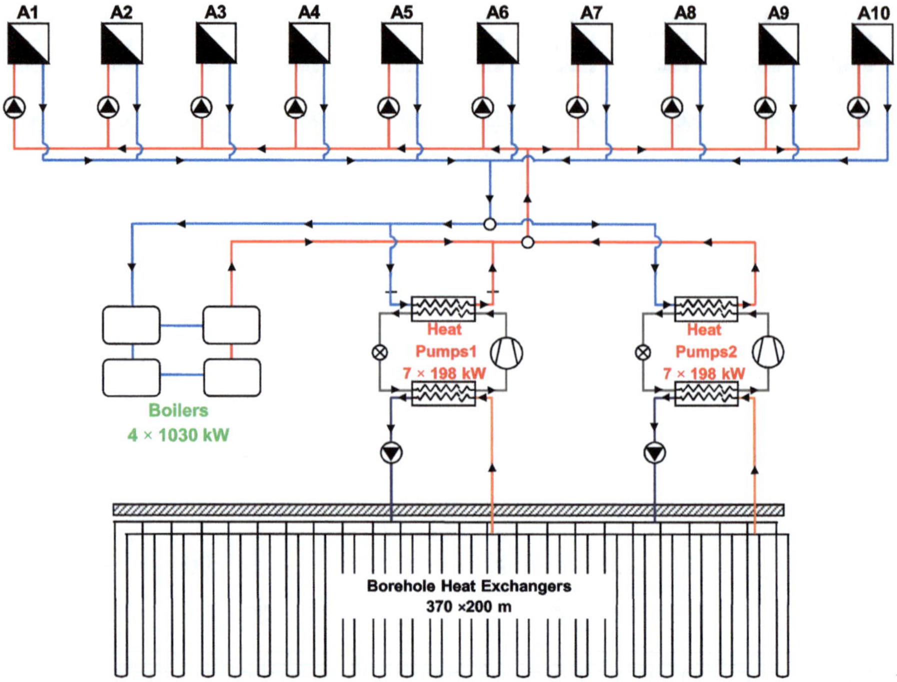


**Fig. 7.87** Flowchart of the BTES system in heating mode. (Modified from Kizilkan and Dincer 2015)

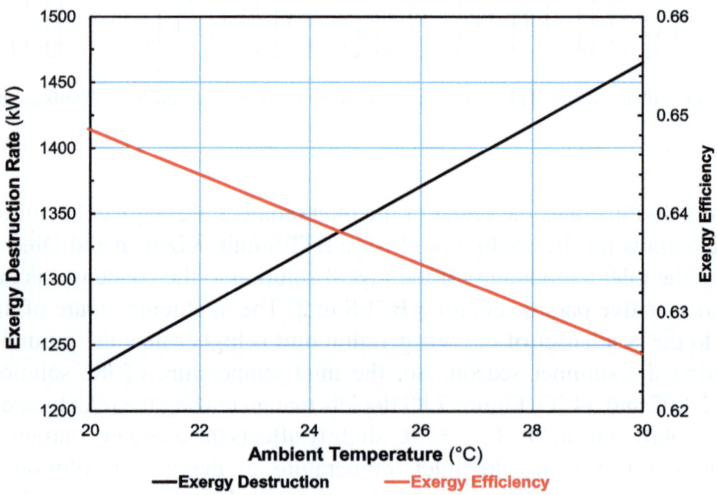


**Fig. 7.88** Influence of ambient temperature on the exergy destruction rate and exergy efficiency. (Modified from Kizilkan and Dincer 2012)

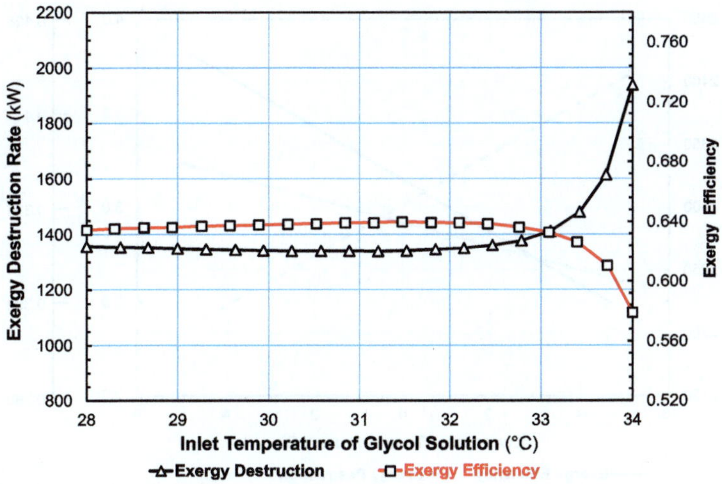


**Fig. 7.89** Influence of glycol inlet temperature on the exergy destruction rate and exergy efficiency. (Modified from Kizilkan and Dincer 2012)

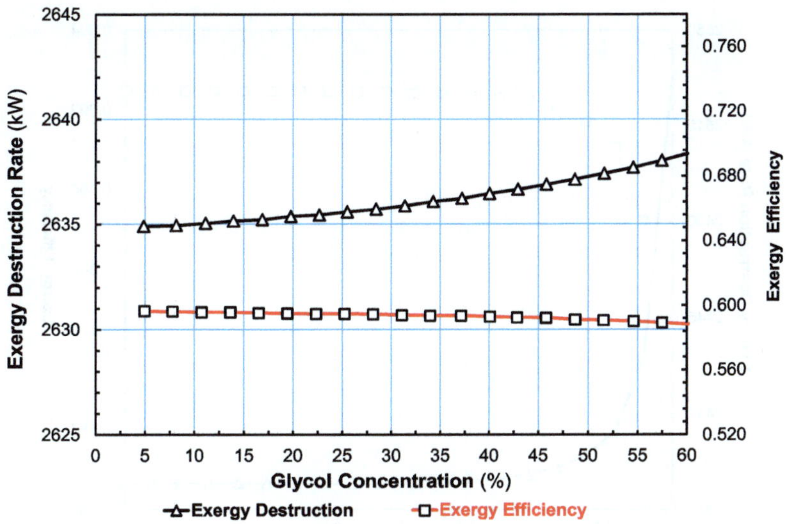


**Fig. 7.90** Influence of glycol concentration on the exergy destruction rate and exergy efficiency. (Modified from Kizilkan and Dincer 2012)

results reveal that varying the glycol concentration from 5% to 60% does not produce a significant change in the exergetic outputs. The maximum variation in the destruction rate remains far below 0.5% for the selected glycol solution concentrations. One can conclude that the influence of glycol solution concentration on the exergetic outputs is negligible. However, it should be noted that glycol with higher concentration implies improved investment costs for a BTES system.

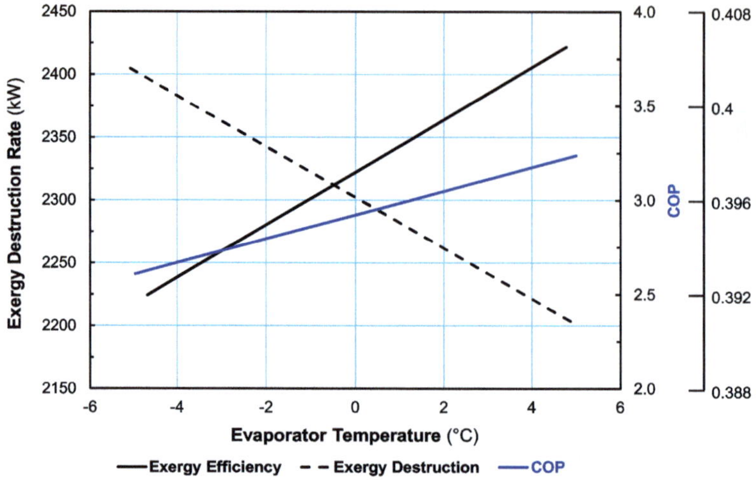


**Fig. 7.91** Influence of evaporator temperature on the performance of the BTES system. (Modified from Kizilkan and Dincer 2015)

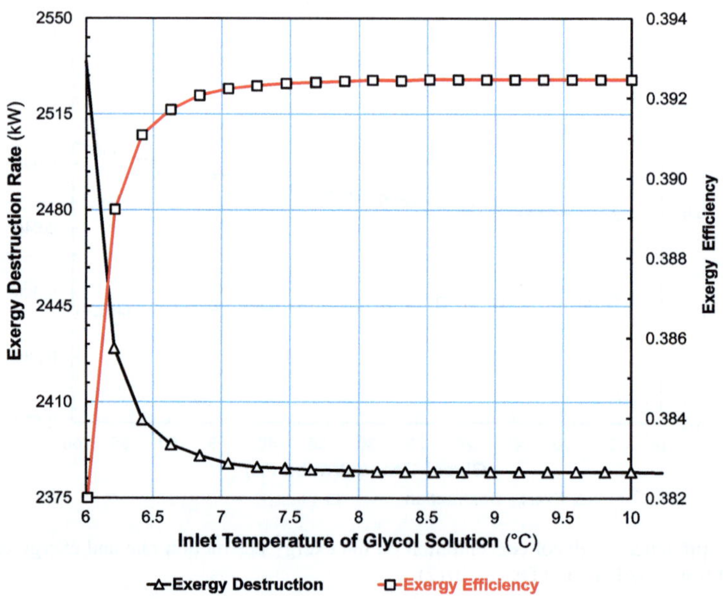


**Fig. 7.92** Influence of inlet temperature of glycol solution on the exergy destruction rate and exergy efficiency. (Modified from Kizilkan and Dincer 2015)

Figures 7.91 and 7.92 show the variations of performance indicators of the BTES unit regarding the evaporator temperature and inlet temperature of the glycol solution for the heating mode, respectively. From Fig. 7.91, one can depict that the COP of the heat pump slightly increases as the evaporation temperature increases. The increment is almost 40% for the selected temperature ranges. Results reveal that the entropy generation decreases as the evaporator temperature increases. Smaller entropy generation leads to a larger COP of the heat pump. Figure 7.91 also shows the exergy destruction rate and exergy efficiency of the system as a function of the evaporator temperature. As in the cooling mode, the variations have a linear pattern for the heating mode. The rate of exergy destruction decreases as the evaporator temperature increases. At higher temperatures, the rate of heat transfer increases, and the difference between the evaporator and condenser decreases. As the rate of exergy destruction reduces with increasing the evaporator temperature, the exergy efficiency of the system increases with increasing the evaporator temperature. For the selected evaporation temperature range, the variation in the exergetic efficiency is almost 4%.

Figure 7.92 shows the variations of the exergy efficiency and the rate of exergy destruction of the BTES system as a function of glycol inlet temperature for the heating mode. The inlet temperature of the glycol solution to the heat pump system is lower than the ambient temperature during winter time. Increasing the inlet temperature of the glycol solution from 6 °C to 7 °C causes significant variations in the exergetic outputs. Beyond this temperature, no signification change is reported. The rate of exergy destruction reduces by 6% in this narrow temperature range. The exergy efficiency, on the other hand, increases roughly from 38.2% to 39.2% in the selected working range. Increasing the temperature of the glycol-water temperature improves the heat transfer from ground to the system. Kizilkan and Dincer (2015) reported that inlet temperature of the glycol-water solution is one of the most influencing parameters on the performance of the BTES unit.

## 7.11 Closing Remarks

This chapter presents the aspects of heat storage technology from micro- to large-scale applications. The case studies that cover the material development, characterization, simulation, thermal management, and implementation of large-scale systems are discussed in detail. One of the most critical steps during the design process of a TES system is the determination of the thermal properties of storage materials. In the characterization section, the cons and pros of two different methods, DSC and T-History, are examined with illustrative examples. Nowadays, there is an increasing trend in the development of nano-enhanced phase change materials to improve the thermal response of the PCMs. In a case study, the influence of GNP (graphite-nano-platelets) on the thermal conductivity and heat storage characteristics of the PCM (arachidic acid) is briefly explained. Another novel approach to produce PCM is the clathrates of refrigerants. The influence of the additives on the thermal properties of the PCM clathrates is represented, and the performance of the candidate materials in a thermal management application is discussed. The chapter also includes several numerical and experimental studies that deal with transient heat transfer and thermal performance investigations, energetic and/or exergetic, of heat storage systems. The performance of aquifers with TES, greenhouse-integrated TES systems, building-integrated PCM blocks, and integration of TES into the multigeneration systems are represented.

## Nomenclature

| | |
|---|---|
| $c_p$ | Specific heat, J/kgK |
| $C$ | Volumetric heat capacity, J/m$^3$K |
| $D$ | Diameter, m |
| $\dot{E}x$ | The rate of exergy, W |
| $F$ | Body force, N/m$^3$ |
| Fo | Fourier number |
| $g$ | Gravitational acceleration, m/s$^2$ |
| $h$ | Convective heat transfer coefficient, W/m$^2$K |
| $h_{sf}$ | Latent heat of fusion, J/kg |
| $H$ | Enthalpy, J |
| $I$ | Incident solar radiation, W/m$^2$ |
| $k$ | Thermal conductivity, W/mK |
| $L$ | Length of the tube, m or latent heat, J/kg |
| Nu | Nusselt |
| $p$ | Pressure, Pa |
| Pr | Prandtl number |
| $r$, $\theta$ | Polar coordinates |
| $R$ | Radius, m |
| $R$ | Aspect ratio |
| Ra | Rayleigh number |
| Re | Reynolds number |
| Ste | Stefan number |
| $s$ | Interface position, m |
| $t$ | Time, s |
| $T$ | Temperature, K or °C |
| $u$, $v$ | Velocity components, m/s |
| $W$ | Mass, kg |
| $x$, $y$ | Cartesian coordinates, m |

## Greek Letters

| | |
|---|---|
| $\alpha$ | Thermal diffusivity, m$^2$/s |
| $\beta$ | Thermal expansion coefficient, 1/K |
| $\varepsilon$ | Emissivity |
| $\Delta$ | Difference |
| $\theta$ | Dimensionless temperature |
| $\rho$ | Density (kgm$^{-3}$) |
| $\mu$ | Dynamic viscosity (kgm$^{-1}$ s$^{-1}$) |

## Subscripts

| | |
|---|---|
| $B$ | Body |
| $conv$ | Convection |
| $d$ | Destruction |
| $i$ | Initial or inner |
| $in$ | Indoor or inner |

| | |
|---|---|
| *l* | Liquid |
| *m* | Melting or maximum |
| *n* | Nucleation |
| *o* | Dead state |
| *out* | Outlet or outer |
| *rad* | Radiation |
| *ref* | Reference |
| *s* | Solidification or solid |
| *sf* | Solid to liquid |
| *sur* | Surrounding |
| *surf* | Surface |

## References

Agyenim, F., Hewitt, N., Eames, P., & Smyth, M. (2010). A review of materials, heat transfer and phase change problem formulation for latent heat thermal energy storage systems (LHTESS). *Renewable and Sustainable Energy Reviews, 14*(2), 615–628.

Almahdi, M., Dincer, I., & Rosen, M. A. (2016). A new solar based multigeneration system with hot and cold thermal storages and hydrogen production. *Renewable Energy, 91*, 302–314.

Al-Sulaiman, F. A., Dincer, I., & Hamdullahpur, F. (2011). Exergy modeling of a new solar driven trigeneration system. *Solar Energy, 85*(9), 2228–2243.

Al-Sulaiman, F. A., Hamdullahpur, F., & Dincer, I. (2012). Performance assessment of a novel system using parabolic trough solar collectors for combined cooling, heating, and power production. *Renewable Energy, 48*, 161–172.

AlZahrani, A. A., & Dincer, I. (2016). Performance assessment of an aquifer thermal energy storage system for heating and cooling applications. *Journal of Energy Resources Technology, 138*(1), 011901.

Arezov, A., & Niyazov, S. K. (1980). *Appl. Sol. Eng., 16*, 430.

ASTM D1481-17. (2017). Standard test method for density and relative density (specific gravity) of viscous materials by Lipkin bicapillary pycnometer. ASTM International, West Conshohocken.

ASTM D1986-14. (2014). Standard test method for determining the apparent viscosity of polyethylene wax. ASTM International, West Conshohocken.

ASTM D2534-88. (2017). Standard test method for coefficient of kinetic friction for wax coatings. ASTM International, West Conshohocken.

ASTM D2669-16. (2016). Standard test method for apparent viscosity of petroleum waxes compounded with additives (hot melts). ASTM International, West Conshohocken.

ASTM E1269-11. (2011). *Standard test method for determining specific heat capacity by differential scanning calorimetry*. ASTM International, West Conshohocken.

ASTM E967-08. (2014). Standard test method for temperature calibration of differential scanning calorimeters and differential thermal analyzers. ASTM International, West Conshohocken.

Baille, J. (1987). In C. Von Zabeltitz (Ed.), *Energy conservation and renewable energies for greenhouse heating*. FAO.

Bakr, M., van Oostrom, N., & Sommer, W. (2013). Efficiency of and interference among multiple aquifer thermal energy storage systems; A Dutch case study. *Renewable Energy, 60*, 53–62.

Balaras, C. A. (1996). The role of thermal mass on the cooling load of buildings. An overview of computational methods. *Energy and Buildings, 24*(1), 1–10.

Balducci, M. (1985). Solar agricultural greenhouse, Hag Brochure, EEC, DG 17.

Barreneche, C., Solé, A., Miró, L., Martorell, I., Fernández, A. I., & Cabeza, L. F. (2013). Study on differential scanning calorimetry analysis with two operation modes and organic and inorganic phase change material (PCM). *Thermochimica Acta, 553*, 23–26.

BASF. (2018). https://www.basf.com/. Accessed 2 Jan 2018.

Bauer, T., Laing, D., & Tamme, R. (2010). Overview of PCMs for concentrated solar power in the temperature range 200 to 350 °C. *Advances in Science and Technology, 74*, 272–277 Trans Tech Publicatioens.

Baytorum, N. (1987). In C. Von Zabeltitz (Ed.), *Greenhouse heating with solar energy*. FAO.

Benli, H. (2006). *Isı depolamalı cam seralarda sıcaklık değişiminin incelenmesi (in Turkish)*. Ph.D. Thesis. University of Fırat, Elazığ, Turkey.

Benli, H., & Durmuş, A. (2009). Performance analysis of a latent heat storage system with phase change material for new designed solar collectors in greenhouse heating. *Solar Energy, 83*(12), 2109–2119.

Bonrehi, F. (1982). *Proc. Conf. La Serre Solaire* (p. 230). Perpignan.

Bouhdgar, A., & Boulbing, A. (1990). In A. Sayigh (Ed.), *Proc. Congress Energy and the Environment* (p. 2325). Reading.

Boulard, T., & Baille, J. (1987). *Agricultural and Forest Meteorology, 32*, 270.

Brandstetter, A. (1987). *Proc. ISES Solar World Congress* (p. 3353). Hamburg.

Bredenbeck, H. (1987). In C. Von Zabeltitz (Ed.), *Greenhouse heating with solar energy*. FAO.

Bricault, M., Jeremie, G. Paris, J., & Jackson, H. A. (1982). *Proc Conf. Energex '82* (p. 564). Regina.

Campiotti, C., Camporcale, J., & Salice, R. (1988). *ACTA Hortic.*

Carling, R. W., Kramer, C. M., Bradshaw, R. W., Nissen, D. A., Goods, S. H., Mar, R. W., et al. (1981). *Molten nitrate salt technology development status report* (No. SAND-80-8052). Livermore/Albuquerque: Sandia National Labs.

Castellón, C., Günther, E., Mehling, H., Hiebler, S., & Cabeza, L. F. (2008). Determination of the enthalpy of PCM as a function of temperature using a heat-flux DSC—A study of different measurement procedures and their accuracy. *International Journal of Energy Research, 32*(13), 1258–1265.

Celik, A., Metin C. S., & Ezan M. A. (2017, September 13–16). Implementation of PCM with Li-Ion Battery, ULIBTK'17 21. Ulusal Isı Bilimi ve Tekniği Kongresi, Çorum.

Cibsejournal. https://www.cibsejournal.com/cpd/modules/2010-04/.. Accessed 12 Nov 2017.

Coleparmer. https://www.coleparmer.com/i/kimble-chase-15123r-50-pycnometer-bottle-50-ml/3459074. Accessed 12 Nov 2017.

Damrath, D., & Von Zabeltitz, C. (1987). In C. Von Zabeltitz (Ed.), *Greenhouse heating with solar energy*. FAO.

Deforche. (1990, May). European Community demonstration programme in solar thermal energy, EEC.

Eggers, H., & Vickermann, E. (1983). *Proceedings of the 1st EEC conference on solar heating* (p. 467). Amsterdam.

Erek, A., Seki, Y., Ezan M. A., Turgut, A., & Cetin, L. (2015). *Investigation of thermal behavior of nano-enhanced phase change materials and application in the encapsulated thermal energy storage systems*. TÜBİTAK MAG Project, Number: 112M164, pp. 1–150.

Esquira, I., Segal, I., & Antler, A. (1987). In C. Von Zabeltitz (Ed.), *Greenhouse heating with solar energy*. FAO.

Ezan, M. A. (2011). *Experimental and numerical investigation of cold thermal energy storage systems*. PhD thesis, Graduate School of Natural and Applied Sciences of Dokuz Eylul University, Izmir.

Ezan, M. A., Ince, S., Seki, Y., Turgut, A., Cetin, L., & Erek, A. (2014). *Improvement the thermal performance of Myristic acid*. International conference on Heat Transfer, Fluid Mechanics and Thermodynamics (HEFAT 2014), Orlando.

Farah, J. (1987). In C. Von Zabeltitz (Ed.), *Greenhouse heating with solar energy*. FAO.

Farid, M., Kim, Y., Honda, T., & Kanzawa, A. (1989). The role of natural convection during melting and solidification of PCM in a vertical cylinder. *Chemical Engineering Communications, 84*(1), 43–60.

Flerking, M. (1981). *Serres Solaires de Production,* PIC edition. ARES.

Fotiades, I. (1987). In C. Von Zabeltitz (Ed.), *Energy conservation and renewable energies for greenhouse heating*. FAO.
Fotiades, I. (1988). *Use of solar energy for heating of greenhouses in Cyprus*. Final report to the Ministry of agriculture, Cyprus.
Fourcy, A. (1981). *Serres Solaires de Production*, PIC edition. ARES.
Fourcy, A. (1982). *Proc. Conf. Solar Greenhouses* (p. 133). Perpignan.
Fraunhofer IFAM. (2018). https://www.ifam.fraunhofer.de. Accessed 2 Jan 2018.
Fraunhofer ISE. (2018). https://www.ise.fraunhofer.de. Accessed 2 Jan 2018.
Garcia, M. (1987). In C. Von Zabeltitz (Ed.), *Energy conservation and renewable energies for greenhouse heating*. FAO.
Gebhart, B., & Mollendorf, J. C. (1977). A new density relation for pure and saline water. *Deep Sea Research, 24*(9), 831–848.
Gong, Z. X., & Mujumdar, A. S. (1998). Flow and heat transfer in convection-dominated melting in a rectangular cavity heated from below. *International Journal of Heat and Mass Transfer, 41* (17), 2573–2580.
Gowan, T., & Black, H. (1980). *Sunworld, 45*, 150.
de Gracia, A., & Cabeza, L. F. (2017). Numerical simulation of a PCM packed bed system: A review. *Renewable and Sustainable Energy Reviews, 69*, 1055–1063.
Grafiadellis, M. (1985). *Proc. 2nd Greek Conf. on Renewable Energy Sources*. Salonika.
Grafiadellis, M. (1987). In C. Von Zabeltitz (Ed.), *Greenhouse heating with solar energy*. FAO.
Groves, D. (1984). *Solar energy in Israel*. Jerusalem: Ministry of Energy and Infrastructure.
Gschwander, S., Lazaro, A., Cabeza, L. F., Günther, E., Fois, M., & Chui, J. (2011). *Development of a test standard for PCM and TCM characterization part 1: Characterization of phase change materials*. IEA Report A,2.
Günther, E., Hiebler, S., Mehling, H., & Redlich, R. (2009). Enthalpy of phase change materials as a function of temperature: Required accuracy and suitable measurement methods. *International Journal of Thermophysics, 30*(4), 1257–1269.
Gupta, A., & Tiwari, G. N. (2002). Computer model and its validation for prediction of storage effect of water mass in a greenhouse: A transient analysis. *Energy Conversion and Management, 43*(18), 2625–2640.
Haghshenaskashani, S., & Pasdarshahri, H. (2009). Simulation of thermal storage phase change material in buildings. *World Academy of Science, Engineering and Technology, 58*, 111–115.
Hendricks, J. H. C., & Sark, W. G. J. H. M. (2013). Annual performance enhancement of building integrated photovoltaic modules by applying phase change materials. *Progress in Photovoltaics: Research and Applications, 21*(4), 620–630.
Hong, H., Kim, S. K., & Kim, Y. S. (2004). Accuracy improvement of T-history method for measuring heat of fusion of various materials. *International Journal of Refrigeration, 27*(4), 360–366.
Huang, B. K., Toksoy, M., & Cengel, Y. A. (1986). Transient response of latent heat storage in greenhouse solar system. *Solar Energy, 37*(4), 279–292.
Ismail, K. A., Henriquez, J. R., & Da Silva, T. M. (2003). A parametric study on ice formation inside a spherical capsule. *International Journal of Thermal Sciences, 42*(9), 881–887.
ISO 12185:1996, Crude petroleum and petroleum products – Determination of density – Oscillating U-tube method.
ISO 3675:1998, Crude petroleum and liquid petroleum products – Laboratory determination of density – Hydrometer method.
Jaffrin, A., & Makhlouf, S. (1987). *Proc. ISES Solar World Congress* (p. 3358). Hamburg.
Jaffrin, A., & Cadier, P. (1982). Latent heat storage applied to horticulture. *Solar Energy, 28*(4), 313–321.
Jaffrin, A., Hallot, J., & Lubrano, M. (1982). La seize solaire architecture agriculture, UNESCO.
Javani, N., Dincer, I., Naterer, G. F., & Yilbas, B. S. (2014). Heat transfer and thermal management with PCMs in a Li-ion battery cell for electric vehicles. *International Journal of Heat and Mass Transfer, 72*, 690–703.

Jelinkova, H. (1987). In C. Von Zabeltitz (Ed.), *Greenhouse heating with solar energy*. FAO.

Karthikeyan, S., Solomon, G. R., Kumaresan, V., & Velraj, R. (2014). Parametric studies on packed bed storage unit filled with PCM encapsulated spherical containers for low temperature solar air heating applications. *Energy Conversion and Management, 78*, 74–80.

Kavin, J., & Kurtan, S. (1987). In C. Von Zabeltitz (Ed.), *Greenhouse heating with solar energy*. FAO.

Khodadadi, J. M., & Zhang, Y. (2001). Effects of buoyancy-driven convection on melting within spherical containers. *International Journal of Heat and Mass Transfer, 44*(8), 1605–1618.

Khodadadi, J. M., Fan, L., & Babaei, H. (2013). Thermal conductivity enhancement of nanostructure-based colloidal suspensions utilized as phase change materials for thermal energy storage: A review. *Renewable and Sustainable Energy Reviews, 24*, 418–444.

Kizilkan, O., & Dincer, I. (2012). Exergy analysis of borehole thermal energy storage system for building cooling applications. *Energy and Buildings, 49*, 568–574.

Kizilkan, O., & Dincer, I. (2015). Borehole thermal energy storage system for heating applications: Thermodynamic performance assessment. *Energy Conversion and Management, 90*, 53–61.

Kyritsis, S., & Mavroyiannopoulos, G. (1987). In C. Von Zabeltitz (Ed.), *Energy conservation and renewable energies for greenhouse heating*. FAO.

Lázaro, A., Günther, E., Mehling, H., Hiebler, S., Marín, J. M., & Zalba, B. (2006). Verification of a T-history installation to measure enthalpy versus temperature curves of phase change materials. *Measurement Science and Technology, 17*(8), 2168.

Levav, N., & Zamir, N. (1987). In C. Von Zabeltitz (Ed.), *Greenhouse heating with solar energy*. FAO.

Lin, D. S., & Nansteel, M. W. (1987). Natural convection heat transfer in a square enclosure containing water near its density maximum. *International Journal of Heat and Mass Transfer, 30*(11), 2319–2329.

Liskola, K. (1987). In C. Von Zabeltitz (Ed.), *Greenhouse heating with solar energy*. FAO.

Liu, J., Xu, C., Ju, X., Yang, B., Ren, Y., & Du, X. (2017). Numerical investigation on the heat transfer enhancement of a latent heat thermal energy storage system with bundled tube structures. *Applied Thermal Engineering, 112*, 820–831.

Liu, M., Tay, N. S., Bell, S., Belusko, M., Jacob, R., Will, G. et al. (2016). Review on concentrating solar power plants and new developments in high temperature thermal energy storage technologies. *Renewable and Sustainable Energy Reviews, 53*, 1411–1432.

Mac Kinnon, D. (1981). Serres Solaires de Production, PIC edition. ARES.

Machida, Y., Kudoh, Y., & Takeda, T. (1985). *Proc. Symposium on Thermal Application of Solar Energy* (p. 503). Japan ISES.

Malik, M., Dincer, I., & Rosen, M. A. (2016). Review on use of phase change materials in battery thermal management for electric and hybrid electric vehicles. *International Journal of Energy Research, 40*(8), 1011–1031.

Mao, Q. (2016). Recent developments in geometrical configurations of thermal energy storage for concentrating solar power plant. *Renewable and Sustainable Energy Reviews, 59*, 320–327.

Marín, J. M., Zalba, B., Cabeza, L. F., & Mehling, H. (2003). Determination of enthalpy–temperature curves of phase change materials with the temperature-history method: Improvement to temperature dependent properties. *Measurement Science and Technology, 14*(2), 184.

Mazria, E., & Baker S. (1981). Serres Solaires de Production, PIC edition. ARES.

Mercier, J. R. (1982). *Design and operation of a solar passive greenhouse in the south-west of France*. Energy conservation and use of renewable energies in the bio-industries: Proceedings of the International Seminar on Energy Conservation and the Use of Solar and other Renewable Energies in Agriculture, Horticulture, and Fishculture, held at the Polytechnic of Central London, 15–19 September 1980. Oxford: Pergamon Press.

Molz, F. J., Warman, J. C., & Jones, T. E. (1978). Aquifer storage of heated water: Part I – A field experiment. *Groundwater, 16*(4), 234–241.

Molz, F. J., Parr, A. D., & Andersen, P. F. (1981). Thermal energy storage in a confined aquifer: Second cycle. *Water Resources Research, 17*(3), 641–645.

Montero, M., Marfa, M., Serrano, T., & Anton, S. (1987). In C. Von Zabeltitz (Ed.), *Energy conservation and renewable energies for greenhouse heating*. FAO.

Moraga, N. O., Xamán, J. P., & Araya, R. H. (2016). Cooling Li-ion batteries of racing solar car by using multiple phase change materials. *Applied Thermal Engineering, 108*, 1041–1054.

Mougou, A., & Verlodt, H. (1987). In C. Von Zabeltitz (Ed.), *Greenhouse heating with solar energy*. FAO.

Nash, R., & Williamson, J. (1978). *Proc. ISES Solar World Congress* (p. 64).

Navarro, L., de Gracia, A., Castell, A., Álvarez, S., & Cabeza, L. F. (2015). PCM incorporation in a concrete core slab as a thermal storage and supply system: Proof of concept. *Energy and Buildings, 103*, 70–82.

Nissen, D. A. (1982). Thermophysical properties of the equimolar mixture sodium nitrate-potassium nitrate from 300 to 600 degree C. *Journal of Chemical and Engineering Data, 27* (3), 269–273.

Öztürk, H. H., & Başçetinçelik, A. (2003). Energy and exergy efficiency of a packed-bed heat storage unit for greenhouse heating. *Biosystems Engineering, 86*(2), 231–245.

Pacheco, M., Marreivoc, S., & Rosa, M. (1987). In C. Von Zabeltitz (Ed.), *Energy conservation and renewable energies for greenhouse heating*. FAO.

Paksoy, H. O., Andersson, O., Abaci, S., Evliya, H., & Turgut, B. (2000). Heating and cooling of a hospital using solar energy coupled with seasonal thermal energy storage in an aquifer. *Renewable Energy, 19*(1–2), 117–122.

Papadopulos, S. S., & Larson, S. P. (1978). Aquifer storage of heated water: Part II – Numerical simulation of field results. *Groundwater, 16*(4), 242–248.

Paris, V. (1981). *Proc. Conf. Solar Greenhouses* (p. 240). Perpignan.

Peck, J. H., Kim, J. J., Kang, C., & Hong, H. (2006). A study of accurate latent heat measurement for a PCM with a low melting temperature using T-history method. *International Journal of Refrigeration, 29*(7), 1225–1232.

Pelay, U., Luo, L., Fan, Y., Stitou, D., & Rood, M. (2017). Thermal energy storage systems for concentrated solar power plants. *Renewable and Sustainable Energy Reviews, 79*, 82–100.

Pielichowska, K., & Pielichowski, K. (2014). Phase change materials for thermal energy storage. *Progress in Materials Science, 65*, 67–123.

Reddy, K. S., Mudgal, V., & Mallick, T. K. (2018). Review of latent heat thermal energy storage for improved material stability and effective load management. *Journal of Energy Storage, 15*, 205–227.

Saidov, S. A., & Akhtamov, R. A. (1987). Experimental greenhouse with solar and geothermal heating systems. *Geliotechnica, 23*(6), 78–80.

Şahan, N., Fois, M., & Paksoy, H. (2015). Improving thermal conductivity phase change materials – A study of paraffin nanomagnetite composites. *Solar Energy Materials and Solar Cells, 137*, 61–67.

Sallanbas, H., Durceylan, E., & Yelboga, K. (1987). In C. Von Zabeltitz (Ed.), *Greenhouse heating with solar energy*. FAO.

Sandnes, B., & Rekstad, J. (2006). Supercooling salt hydrates: Stored enthalpy as a function of temperature. *Solar Energy, 80*(5), 616–625.

Santamouris, M., Balaras, C. A., Dascalaki, E., & Vallindras, M. (1994). Passive solar agricultural greenhouses: A worldwide classification and evaluation of technologies and systems used for heating purposes. *Solar Energy, 53*(5), 411–426.

Seki, Y., Ince, S., Ezan, M. A., Turgut, A., & Erek, A. (2015). Development and evaluation of graphite nanoplate (GNP)-based phase change material for energy storage applications. *International Journal of Energy Research, 39*(5), 696–708.

Solé, A., Miró, L., Barreneche, C., Martorell, I., & Cabeza, L. F. (2013). Review of the T-history method to determine thermophysical properties of phase change materials (PCM). *Renewable and Sustainable Energy Reviews, 26*, 425–436.

Sørensen, B. (1989). Experiments with energy storage in a high-latitude greenhouse. *Solar Energy, 42*(4), 293–301.

Stanković, S. B., & Kyriacou, P. A. (2012). The effects of thermistor linearization techniques on the T-history characterization of phase change materials. *Applied Thermal Engineering, 44*, 78–84.
Straub, D., Coppinger, E., & Bylon, D. (1978). *3rd Conf. on Solar Energy filr Heating of Greenhouses Residence Combination* (p. 101).
Sykes, J. F., Lantz, R. B., Pahwa, S. B., & Ward, D. S. (1982). Numerical simulation of thermal energy storage experiment conducted by Auburn University. *Groundwater, 20*(5), 569–576.
Tan, F. L. (2008). Constrained and unconstrained melting inside a sphere. *International Communications in Heat and Mass Transfer, 35*(4), 466–475.
Tan, F. L., Hosseinizadeh, S. F., Khodadadi, J. M., & Fan, L. (2009). Experimental and computational study of constrained melting of phase change materials (PCM) inside a spherical capsule. *International Journal of Heat and Mass Transfer, 52*(15–16), 3464–3472.
Tavman, I., & Turgut, A. (2010). An investigation on thermal conductivity and viscosity of water-based nanofluids. *Microfluidics Based Microsystems*, 139–162.
Tsang, C. F., Buscheck, T., & Doughty, C. (1981). Aquifer thermal energy storage: A numerical simulation of Auburn University field experiments. *Water Resources Research, 17*(3), 647–658.
Turgut, A., Sauter, C., Chirtoc, M., Henry, J., Tavman, S., Tavman, I., & Pelzl, J. (2008). AC hot wire measurement of thermophysical properties of nanofluids with 3ω method. *The European Physical Journal-Special Topics, 153*, 349–352.
Ukrainczyk, N., Kurajica, S., & Šipušić, J. (2010). Thermophysical comparison of five commercial paraffin waxes as latent heat storage materials. *Chemical and Biochemical Engineering Quarterly, 24*(2), 129–137.
Von Zabeltitz, C., & Rosocha, C. (1987). In C. Von Zabeltitz (Ed.), *Greenhouse heating with solar energy*. FAO.
Vonarburg, J. J., & Gallacher, G. (1982). *Proc. Conf. Solar Greenhouses* (p. 156). Perpignan.
Wang, X., Niu, J., Li, Y., Zhang, Y., Wang, X., Chen, B., et al. (2008). Heat transfer of microencapsulated PCM slurry flow in a circular tube. *AICHE Journal, 54*(4), 1110–1120.
Woodston, D. G. (1982). *Proc. Conf. Solar Greenhouses* (p. 156). Perpignan.
Wu, M., Xu, C., & He, Y. (2016). Cyclic behaviors of the molten-salt packed-bed thermal storage system filled with cascaded phase change material capsules. *Applied Thermal Engineering, 93*, 1061–1073.
Xia, L., Zhang, P., & Wang, R. Z. (2010). Numerical heat transfer analysis of the packed bed latent heat storage system based on an effective packed bed model. *Energy, 35*(5), 2022–2032.
Yamagishi, Y., Takeuchi, H., Pyatenko, A. T., & Kayukawa, N. (1999). Characteristics of microencapsulated PCM slurry as a heat-transfer fluid. *AICHE Journal, 45*(4), 696–707.
Yavuz, F. E. (2017). *Natural convection driven phase change inside spherical capsules*. MSc thesis, Graduate School of Natural and Applied Sciences of Dokuz Eylul University, Izmir.
Yiannoulis, P. (1990). *ACTA Hortie.*
Yinping, Z., Yi, J., & Yi, J. (1999). A simple method, the-history method, of determining the heat of fusion, specific heat and thermal conductivity of phase-change materials. *Measurement Science and Technology, 10*(3), 201.
Yoshioka, T. (1989). *Proc. ISES Solar World Congress* (p. 3070). Kobe.
Yu, J., Chen, X., Ma, X., Song, Q., Zhao, Y., & Cao, J. (2014). Influence of nanoparticles and graphite foam on the supercooling of acetamide. *Journal of Nanomaterials, 2014*, 214.
Zafar, S. (2015). *Experimental and theoretical investigation of novel phase change materials for thermal applications*. PhD dissertation, Faculty of Engineering and Applied Science, University of Ontario Institute of Technology, Oshawa.
Zafar, S., Dincer, I., & Gadalla, M. (2017). Evaluation of thermophysical properties of refrigerant clathrates with additives. *International Communications in Heat and Mass Transfer, 89*, 165–175.
Zeroni, M. (1990). *ACTA Hortic.*
Zhou, Z., Zhang, Z., Zuo, J., Huang, K., & Zhang, L. (2015). Phase change materials for solar thermal energy storage in residential buildings in cold climate. *Renewable and Sustainable Energy Reviews, 48*, 692–703.

FSC
www.fsc.org
MIX
Papier aus verantwortungsvollen Quellen
Paper from responsible sources
FSC® C105338

If you have any concerns about our products,
you can contact us on
**ProductSafety@springernature.com**

In case Publisher is established outside the EU,
the EU authorized representative is:
**Springer Nature Customer Service Center GmbH**
**Europaplatz 3, 69115 Heidelberg, Germany**

Printed by Libri Plureos GmbH
in Hamburg, Germany